Ecological theories and hypotheses are usı
variability in space and time, which often makes the design of exper
difficult. The statistical tests we use require data to be collected carefully and with proper regard to the needs of these tests. This book describes how to design ecological experiments from a statistical basis using analysis of variance so that we can draw reliable conclusions.

The logical procedures that lead to a need for experiments are described, followed by an introduction to simple statistical tests. This leads to a detailed account of analysis of variance, looking at procedures, assumptions and problems. One-factor analysis is extended to nested (hierarchical) designs and factorial analysis. Finally, some regression methods for examining relationships between variables are covered.

Examples of ecological experiments are used throughout to illustrate the procedures and examine problems. This book will be invaluable to practising ecologists in addition to advanced students involved in experimental design.

EXPERIMENTS IN ECOLOGY

EXPERIMENTS IN ECOLOGY

Their logical design and interpretation using analysis of variance

A. J. UNDERWOOD
Institute of Marine Ecology, University of Sydney

PUBLISHED BY THE PRESS SYNDICATE OF THE UNIVERSITY OF CAMBRIDGE
The Pitt Building, Trumpington Street, Cambridge CB2 1RP, United Kingdom

CAMBRIDGE UNIVERSITY PRESS
The Edinburgh Building, Cambridge CB2 2RU, United Kingdom http://www.cup.cam.ac.uk
40 West 20th Street, New York, NY 10011-4211, USA http://www.cup.org
10 Stamford Road, Oakleigh, Melbourne 3166, Australia

First published 1997
Reprinted 1998 (twice)

Printed in the United Kingdom at the University Press, Cambridge

Typeset in 10/13pt Times [wv]

A catalogue record for this book is available from the British Library

Library of Congress cataloging in publication data

Underwood, A. J.
Experiments in ecology : their logical design and interpretation using analysis of variance / A. J. Underwood.
p. cm.
Includes bibliographical references and index.
ISBN 0 521 55329 6 (hardback). – ISBN 0 521 55696 1 (pbk.)
1. Ecology – Experiments. I. Title.
QH541.24.U54 1997
574.5′072 – dc20 96-15180 CIP

ISBN 0 521 55329 6 hardback
ISBN 0 521 55696 1 paperback

The wide subject of experimental design was opened up, aimed at . . . avoiding waste of effort in the accumulation of ill-planned, indecisive, or irrelevant observations.
(R. A. Fisher, 1955)

The task now is to prevent it from being closed!

Contents

Acknowledgements

The Australian Research Council have generously supported my research for 20 years. My long-time friend Howard Choat (University of Auckland and James Cook University), the World Aquaculture Congress (Halifax, Nova Scotia), Jim Clegg (Bodega Bay), Chris Battershill (NIWAR, New Zealand), Bob Kearney (Fisheries Research Institute, Sydney), Peter Scanes (EPA, Sydney), South Australian State Fisheries, the Tjärnö Marine Laboratory (twice) and the University of Stockholm (Sweden), Bob Vadas (University of Maine), Macquarie University (Sydney), the University of Évora (Portugal) and the Black Sea Environmental Program (Odessa, Ukraine) have all invited me to their institutions to learn how to teach. Bob Clarke (Plymouth), Mike Foster (Moss Landing), Roger Green (Western Ontario), John Gray (Oslo), Pete Peterson (North Carolina), Wayne Sousa (Berkeley) and Bob Vadas (Maine) encouraged discussion. David Fletcher (Otago) was particularly helpful with mathematical and statistical advice. My postgraduate students and postdoctoral researchers taught me about complex and interesting experiments.

My family, Anne, Clare and James were remarkable in putting up with this book. The mechanics were greatly aided by Sylvia Warren and Jenni Winzar (typing), Gee Chapman and Vanessa Mathews (figures), Tim Glasby (tables), Karen Astles (references), Greg Skilleter (keeping computers functioning). John Lawton and Susan Sternberg (Blackwells) initiated the book, but did not wish to publish it. Without Alan Crowden's and CUP's enthusiasm and encouragement, it would not have been finished.

More than for anyone else I know, my work has profited from the talents of several gifted women scientists. I am indebted to my friends Anne Underwood, Patsy Armati, Ros Hinde and Mary Peat for all of their help over many years. I am particularly indebted for help with all matters experimental and for reading and commenting on the book in her uniquely critical way to my colleague Gee Chapman, with whom I teach this topic and have most wrestled with the problems. Without her help, this book would not have been written. Neither she nor anyone else should be blamed that it has! The errors are mine. There are many fewer because of the efforts of Sandi Irvine, who read this with great care.

1

Introduction

'Not another book attempting to explain statistics in a simple-minded way!' could well be the reaction of an ecologist coming in from a hard day's work in the field and finding this book. It is a reasonable comment. There already exist numerous books on introductory statistics. There are, however, few books that focus on the nature of biological experiments from the perspective of biological hypotheses and the practical problems that biologists routinely encounter.

For many years, classes I have taught in ecology have struggled with mathematical and statistical concepts.

Many professional biologists also have problems with issues of logic, mathematics, statistics and formal structures. We do, however, collect data, test hypotheses and do experiments for a biologically driven mixture of reasons of curiosity (when we are young), compulsion (when we are trying to get a job), duty (when we've got one) and habit (when we're older!). Whatever the motive, all of this activity ought to result in clear and coherent advances in our understanding of nature. The challenge is to ensure that we are not lost in an overwhelming amount of indigestible quantitative information. For years, many of us have been trying to understand a logically rational framework for ensuring that the *biology* stays as the primary focus of the design of sampling programmes and experiments and of the collection, analysis and interpretation of data.

Unfortunately, the relationships between the logical structure of hypotheses, the variability of biological systems and the design of experiments are complex and impose the use of statistical procedures. So, it is inevitable that the foundations of statistical procedures should be laid long before the towering edifices of complex experimental designs and grand biological visions are erected. There is, sadly, compelling evidence that this is not always the order of events in ecology. For example,

Underwood (1981, 1986) and Hurlbert (1984) documented numerous instances of serious widespread problems in the use of statistical procedures and in the logical relationships between the hypotheses proposed and the experiments actually done.

Many of us are inevitably compelled to have to deal with statistical concepts and constraints before we can sensibly plan and interpret the experiments we want or need to do (or have promised grant agencies we would do). This book therefore considers three themes. First, there must be some structure in the logic of experimental procedures. If there are no clear goals in a research programme, there will be no useful results. The logical structure of hypotheses, of tests of hypotheses and the interpretation of the outcome of such tests are considered in an introductory way.

Second, the basic concepts of relevant, simple statistical procedures are considered. This is aimed primarily at those in earlier stages of research careers. For many readers, this will be revisionary rather than visionary, but experience suggests that many readers will welcome some revision of basic statistical concepts. It is always possible to develop greater understanding, to jog the memory or to learn something from a different viewpoint. Of course, like most books discussing statistical themes, the notation of later sections will be completely baffling unless the earlier bits are read, so there is some coercion to revise!

Third, there is the much more interesting topic of the design of real experiments capable of helping us to understand the real world, where natural variation is large. Realism requires that experiments are practical and include appropriate attention to spatial variability (i.e. different outcomes of ecological processes in different places), temporal change (i.e. different patterns and time-courses of biological processes in different seasons, at different ages or at different stages of development of an organism or an assemblage) and complex interactions of space and time. To retain the logic of any test of an hypothesis requires close attention to issues of replication, independence of data and controls for experimental procedures. These make the design of most experiments necessarily complex.

The approach used here is, unashamedly, based on the framework provided by analysis of variance. The general approach of analysis of variance can handle much of the complexity identified in the previous paragraph as a necessity in most experiments. This is true, even if the data collected cannot be made to conform to the requirements of the procedures. So, it forms a very valuable conceptual framework for experimental design. Most ecological experiments will be complex and difficult. There are no obviously visible alternatives that have the same property of helping to translate the logical structures required by complex hypotheses into the experimental layout needed to complete valid tests. Some alternative approaches are not well suited to the complex structures necessary in many ecological experiments (e.g. GLIM – Generalized Linear Modelling – cannot handle complex experimental designs involving mixtures of nested and orthogonal factors; Crawley, 1993).

Some reviewers of an early draft of this book were extremely hostile to the concept that biological (in general) and ecological (in particular) research requires experiments to test hypotheses. They expressed dismissive, but oversimplified, views on the role of hypothesis testing in biology. Nevertheless, they provided no alternative framework in which to evaluate alternative, competing conceptual models about the way some part of the world works. They therefore provided no framework for progress except the view that estimation (i.e. description of observations) would somehow be satisfactory. So, it is worth a clear statement here that experiments have particular roles and these cannot be circumvented by alternative approaches.

Many scientists will continue to do experimental science – progressing by testing hypotheses in a logical framework and using the outcomes of experiments to evaluate theories and hypotheses in an empirical manner. We shall continue to do this despite the views raised by people who object to hypothesis testing. Much of the objection appears to derive from a naïve (or perhaps arrogant?) view that the nature of hypothesis testing is based solely on the use of a probability statement. For example, the article by Perry (1986) has been suggested to me as a clear statement of why statisticians object to tests of hypotheses. Perry (1986) and other authors on the topic (such as Wonnacott, 1987) make a number of cogent arguments concerning the inadequacies of hypothesis testing without proper examination of the data and without regard to the importance of patterns or differences revealed in a set of data. In particular, Perry (1986) stated that 'the primary interest in ecology is to estimate the magnitude of treatment effects' and listed several reasons why tests of hypotheses would be misleading. Yet, he discussed no rationale for wanting or needing to *describe* (i.e. estimate the magnitude of) a particular biological phenomenon, no alternative framework in which a description can be used or interpreted and no understanding of the use of data to evaluate theories. Even Wonnacott (1987), despite considerable antipathy for tests of hypotheses, concluded that there was convergence between the use of confidence limits to provide well-described estimates of differences between means of variables and the use of a probabilistic outcome of a statistical test of the relevant null hypothesis. Wonnacott (1987) further advocated the use of probabilistic tests of hypotheses in situations where there are several parameters to be estimated. As will be seen in this book, such are the norm for ecological investigations – numerous experimental populations are required by the hypotheses or because of the nature of the experimental design. These authors and other supposed critics of testing hypotheses are, however, remarkably coy about why any experimental investigation is being done. I have no quarrel with their insistence that the data should be examined properly instead of dismissed and summarized by the probability associated with a particular test. I am, however, firmly of the view that data are gathered to examine some world-view in order to arrive at a decision in terms of discriminating among various ideas or theories.

The issue that matters is to identify the hypotheses clearly before the experiment is done. Hence, testing hypotheses is the approach to be emphasized here. In my experience, the problem with any aspect of quantitative science and, explicitly, with experimental tests to evaluate models and theories is that the structures in which the data are to be interpreted are often vague, imprecise or opaque. Hence, my insistence that there should be a clear structure. I provide a summary of one – well used and of long standing – to make a framework explicit. People who are not yet clear about their own views and philosophies may find the structure helpful and will see the relationship to empirical data in ways not explicit in many statistical texts. Anyone who disagrees with it should identify for him/herself what alternative framework is preferred. Then, everyone can progress with his/her experiments and may find some of my attempts to clarify issues helpful.

Those who deny the need for any structure, or who believe that quantitative data are gathered for the sole purpose of description, will see no point in a book such as this. Indeed, there would be none for such limited purposes. Gathering data without clear, pre-defined purposes (which I identify as hypotheses to be tested) will continue to be the bane of ecological and environmental science. My response to the view that philosophical or other frameworks are irrelevant to analysis of empirical quantitative data is that statistical treatises abound, but have not yet solved many of the problems biologists have with experimental design, because the nature of the investigation has not been a primary feature of discussions of how to analyse data.

The statistical treatment here is very basic. I am not a statistician and I don't like maths. Some readers will object, for example, to the lack of attention to so-called non-parametric procedures (but they cannot be extended to handle many complex experimental designs). Others will probably object to the lack of treatment of issues such as models for regression. These have been handled better elsewhere.

Other readers would rather that the book identified and explained how to use current computer software that can do the analyses. My response to this is that most software will be superseded and much of it is idiosyncratic. It is better to discuss how the experiments are designed and the data to be gathered, analysed and interpreted, rather than the mechanics of how to do the analyses.

The principle on which this book was written is simple. All statistical procedures must be subservient to the bio- or eco-logical concepts being examined. Hence, the mixture and progression of themes. The three themes (logic, statistics, experimental design) ebb and flow through the book. The aim is to provoke discussion and appraisal of the relevant issues before the experiments are done. Then, more focus can be brought into seeking advice from statisticians.

What biologists, particularly ecologists, do is important for numerous aspects of social welfare, for example for management of human impacts on our environment. How we do what is important is therefore, itself, very important. Science is necessary for the planet. Science must be experimental. So, experimental design is a crucial underpinning to what we do and how we do it. The purpose of this book is to explore improvements to our abilities to do science and to provide some firm, stable stepping-stones through the quagmire of biological complexity.

I have chosen to start with an introduction to statistical notions as they relate to tests of hypotheses. I am not convinced that most biologists of my acquaintance have fared well in statistics courses until they are involved in their own research problems and must deal with data of specific relevance to themselves. Therefore, I have started where my courses to professional and postgraduate biologists start.

I make no apology for treating concepts of planning, analysis and interpretation without reference to some modern ideas in statistical theory and practice. The ecological literature in particular and biological experimentation in general abound with problems of replication, controls, confounding and scale. Getting the fundamentals right seems more valuable than exploring the most advanced concepts. It is, however, the case, that a well-conceived experimental procedure based on realistic fits to fundamental principles will allow more straightforward consultation of expert statisticians and a satisfactory outcome for all concerned.

Any reader who is confident about sampling, simple descriptive statistics, Type I and Type II errors, power, the construction of alternative hypotheses, etc., can usefully proceed from Chapter 2 directly to Chapter 7. In teaching, I have found a useful self-test to be: can you draw a graph of the relationships between each of the sample mean, sample variance and standard error and the size of samples – including the scatter around any trend? If so (and you can check against Figure 5.7), after Chapter 2, proceed to Chapter 7. Otherwise, you will find examination of Chapters 3 to 6 helpful.

2

A framework for investigating biological patterns and processes

2.1 Introduction

Progress in understanding biological phenomena requires a formal procedure to identify what are the phenomena and what constitutes understanding. Unless the steps and sequences of a research programme are clear to others, there will be argument and debate about the problems, the data, their interpretations, etc. This is usually a very unprofitable debate because it is not about our understanding of natural systems, but about our inadequacies in dealing with them. The framework used here is not unique; nor is it widely agreed by others (e.g. Diamond, 1986). It is, however, a logical procedure. Failure to conform to some coherent framework or failure to provide any rational justification for the way some study is done only adds to the confusion already widespread in the science of ecology (Peters, 1991). As will be described below, there is always a need for some logical procedure to identify the most likely explanations for biological phenomena.

The procedure described here is a version of one in widespread use in ecology. This version of it was described in full by Underwood (1990), with detailed ecological examples discussed by Underwood (1991a). It is not important, at this stage, for you to agree with the components and their linkages. What is important is to realize the need for all experimental investigations to be viewed as a step in some sequence of logical steps. Each experimenter must be clear about the sequence so that the conclusions reached from an experiment can be justified in terms of the logical structure in which it was done.

We need some procedure to allow discrimination among alternative explanations (or theories) for observed phenomena (usually biological patterns or other biological observations). 'Experimental' procedures, with which this book is solely concerned, are formal tests of the claims (predictions or hypotheses) derived from alternative explanations. Thus, to interpret them requires clear understanding of the original observations, the various explanations of them, the construction of predictive hypotheses and the logical validity of the tests. If these are all clear, the rest (i.e. the interpretation of the results, the publication of the findings and resultant fame and glory) is relatively easy!

2.2 Observations

Biological research programmes start with observations. We see patterns or phenomena. We are sometimes confused by the variety and scale of our observations and sometimes we are deluded by the things we see, so that we believe certain patterns exist that are not supported by subsequent, more detailed investigation. Ecological observations vary from widespread ones (trees do not extend to the tops of high mountains) to very local ones (there is a great variety of animals in the localized habitat formed by a hole in a tree). Observations may be of very long-term phenomena (deciduous forests in the north-eastern USA have changed in composition of species over the last few thousand years) to very short-term ones (rabbits run away when disturbed).

Whatever the scale, scope or time-course of our observations, the work of a biologist is to provide some convincing account of why these observations were made. This involves several types of explanation, including social and behavioural ones about the observer (e.g. Hanson, 1959; Koyre, 1968; Feyerabend, 1975). Why are particular things observed and others ignored? What 'rules' determine what we see, or choose to describe?

A second type of explanation about observations is to determine whether the particular observations were made because they do, indeed, represent reality. Alternatively, the observer may have been mistaken (see the discussion by Underwood, 1990). These models are about the structure of biological patterns and lead to experimental tests of the patterns to validate or corroborate quantitatively the existence of the claimed observations (see Section 5.10).

Finally, if the observations have been demonstrated to represent real patterns, i.e. to relate to real natural phenomena, the most interesting biological models are relevant. These are models to account for the existence of the phenomena observed. They are explanations of *biological* processes that cause the observed biological patterns. Much of modern biology is concerned with increasingly sophisticated models about processes,

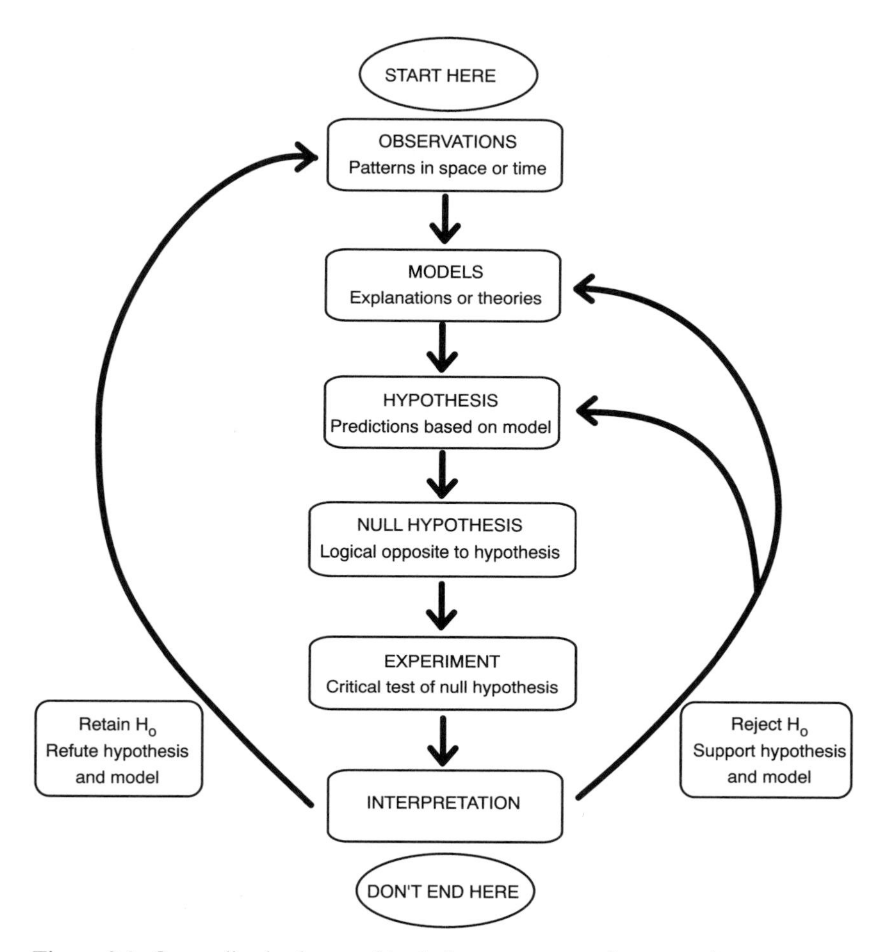

Figure 2.1. Generalized scheme of logical components of a research programme. Relationships are discussed in the text and described in detail by Underwood (1990, 1991a). H_0 is the symbol used to represent the null hypothesis.

although a large proportion of these are purely retrospective and provide little predictive capacity (for a review, see Peters, 1991).

So, a framework for biological research leading to experiments must begin, as in Figure 2.1, with quantitatively validated observations. As an example, consider the observation that a species of tree is distributed only up to a certain height on mountains in some region. At higher levels, there are numerous goats. Further, consider that these observations have been made by many people at numerous times in many places and are therefore unlikely to be illusory (unless mass hysteria has caused widespread delusion!). The observations have been subjected to adequate preliminary tests to demonstrate that they are adequate representations of reality.

2.3 Models, theories, explanations

There are numerous potential explanations for the existence of the upper boundary of distribution of the trees. These can be summarized in five groups.

2.3.1 Models of physiological stress

The physiological and other resource requirements of the trees cannot be met above their upper limit because their resources are absent above a certain level. Detailed forms of this model would include identification of specific resources (e.g. water, inorganic chemicals, soil, etc.) that are missing above the tree-line.

2.3.2 Models based on competition

These propose that the upper limit of the trees is set by competition from other species. The physiological and other resource requirements of the trees cannot be met above their upper limit because some other species or groups of species are occupying, using, pre-empting, or otherwise removing the resources. At higher levels on the mountains, the other species are competitively superior in acquisition of the resources. The other species are, however, either absent from, or less effective at acquiring resources in, the areas where the trees occur. Alternatively, the putative competing species are present lower down the mountain, but in insufficient numbers to acquire all the resources needed by the trees. Therefore, the trees are observed not to extend higher than the lower limit of effective competition from other species.

2.3.3 Grazing models

The resources of the trees *are* available above their current upper limit, but the trees cannot survive there because of the observed grazers (the goats). Any young tree that becomes established in the area above where trees are observed to survive is consumed by grazers. For whatever reason, the goats do not extend, or are less effective, below the upper limit of distribution of the trees.

2.3.4 Models to do with hazards

Here, the trees could survive above their current upper limit were it not for accidents or physical stresses associated with the higher environments. Thus, the resources they require are available and there are insufficient grazers to prevent them from surviving higher on the mountains. Nevertheless, they cannot live there because some feature of the environment above their current limit kills them. Such hazards are less frequent, or are absent, or less intense, at lower levels on the mountain. The feature of the environment that kills trees might be, for example, avalanches during the spring when snow starts to melt. Or, perhaps, winter storms are more severe in the highest parts of the mountains.

Although this might be considered a subset of physiological requirements (if physiological requirements include suitable shelter from storms), it has often been useful to keep accidents and hazards separate from resources in analyses of processes that create or maintain patterns of distribution of animals and plants (Andrewartha & Birch, 1954).

2.3.5 Models of failure of recruitment

It is possible that the trees cannot extend to higher levels because, even though there is no reason for the trees not to survive and do well in the higher areas, their propagules cannot get there. Seeds are not transported to the highest levels of the mountains by wind (or water, which runs downhill from the adults). Other animals that might carry them do not extend to the higher levels.

All of these models share the common definitional feature of a *model* or *theory*. Each *must be capable of explaining the original observations*. In each of the five models listed above, if the model were true, it would explain why the observations were made. Any one of these models could account for the existence of the observed upper limit of the trees.

This is the only rule about models. A valid logical model must be able to account for the observations that it is proposed to explain. This is important, but beyond the theme of this book. As a contrast, consider the competitive model, if it were stated less precisely as: 'some species outcompetes the trees at higher levels'. On its own, this cannot explain the existence of the observed upper boundary of distribution of the trees, because it does not include any statement about why the trees are able to survive lower down. Thus, it does not explain the existence of a *boundary*. As stated previously above, the competition model can account for the observations because it includes some statement about why competition is not also occurring at lower levels on the mountains. Without this, the model can explain only the absence of trees from the highest parts of the mountains, but not the existence of the observed boundary. Such a model is, by definition, useless.

Without exploring all of these further, resulting in a text on plant ecology (which has been done much better by Harper, 1977), it is clear that there are numerous possible and contradictory explanations for the absence of trees above the observed limit.

Furthermore, numerous other models are combinations of the five kinds already specified. Thus, it could be a combination of physiological stresses (perhaps due to the colder temperatures prevailing at higher levels) and competition from more cold-hardy species that explains why the trees are absent above a certain observed limit.

2.4 Numerous competing models

This example demonstrates the most important point. It is extremely unlikely, particularly near the beginning of a research programme, that there will be only one, or even one or two, realistic models that can potentially explain some observed phenomenon. Here, there are five quite distinct types of explanation. Some may be readily eliminated from further contention, but only by some further investigation outside the observations already made. The acquisition of further observations will inevitably therefore constitute some form of test of the validity of these models. Such tests are experiments.

Because there are always going to be alternative proposals to account for most, if not all, biological observations, it is crucial to have some procedure that allows us to discriminate among the models. If we could eliminate any that are known to be wrong, we could make progress by simplifying the situation. If we could find some way to demonstrate that only one of the models was correct, we would have solved the problem of explaining our original observations.

Usually, however, we must consider all the models proposed so far and these may not yet include the true explanation for our observations. If investigated properly, all these models would be shown to be wrong and much more effort would go into trying to propose new ones.

In contrast, sometimes a model is well supported by the available evidence and, in contrast to alternative models that can be shown to be wrong, is concluded to be correct. Yet it may still be wrong, although capable of explaining most or all of what has so far been observed. This is what happened with Newtonian mechanics. The theory (the 'laws' of motion) was, for hundreds of years, successful in explaining and predicting patterns of movement of objects. There was, however, an accumulation of observations that were inconsistent with the predictions of Newton's theories (Chalmers, 1979). Ultimately, this led to a realization of the need for a better (and different) theory that could explain all the observations. Hence the need for Einstein's theory of relativity. Newton's theories had, in fact, always been wrong, but were successful as explanations of the phenomena observed over many years.

We need a framework like that in Figure 2.1 to distinguish among alternative explanations. We need a framework to encourage and promote constant probing of established models so that we do not fall into the trap of accepting them just because they have worked so far.

2.5 Hypotheses, predictions

To discriminate amongst alternative models, different logical hypotheses can be proposed from them. These have only the property that they must be logically derivable from the model and must be about some observation that has not already been made. Otherwise, they are not predictive. If the 'predicted' event were already known to occur, the set of observations being investigated was obviously not described completely. If the predicted event has already been seen, there must already exist the observations to demonstrate it. All of these observations were available at the beginning, so cannot be used as attempts to verify some theory about the system being investigated. So, an hypothesis is some prediction about unknown events based on (derived from) a model to explain the original observations.

It is not necessary to go through all of the hypotheses derivable from the above models. Just consider the one about grazing: the model states that trees are prevented from extending further up the mountain because the goats at higher levels eat them. It is simple to propose that, if goats are kept out of some higher areas, say by shooting them, or by putting up goat-proof fences (if such things exist!) around experimental plots, trees should begin to grow in the grazer-free areas. Above the upper level of distribution of the trees, control areas (i.e. where shots have been fired but no goats killed, or where partial fences have been set up to disrupt the habitat in the same way, but not to keep out the goats) will continue to be free from trees. In contrast, if any of the other models is correct, this proposed difference between experimental and control areas cannot happen, because the seeds will not grow up due to physiological stress, competition, or accidents, or will not arrive due to failure of recruitment. The prediction about the effect of removing goats stems directly from the model and is definitely about observations that have not yet been made (i.e. the prediction is about what will happen when goats are experimentally removed from some places).

Much more biology would need to be included in the development of the hypothesis, to ensure that the experiment was done in the appropriate season and with appropriate timing. There is, for example, no point in doing the experiment only over a period of time when seedlings cannot grow because of the weather. Such an experiment would indicate that the hypothesis was wrong, even though trees might have grown in the absence of goats if they had been examined during the appropriate season.

Considerable thought must go into the definition of how long the experimental areas would need to be observed. Thought must be given to how old a seedling or sapling would need to be able to grow before it could be concluded that the upper limit of the trees had been extended upwards by removing goats. If adult trees must be established for 20 years to demonstrate that competition, hazards and physiological stresses are not going to kill young trees, this must be known in advance so that the experiment is monitored for 20 years.

The answer to this and many other questions would presumably be available from making sure that the original observations were very precisely stated (e.g. the age of trees extending to the upper limit could be noted to help determine how old an experimental tree would have to be in order to be old enough to show an extension of distribution).

Hypotheses could equally be developed for each of the other models, although some of them would be logistically much more difficult to imagine testing. Some would be very difficult except in the most general terms, because there is not enough information about potential resources, competitors and physical hazards to be able to make coherent and specific statements. Nevertheless, in principle, hypotheses can be constructed from any model that has been advanced to explain phenomena.

2.6 Null hypotheses

Having established an hypothetical prediction about what would happen under some stated (experimental) conditions as a consequence of a particular model, it is essential to test it. This involves creating the conditions required by the hypothesis (removing the goats as specified) and observing the results. There is, however, a problem in logic to do with the nature of proof. Proving that some prediction has come true requires what is known as an inductive argument (Popper, 1968). Technically, proof requires every possible observation to be available. Or it requires the assumption that what happens in all possible circumstances not actually observed can be inferred from the cases actually available in the experiment. The details of this are not necessary here (for a more complete account, see Underwood, 1990, and the references therein). As a result, however, it is not practicable and, in many cases, not possible to *prove* that something has occurred. There is always the possibility that insufficient evidence has been obtained and the contrary observation has not yet shown up.

Therefore, it has become customary to use a logical piece of sleight-of-hand to help us past this problem. The procedure is known as a falsificationist procedure (after Popper, 1968), so-called because no attempt is made to prove anything. Instead, something is to be disproved, which is obviously more straightforward. Once disproved, there is no danger that some later observations will alter your conclusion. There is no need to assume anything about observations that have not yet been made. Once an observation has been made that does disprove something, that thing stays disproven. Something cannot become non-disproved!

The trick consists of constructing what is termed the logical null hypothesis – the opposite statement to the hypothesis – and then attempting to disprove it. The argument goes that disproof of the null hypothesis, by definition, leaves the original hypothesis as the only alternative. Thus, falsification of the null hypothesis is a procedure not fraught with the problems of proof. When it occurs, disproof is unarguable. Here, the logical hypothesis is in the form that seeds will germinate and trees will grow more frequently (i.e. the numbers will be greater than zero) in areas where goats are removed than in control areas with goats (where the numbers of trees growing will be zero). If the number in the control areas is not zero, the trees have also extended up the mountain in regions where there are goats, so something must be wrong with the observations, not the model and its hypothesis.

The null hypothesis consists of all the alternatives to the hypothesis. Thus, the null hypothesis is:

1. that an equal number of trees will grow in areas where there are goats and in areas where goats are removed, or
2. there will be fewer trees where there are no goats than where there are goats.

In the case discussed here, point (2) is impossible because there are already supposed to be zero trees where there are goats. In this example, therefore, the hypothesis is that removal of goats will lead to the appearance of trees in experimental plots. The null hypothesis is that removal of goats will make no difference to the lack of trees – experimental areas will continue to have no trees, just like control plots where goats are present.

We shall return to null hypotheses at various points in this book. This is a logical null hypothesis – the opposite statement to that contained in a logical hypothesis. This can be tested by doing the experiment.

2.7 Experiments and their interpretation

Once a logical null hypothesis has been proposed, it should be tested by an experiment. The conditions specified in the hypothesis – the conditions under which some prediction(s) have been made – must be created in order to determine whether the predictions of the null hypothesis can be shown to be wrong. Any coherent and rational method of achieving this is a valid experimental test of a null hypothesis. For some studies, the test would be to measure some variable or to ascertain a pattern of dispersion of distribution or sizes of organisms. These are sometimes called *mensurative* experiments (e.g. Hurlbert, 1984) to contrast them with *manipulative* or intrusive experiments. The latter involve changing (manipulating) components of the habitat or the organisms to examine whether they change or behave or recover, etc., as predicted. This type of experiment is commonly used to test hypotheses about the *processes* causing patterns in nature.

The distinction between types of experiment is a distraction. It does not matter whether the system is measured or manipulated and measured. Each is appropriate for different circumstances and different models. What matters is that the experiment is clearly related to the need to test a logically defined null hypothesis. The experiment must then be done so that it preserves the logical structure and allows a logical conclusion. The design of experiments is therefore a crucial matter – to avoid illogicalities creeping in when the experiment is done. These are discussed extensively throughout this book (see particularly Chapter 3).

Once the experiment is done, it should be possible to arrive at one of only two possible conclusions (Figure 2.1). One is that the null hypothesis is wrong (trees did grow where goats were removed at the higher levels on the mountain, but not in nearby control areas with goats). Rejection of the null hypothesis therefore supports the hypothesis and thereby the model on which it was based. The only possibility left once the null hypothesis has been rejected is the hypothesis. So, the hypothesis is corroborated and the model is therefore supported (its predictions have been corroborated).

In contrast, the only other outcome is that the null hypothesis must be retained (trees did not grow in the areas from which goats were removed, or grew in similar numbers in those areas and in the grazed controls). In these cases, the hypothesis is clearly wrong. It is disproven and therefore the original model on which it was based must also be wrong (its predictions did not turn out to be correct). Thus, the model is falsified.

Note that, in this scheme of things, either outcome leads to disproof of something. If the null hypothesis is disproven, the hypothesis is supported by falsification of its alternatives. If the null hypothesis is retained, the hypothesis and model are themselves falsified.

So, the only appropriate outcome of an experiment or test of a logically derived null hypothesis is a disproof of either the null hypothesis or the hypothesis. There are many problems with this in practice (for discussions, see Connor & Simberloff, 1986; Underwood, 1990). One problem is that careless thought may have allowed proposal of an hypothesis that could equally well have been derived from one (or more) other model(s). There are also, as will be developed later, problems of determining whether or not the null hypothesis should, in fact, be retained or rejected (see the later discussion of Type I and Type II errors and see Underwood, 1990). There are also cases (such as goodness-of-fit procedures) where the logical null hypothesis can neither be proposed nor tested because of the form of the hypothesis and the needs of statistical procedures (see Section 5.10).

Nevertheless, there is considerable logical structure in this set of steps and the falsificationist procedure is an extraordinarily powerful tool in the pursuit of biological knowledge.

2.8 What to do next?

Once a conclusion has been reached to reject or retain the null hypothesis, what should happen next? One possibility is to publish the results as some advance in our understanding – either because some model has been rejected, simplifying subsequent research programmes, or because it has been supported, suggesting an explanation for some observed phenomenon. This not only advances the career(s) of the research team, but also lets others know that some progress had been made and should (although sadly does not always!) alter their perception of the way the world works. Publication also provides about the only public accounting of what the research grant or time of the investigators was actually spent doing. So, it is an obligation, not a choice. It also carries with it certain social components to do with the very definition of a piece of science or a scientist (Merton, 1977).

This is, however, not the end-point of the science. Having established that a certain model's predictive hypothesis has been supported by rejection of the relevant null hypothesis, the model should be probed again. This time, the model should be made more general or more specific and subject to increased harshness and rigour of test. Thus, the model should now be made more general to account for areas where the conditions at the tops of mountains are otherwise different. From the model it should be possible to predict what will happen to the upper limit of the trees in other circumstances, such as where there are naturally no goats, or where there are variations in habitat at the tops of mountains so that there are refuges from goats. The model could be made more refined by increasing its content to include an explanation of why the goats do not graze as efficiently or effectively at lower levels on the mountain, i.e. why are there any trees at all? Or, the model could be refined to include explanations of the variations in height of the upper limit of the trees in response to different densities or rates of grazing by the goats. There are numerous possibilities. Complacency that the problem had been solved is misplaced. The model may not be correct – it just may be correct for the one prediction made from it. Further examination of the model should be an on-going occupation, so that the model does not linger around in the literature even though inadequately examined.

Far too often, conclusions are reached from single, small-scale experiments in only one site, with little replication (for a plea to revisit experimental sites, see Connell & Sousa, 1983). In the logical framework advocated here, models should be more widely examined and should not just be accepted as part of the background lore of old ecology – handed on from one generation of ecologists to another.

Alternatively, the model may have been thrown out because the null hypothesis was retained and the hypothesis was clearly wrong. Under these circumstances, there is no current explanation for the original observations. This is not a nihilist state to be avoided (*sensu* Paine, 1991), but rather a state in which a new model is required. The experiment has revealed that the previous model is no good. This is clearly an advance brought about by the elimination of falsehood rather than by the discovery of truth. It is still an advance.

Now, however, there are the original observations and the new ones obtained during the experiment (there are no trees above the natural upper limit observed, but, if goats are removed, there are still no trees above the upper limit). A model must now be proposed to account for the combined set of observations. You may fall back on one of the other models originally proposed above (but, if possible, you should have been doing experiments simultaneously to attempt to examine the predictions of all the models you can think of). At this point, ingenuity, discussion and advice, plus more consultation of existing literature may help with the proposal of a new model to explain all of the available observations.

Either way, the next step will eventually lead to new hypotheses and experimental tests. Thus, in Figure 2.1 there is no way out of the procedure once you have started it, until you die or change research fields. This is comforting in terms of eventual longevity of employment. It also reminds us that the models we have and routinely use should always be subjected to new improved tests so that we do not become complacent about them and assume that failure to disprove them is the same thing as demonstrating that they are correct. Any new alternative model that is proposed to account for the available observations should be accorded the same serious consideration as some established one (usually known as a paradigm; Kuhn, 1970).

Re-examination, novel testing, more rigorous experimental analysis are part of the framework of investigation. No single study – whether it rejects or retains some null hypothesis – is sufficient to declare a problem solved.

2.9 Measurements, gathering data and a logical structure

The preceding account of a logical framework for investigating phenomena in nature is oversimplified in one important aspect. It starts with observations; yet observations are, in fact, rarely made as if no previous biology or ecology were known. This has led to severe criticism of observations (e.g. Hanson, 1959; Koyre, 1968; Popper, 1968; Feyerabend, 1975) because they are usually 'theory-laden' – i.e. they are taken only because some previous model, theory, world-view, paradigm (call it what you will) makes you incapable of making different observations. We see what we are trained to see or what we think should be there. This notion would obviously cause considerable bias in the types of observation we make, leading us to operate entirely within a constrained view of the world. The limitations of this would be largely unrecognized by ourselves or others.

Biologists and ecologists often describe their observations and measurements as though there were no underlying theory. Data are gathered 'to find out something'. Thus, rates of growth of lizards may be measured because 'it will be useful to understand why they reach different adult sizes in different years'. At one simple level, this reasoning must be true, but this is not usually what is happening. Usually, the observations are being collected to demonstrate the correctness of some unstated world-view (e.g. Dayton, 1979). The notion that gathering observations will somehow, on its own, increase rational understanding is an old one (Bacon, 1620) that has not really stood up well to the ravages of close scrutiny over time.

The unthinking, uncritical gathering of data certainly reinforces the view that scientists continually operate with theory-laden observations, but are unaware of the difficulties imposed by not recognizing the underlying theories. Consider the following simple examples concerning numbers of eggs in nests of a particular species of bird. Assume that the species has already been studied in some part of the world. This is not an unreasonable assumption. Even if this species has not been studied elsewhere, other birds or other organisms have. We already have theories about breeding of birds. Some of these have been well established for years (Lack, 1954). No-one can realistically claim to be making observations about birds without considering the underlying theories. It would, in fact, be ignorant and unprofessional not to find out about the existing theories and models at some point in the investigation. The previous theories usually create expectations about the measurements we might make. The measurements and observations are useful to evaluate the theories. The two processes go together, inexorably and intimately.

A biologist studying the bird in southern Europe may wish to measure the numbers of eggs per nest because 'it will be useful', 'it will advance our knowledge', 'it will tell us something'. These statements do not clarify what is meant by 'useful', nor what is the 'something' that we might be told. The only 'advance' will be the acquisition of new knowledge – the number of eggs per nest.

There will always be some descriptive studies in which all that is required is a measurement. For example, in taxonomy, the description of a new species of fly may include measurements of the lengths of the wings of the individuals being described. But, in most areas of biology, observations are almost invariably made in relation to some existing theory and should be collected to test explicit predictions of such theory. Then, observations and measurements will rarely be made as if gathering them were the only point of the exercise.

2.10 A consideration: why are you measuring things?

Despite claims to the contrary, most biologists measure things in order to determine whether the measurements fit with preconceptions or predictions or hypotheses. The decision about what to measure is rarely taken in a vacuum. A biologist in southern Europe may know from the literature in other parts of the world, where measurements have already been taken, that the continued persistence of populations of a particular species is dependent on them producing, on average, three eggs per nest. Any measurement of the number of eggs per nest in southern Europe must, in some form, be a test of whether such a condition also holds there. As described above, the observations are that the species lives in southern Europe in habitats similar to those it occupies elsewhere.

The model being proposed is that the birds in southern Europe are like those elsewhere. This is, of course, part of a more general model that members of a species, even in different geographic areas, have similar ecologies and life-histories. If this model is correct, then it is logical to propose that the average number of eggs you would find per nest is three. Measurement of the actual number would be needed to test this hypothesis. Note that the measurements being made are now clearly identified as being useful to test the stated hypothesis. The measurements are no longer simply being made because they can be and because some biologist wishes to make them.

This procedure (gathering data to determine whether they conform to some stated quantitative prediction) has been called a mensurative experiment (Hurlbert, 1984), but is more commonly known as a 'goodness-of-fit' test. A goodness-of-fit test is one where the value of the defined variable is specified by the hypothesis. The number of eggs is then measured in order to determine how well it conforms to, or 'fits', the value specified from previous theory.

In more complex biological studies, the measurement is not singular and is not made as part of a goodness-of-fit test but in order to compare it with another measurement (or several other measurements). For example, birds of a particular species may be nesting in trees in thick woodland or in trees in scattered groups. Previous ecological research and knowledge of the natural history of the birds may suggest that the eggs in nests in dense woodland are more likely to fall prey to a predator than are those in scattered clumps of trees (perhaps because the predators will not cross open ground between trees and need more food than is normally available in a small patch). In contrast, where predators are absent, there are greater supplies of food for the birds in dense woodland and they should therefore tend to lay more eggs. Here, the theory or model is that the pattern of reproductive success is different in the two habitats and it is proposed that this model applies to the present populations of birds. No information is yet available about the number of eggs per nest in either habitat, but the model leads to the predictive hypothesis that there should be fewer or more eggs per nest in nests in scattered clumps of trees than in nests in dense woodland, depending on whether predators are or are not active in the area studied. Measurements would then be taken in the two populations of nests in order to determine whether the prediction was likely to be true or false. Here, the model and hypothesis do not specify the magnitude of the measurements; they specify a pattern of difference between the measurements. The measurements are used to test whether the predicted pattern has been found.

2.11 Conclusion: a plea for more thought

Biological science in general and ecology in particular would be well served and make more progress if the underlying models (world-views, paradigms, biases, constraints, etc.) were explicit. Then it would be possible to determine their influences on what and how measurements are taken (see e.g. Underwood & Denley, 1984). The procedure described in this chapter identifies the components. These should be identifiable in any study. Then we could identify the relationships between the logical structure, the observations made and the data collected, the use of the data in tests of their match to hypothetical predictions and the conclusions reached.

From this account, it is clear that an experiment is a test of a logically stated null hypothesis. A proper test involves having only one of two possible outcomes – rejection or retention of the null hypothesis and therefore support for or rejection of the model from which it was derived. It is an important part of the whole process that the experiment is planned, done, analysed and interpreted without causing new forms of illogicality and misleading interpretation because of faults in its design and execution. Preservation of the logical structure in which an experiment is done must be an important and crucial goal of research. By being explicit about the steps in the procedures, it should be easier to learn and understand where things go wrong and thereby to avoid them. One major objective of this book is to avoid some of the common traps of poor experimental design, so that there is a greater chance of the hard work of experiments in biology resulting in logical rather than nonsensical conclusions.

If the framework proposed here is unacceptable, you have some obligation to develop another one. The alternative you choose must have some structured underpinning for the roles of experiments, evidence and conclusions from experiments and the relationship between experimentally acquired data and their interpretation in relation to the original biological problem or question. Without this, it is not possible for anyone else to determine the purpose of a study. It is also not possible to evaluate how effective a study has been in terms of the relationships between acquisition of data and reliability of any conclusions or interpretations about them.

Above all, unless the framework in which data are to be interpreted is clearly identified, there are numerous dangers in analysing the data. There is a potentially vast difference between a coherent and unarguably correct statistical analysis of data and a coherent and unarguably correct interpretation of data in relation to some specific biological problem.

3

Populations, frequency distributions and samples

3.1 Introduction

The preceding account of a logical framework for investigating biological observations is not as easy to use as might be hoped. The major problem is that neither the original observations nor the results of tests are unambiguous. Biological observations (in whatever form the data are collected) are not constant, fixed truths, but vary from place to place and from time to time. So, of course, do measurements in other sciences, such as chemistry and physics. Even where intrinsic variability is small or absent (i.e. the measurement is of some physical or chemical constant), both the machinery used to make measurements and the observers themselves are not constant.

3.2 Variability in measurements

There are intrinsic and extrinsic reasons why biological observations are variable. Intrinsic reasons include, first, fundamental properties of biological systems. For example, the size of an animal, the rate of growth of a plant, the speed of movement of a predator are all subject to genetic variability from one individual to another. Thus, unless the individuals are identical in all genetically determined processes of growth or speed (respectively), there is no possibility that their sizes (or speeds) will be identical.

Second, in addition to innate properties of systems, there are intrinsic processes causing variability. For example, the diameters of oocytes in a gonad are influenced by stresses, pressures, rates of cellular division, position in the gonad, etc. Not all cells (even if genetically identical) will finish up the same size. Even genetically identical individuals will not grow at the same rate (and therefore finish up the same size) because they do not encounter nor process identical amounts and quality of food throughout their lives.

The numbers of seeds or barnacles in different parts of some habitat are not identical because of the processes that disperse them from their parental sources. Seeds blown around by the wind and barnacle larvae transported along a coastline by currents, waves and tides cannot possibly arrive in their final destinations in uniform numbers.

Over and above the variability caused by the intrinsic properties and processes affecting natural systems, there are extrinsic causes of variation. Methods of measurement introduce new sorts of variability. The rate of movement of animals cannot easily be measured without error. The amount of chlorophyll per leaf is not measurable without using machinery which requires processes of extraction and preparation that are not constant in their effects on different leaves.

The combined effects of intrinsic and extrinsic causes of variation mean that no measurement we can take or event we can observe will be a constant, fixed representation of the true value. Measuring something on several individuals will result in different values of the measurement. Measuring something on the same individual several times will often result in different values of the measurement because of variations in the methods of measuring and changes induced or naturally occurring in the individuals during the period when the measurements are made.

Apart from variability introduced by systems of measurement, the intrinsic and extrinsic processes operating biologically to create variability are, of course, the whole point in studying the subject. Dealing with variability is therefore the core of our collective discipline of science.

3.3 Observations and measurements as frequency distributions

The simplest way to portray the variability of observations is to use a graphical plot of the frequency distribution of the variable being measured. As an example, consider a very simple distribution of the number of eggs per nest of a particular species of bird in a large forest. There are 730 nests in the forest of which, at the time examined, 90 have no eggs, 164 have one egg each, 209 have two eggs each and so on up to 12 nests that each have five eggs. No nests have more than five eggs. The variation in number of eggs per nest ranges from zero to five eggs. The frequency distribution is shown in Figure 3.1a, which depicts the frequency (as number) of nests with each number of eggs (the x-axis). It is a *discrete* distribution, meaning that there are no fractional

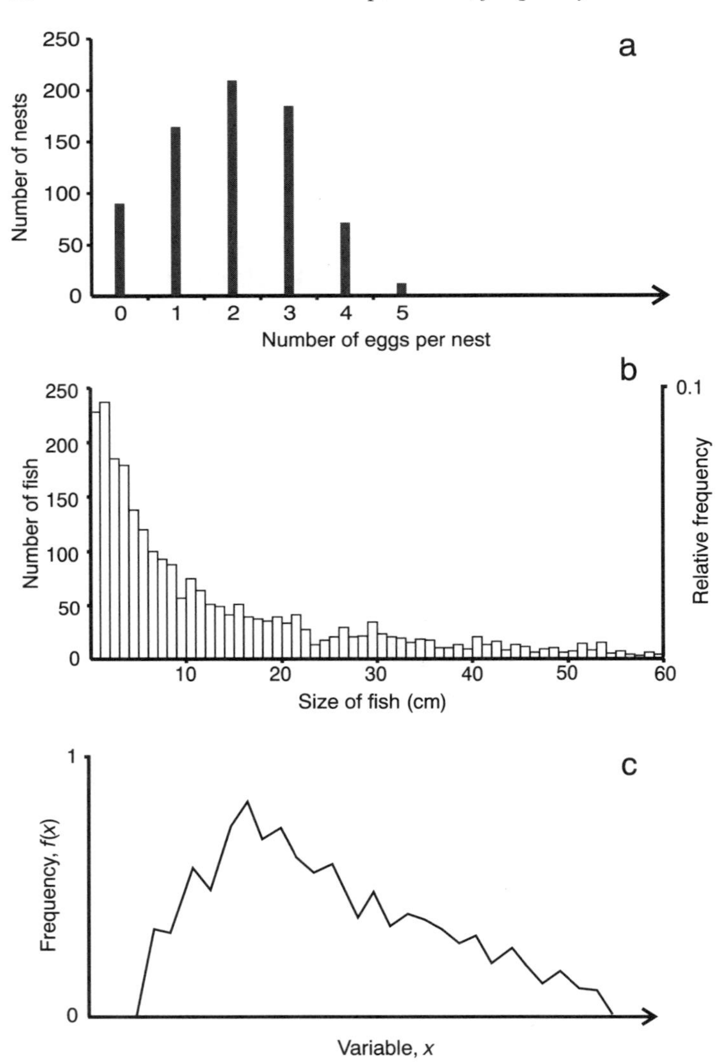

Figure 3.1. Frequency distributions. (a) The number of eggs per nest in a population of 730 nests in a particular location. (b) The sizes of fish in a population of 3000 individuals. (c) The general case of a variable X in a population of values, plotted as the proportion of the population having each possible value of X.

numbers of eggs (there cannot, for example, be 4.2 eggs in a nest). Only whole eggs (and therefore integers) are possible. The reasons for the different numbers occurring in various nests are many and complex (including the genetical make-up and size of the parents; the position of the nest in the forest relative to the weather, availability of food, etc.; the presence, numbers and behaviour of predators that remove eggs, and so on).

A second example is a *continuous* distribution (Figure 3.1b) – the lengths of fish of a species in an aquaculture pond. There are 3000 fish and all of them are measured (there being more money in aquaculture than in many other types of investigation!). Obviously, lengths can vary by infinitesimally small increments, limited only by the smallest subdivisions on the ruler used to measure them. It is thus absurd to attempt to plot the data in the same way as for nests, because there is now essentially an infinite number of sizes along the x-axis. Instead, the data are grouped into intervals of sizes, arbitrarily chosen for convenience of plotting the graph (Figure 3.1b). The result is the histogram of frequency (as number) of fish found to have lengths in the various size-classes chosen.

The general representation of a frequency distribution is also shown in Figure 3.1c, which shows the features of this way of describing observations. The range of values of the variable, X, is plotted with the frequency, $f(X)$, with which each value of X occurs. For convenience, these may be grouped into a histogram (as for the sizes of fish in Figure 3.1b), where the frequencies are the numbers of fish in each centimetre interval. Or the data may be plotted continuously as for the generalized hypothetical data in Figure 3.1c.

Usually, frequency distributions are considered with the frequency plotted as the relative frequency, i.e. the proportion of the entire distribution having each value of the variable (X), rather than as the number of members of the population. This is illustrated for the sizes of fish by the right-hand axis (Figure 3.1b). This is a convenient method of portrayal because it is equivalent to the entire population's relative frequencies summing to one for any population, which makes mathematical manipulations of the data simpler. It also makes relative frequency distributions from several populations directly comparable, regardless of the sizes of the populations. Regardless of how many nests there are in different forests, the frequencies as proportions are directly comparable.

So, to deal with quantitative biological observations and measurements is to deal with frequency distributions of variables. There is no choice!

3.4 Defining the population to be observed

The population to be measured or observed must be defined very carefully. This may be simple. For example, measurements of the chlorophyll content of the leaves may be needed for a population of leaves in experimental pots in a greenhouse. It may, however, be more widespread, such as in the population of leaves in a forest. Or it may need more information because what is wanted is the chlorophyll content of leaves above 2 m above the ground on trees of more than 1 m diameter in north-east facing parts of a forest. The precision of the definition is important for any use of the information gathered.

Suppose that, for whatever reason, a measurement is needed of the size of fish in a population of fish in a particular bay. The fish can be caught only in traps or nets, such that the smallest and fastest-swimming ones will not be caught. The measurements are therefore no longer going to be from the defined population. What is needed is some means of catching them that will obtain the small and fast ones. Perhaps a poison could be used?

If the measurement were, however, supposed to be on the population of fish available by commercially used methods of catching, then the measurements taken would have been appropriate for that population. The difficulty comes when the problem requires measurements on one population (defined in a particular way) and the measurements are actually possible only in (or are thoughtlessly taken from) a different population.

The population being observed must therefore be defined very precisely to avoid confusion and misunderstanding. If the measurements are taken from a population different from that specified in the model being tested, they will not be useful for the problem being addressed. It is, at the least, misleading to describe the data as though they came from a population different from the one actually measured.

Consider the numbers of eggs in the birds' nests. Where the birds are must be defined (here, in a particular forest). When the nests are to be examined is also important. Presumably a measure is wanted in some portion of the breeding season (say, for the sake of example, in April). Otherwise, there are unlikely to be any eggs and the measure would be zero. This would be a perfectly acceptable measure for a time of the year irrelevant to the model.

When the relevant and appropriate population cannot be observed (as with the fish), the measurement should not be taken. Clearly, if it makes no difference what population is being observed, any observation about anything will do. So, if the population relevant to the null hypothesis is not the one being measured, there is no point in the measurement.

If measurements must be taken on a population that is *not* the one specified by the model, hypothesis and null hypothesis, the definition of what was measured becomes very important. A complete description of what was measured is mandatory, particularly for those ecologists thousands of kilometres away reading the results, who are going to be most misled without the proper information. So, if the inappropriate population is being measured, say so.

You then run the risk that no-one will want to know about the study, because the population being measured is uninteresting or irrelevant. At least, though, no-one will have been misled by claims that something different has been measured.

Now, the whole notion of relevance of the population being measured depends on the purpose to which the information is being put. The population being measured must be defined. Making observations and measuring things is inevitably in relation to some stated hypothesis and null hypothesis. Otherwise, the uses of the information are not clear. Without the uses being clear, it is not realistic to attempt to define the population of possible relevant measurements or observations. Before any data are gathered, there must be a definition of a population. This must be based on some clear representation of the relevant null hypothesis and therefore the underlying hypothesis, model and previous observations. If there is no such basis, any type of data can be gathered, in any manner and interpreted in any way to mean anything chosen by the observer.

Such Baconian data-gathering (in the hope that once information is gathered it will turn out to be useful for something) is not a profitable or useful exercise (see the entertaining account of this in Medawar, 1969). Despite the widespread prevalence of unstructured data-gathering, none of it is scientific, using the definitions in Chapter 2. Defining the context in which data are to be interpreted is crucial for ensuring that appropriate data are gathered.

Once the relevant population is defined, it must be observed. There is no known procedure for observing a population that cannot be made available. This often means that the populations in which measurements are made are those in which the measurements can be made. Such populations may not be as large scale or as grand as some ecological and biological theories might require and this obviously sets boundaries on the sorts of question or problem that can be solved using quantitative procedures. It also means that the populations being examined must always be specified as completely as possible. If the females, or the old individuals, or the ones that bite or are too tall to climb are not included in the measurements, they are obviously not included in the population being measured. If they should be (as defined by the models and hypotheses being examined), the measurements are useless and more care must be taken to ensure that those individuals are included. If, however, these particularly difficult subsets do not have to be included with respect to the hypotheses proposed, the population can and should be defined to omit certain types of organism or parts of the habitat.

3.5 The need for samples

Having defined the relevant and appropriate population to be observed, we now hit the next snag. It is extremely unlikely that the *entire* population can actually be observed. In the case of determining the chlorophyll content of leaves in a forest, there are so many leaves that it is extremely improbable that you would have the money or the time to take measurements from all of them. In the case of the birds' nests, you may, in theory, be able to find, climb up to and count all the nests, but, even at 20 per day, this would take 37 days. During this lengthy period you would expect many birds to hatch (thus removing eggs from nests) and, anyway, the period is too long for the defined population (April – see earlier). Even if you could get help to speed things up, you may have ethical problems in causing disturbance to every nest in an area, if your visits were to cause a deleterious change in the nesting behaviour of the parents.

Thus, the problem becomes that of being able to measure some variable in a population without knowing the values of the entire population. The problem, as will be developed below, is to determine what sort of subset or sample of the population, out of those elements actually available to you, will best measure the defined variable on behalf of the entire population. Measurements from samples are known as *statistics* or *statistical estimates*. Thus, measurements require statistics.

The simplest biological study (making a simple quantitative observation or measurement) is therefore intimately related to the procedures for designing a sampling programme that will provide an appropriate measurement from a specified variable population of potential measurements, not all of which can be examined.

3.6 The location parameter

The previous sections have introduced the nature of a measurement of a variable. It is clear that to measure something (a variable), you need an estimate of the magnitude of that variable for all elements in the population of interest. The location parameter is the most useful of such measures. At this point it is worth considering the terms variable and parameter in more detail. A variable, as used here so far, is some measurable quantity that differs in magnitude from one member to another in a population, or defined set of such measurements. All the values of the variable in the entire population can be described graphically as a frequency distribution. The shape and range and all other properties of this frequency distribution are determined by its *moments*. These are mathematical constants that define a frequency distribution. For many distributions (but not all), the moments – the constants defining a distribution – are equal to the *parameters* (or are simple functions of the parameters) of the distribution (as in the next few sections, particularly Sections 3.7, 3.8 and 3.12). Parameters are constants that define moments; moments define the frequency distribution (see Winer *et al.*, 1991). In this sense, we are concerned with parameters and their estimation, because these often have empirical interpretative use in biology.

Confusing the two terms variable and parameter has not aided ecological understanding. Phrases such as 'the parameter salinity was measured in the river . . . ' serve only to confuse things. The variable 'salinity' can be measured, i.e. its magnitude observed. The parameters of a distribution of measures of salinity can be measured only if the entire population of all relevant measurements is made. This seems inherently unlikely and, almost certainly, impossible. Common misusage of 'parameter' is not improving comprehension. We *need* the two separate terms 'variable' and 'parameter'.

The location parameter can be defined operationally as the value which is, on average, as close as possible to all values of variables in the population. Therefore, by definition, it must be in the range of all values in the population and must represent the magnitudes of all members of the population. Consider a population of variables. Each member (X) of the population (1, 2, etc.) is a particular value (X_1, X_2, etc.); any one can be denoted as X_i and there are N members of the population:

$$X_1, X_2 \ldots X_i \ldots X_{N-1}, X_N \tag{3.1}$$

The location parameter (say, arbitrarily for illustration, L) is the number which is 'closest' to all members of the population (i.e. most similar to all X_i values). This can be described by the following equation:

$$\sum_{i=1}^{N}(X_i - L) = 0 \tag{3.2}$$

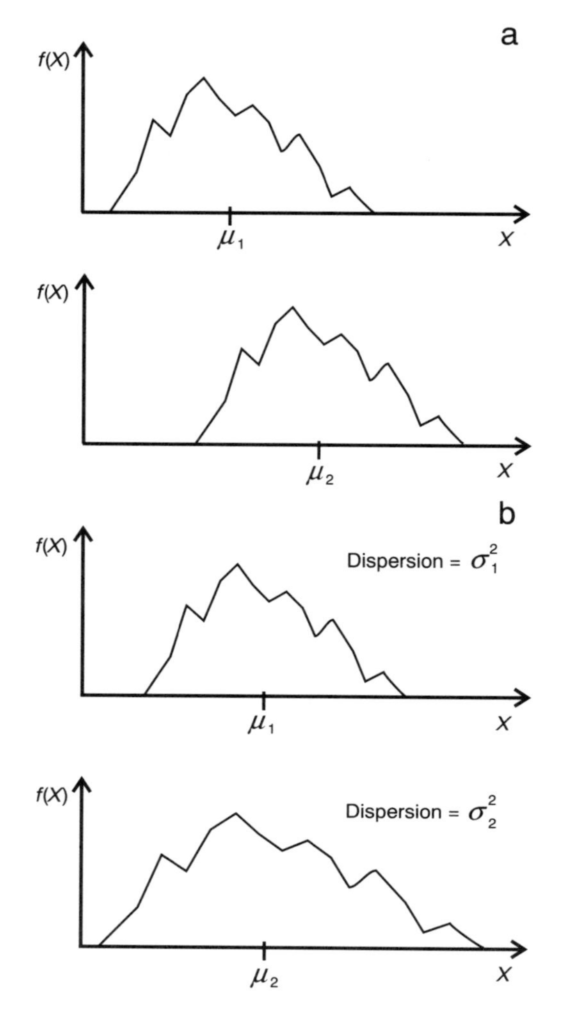

Figure 3.2. Effects of differences in location and dispersion parameters. (a) Two populations of similar frequency distributions, but different location parameters (μ_2 is larger than μ_1). (b) Two populations with similar frequency distributions and the same location parameter, but different dispersion parameters (σ_2^2 is larger than σ_1^2).

From this, it follows directly that:

$$\sum_{i=1}^{N} X_i = NL \tag{3.3}$$

L is usually described as the arithmetic average or mean value of the variable in the population and is traditionally denoted by μ. Therefore, replacing L by μ:

$$\mu = \sum_{i=1}^{N} X_i/N \tag{3.4}$$

This roundabout way of introducing μ as the location parameter will be useful later. The mean of the population encapsulates its location (along the abscissa of the frequency distribution). It has appropriate features of a measurement that can describe the magnitude of the variable of interest. For populations with different magnitudes, the location parameters must be different and therefore, as a single measure, the mean differs (Figure 3.2a). Knowing the mean 'summarizes' the population. For many populations (but not all), the mean is a parameter.

3.7 Sample estimate of the location parameter

So, how would the location parameter (the mean) of a population be estimated, given that the entire population cannot be measured? The mean of an unbiased, representative sample must be an accurate measure of the mean of the entire population. The only really representative samples are those that have the same frequency distribution as that of the population being sampled. If the sample reflects the population properly (i.e. has the same frequency distribution and is therefore representative), its location must be that of the population.

For a population of N elements, from which a representative sample of n elements is measured:

$$\mu = \sum_{i=1}^{N} X_i/N \tag{3.5}$$

is the location parameter (the mean) and:

$$\bar{X} = \sum_{i=1}^{n} X_i/n \tag{3.6}$$

is an unbiased estimate of μ and is the sample mean.

Now it is clear what will happen (formally) if a sample is not representative of the population. If, for example, larger individuals in a population are more likely to be sampled, $\bar{X} > \mu$. The location parameter will be overestimated because of the bias (inaccuracy) of the sample.

Some specific unrepresentative samples may still produce accurate, unbiased estimates of the mean of a population (Table 3.1), but, in

Table 3.1. An unrepresentative sample ($n = 16$) that happens, by chance, to estimate accurately the mean of a population ($N = 100$; $\mu = 3$)

Variable (X)	1	2	3	4	5	6
Frequency in population	20	20	30	10	10	10

$$\mu = \frac{\sum^{N} X}{N} = \frac{((20 \times 1) + (20 \times 2) + (30 \times 3) + (10 \times 4) + (10 \times 5) + (10 \times 6))}{(20 + 20 + 30 + 10 + 10 + 10)} = 3$$

Frequency in sample	4	4	0	4	4	0

$$\bar{X} = \frac{\sum^{n} X}{n} = \frac{((4 \times 1) + (4 \times 2) + (4 \times 4) + (4 \times 5))}{(4 + 4 + 4 + 4)} = 3$$

Table 3.2. Measures of location of a frequency distribution

Consider a population of a variable (X), with a total of $N = 11$ members and values as: 1, 2, 2, 2, 2, 3, 4, 5, 5, 8, 10

Average value	=	Mean	=	4
Central value	=	Median	=	3
Most frequent value	=	Mode	=	2

general, the mean of a sample will not estimate the mean of a population unless the sample is unbiased.

There are other measures of location used in experimental biology (Table 3.2). Two are the *median* and the *mode*. The median is the 'central' value of a variable. It is the magnitude that exceeds and is exceeded by the magnitudes of half the population. It is thus the symmetrical centre of a frequency distribution. Some statistical procedures use the median as the parameter to be estimated. This is particularly common in tests using ranked data. In distributions that are symmetrical around the mean, the median and the mean are equal. The mode is the most frequent value of a variable in a population. This may or may not be the same as either the mean or the median depending on how other parameters in the population change the shape of the frequency distribution.

3.8 The dispersion parameter

The second most important parameter that dictates the shape or form of a frequency distribution is the dispersion parameter. This determines the degree to which the population is scattered or dispersed around its central location (or mean). To illustrate this, consider two populations with the same frequency distribution, except for dispersion (Figure 3.2b). The one with the larger dispersion parameter is more scattered. Practically, this means that measurements in a population with a large dispersion parameter will be much more variable than in a population that is less dispersed.

There are two reasons for needing information about the dispersion of a population. First, to describe a population (as part of a sampling exercise where description of the biological system is the purpose of the study), its dispersion is an important attribute. A much clearer picture of the population of interest can be gained from knowledge of its dispersion than from the simple bald fact of its location.

The second reason is much more important. As will be developed later, in many cases of practical sampling of biologically interesting populations, an estimate of the dispersion parameter may be used to measure the precision of a sample estimate of the mean of a population. Thus, estimation of the population's dispersion parameter will provide a measure of how close a measured sample mean is to an unknown population mean.

There are several possible measures of the dispersion parameter. For example, the range of the population (smallest to largest value) is one measure. It is not, however, a practically useful one, because it is difficult to sample reliably. Only one member of the population has the largest value of any measurable quantity; another single member has the smallest value. They are not particularly likely to be contained in a sample (unless the sample is very large compared to the whole population). The next largest, third largest, etc., and next smallest, third smallest, etc., members of a population may not give an accurate measure of the range.

The most commonly used parameter for dispersion is the population's variance, σ^2. It is preferred for several reasons, largely because its mathematical properties are known in terms of statistical theory (some of this will be used later). The arithmetic definition of a population's variance is not intuitively interpretable. A logical choice of measure of dispersion would be a measure of how far each element in a population is from its location (i.e. its mean). In a widely dispersed population, these distances (or deviations) from the mean would be large compared to a less dispersed population. So, for a population, the average deviation from the mean is:

$$\sum_{i=1}^{N}(X_i - \mu)/N \tag{3.7}$$

and this, in theory, should form the basis for measuring variance.

A little thought makes it clear, however, that this measure cannot be useful because of the definition of location. The mean is the value that is, on average, zero deviation from all the members of a population (see Equation 3.2 earlier). Thus, by definition of the mean, the average deviation from the mean must be zero.

This attempted definition of dispersion is, however, useful as the basis for construction of an index of the magnitude of deviations from the mean in a population. Where the population's dispersion is large (and the population is widely scattered over a broad range of values of the variable), the deviations of many individual members of the population from the population mean are large. If these deviations are *squared*, they are still large relative to those in a population with smaller dispersion, but they no longer add up to zero. The negative values $(X_i - \mu) < 0$ no longer cancel the positive ones $(X_i - \mu) > 0$ because all the squared values are positive. Thus, the variance, σ^2, can be defined as the average *squared* deviation from the mean:

$$\sigma^2 = \sum_{i=1}^{N}(X_i - \mu)^2/N \tag{3.8}$$

This increases as deviations become larger and is no longer constrained to be zero by the definition of the mean.

The variance of a population is, however, in a different scale from the mean of the measurements. For example, in a population of lengths of beaks of birds, the variable (X_i) beak-length is in millimetres, as is the mean (μ). The variance is in square millimetres – a measure of area, not length. To solve this problem, by putting the measure of variance in the same scale as the mean and the variable, we can use the standard deviation – the square root of the variance. σ is also a measure of dispersion but is in the same scale as the variable being measured.

3.9 Sample estimate of the dispersion parameter

As with the location parameter, common sense suggests that the variance of a population can be estimated from a sample by use of the formula that defines the variance. This turns out not to be exactly correct and the appropriate formula for the sample estimate of variance, s^2, is:

$$s^2 = \sum_{i=1}^{n}(X_i - \bar{X})^2/(n - 1) \tag{3.9}$$

for a sample of n elements and for which the sample mean is $\bar{X}$. This is similar to the definition of σ^2 (Equation 3.8), replacing μ by its estimate $\bar{X}$. The difference is the divisor, $(n - 1)$, which is known as the *degrees of freedom*. It is instructive to attempt an explanation of the concept of degrees of freedom, because they are important for much of what follows in later sections of this book.

A demonstration of why the divisor must be changed from N for the population's variance to $(n - 1)$ for the sample estimate must await development of the definition of the 'standard error'. This will be done later (Table 5.1). Here, an operational ('arm-waving') definition of degrees of freedom is given. This is not a mathematically, nor statistically correct view – so don't use it on your statistically-trained friends. It does, however, work!

3.10 Degrees of freedom

The degrees of freedom for a calculation on a set of numbers is the number of elements in the set (i.e. how many numbers there are) minus the number of different (independent) things you must know about the set in order to complete the calculation. An illustration of how this works is given in Table 3.3. Consider a set of $n = 5$ numbers. Obviously, in the absence of any information about them, all five are 'free' to be any value between minus infinity and plus infinity. Suppose, however, you are also told that the sum of the set is 20. Now, only four of the numbers are 'free' and the last one is fixed by your knowledge of the total (Table 3.3). Hence, there are four degrees of freedom. Note that it does not matter which four numbers are discovered first, the final one is always known from the total.

Similarly, if there is a set (or population) of $n = 5$ numbers that have a mean, $\mu = 5$ and a variance, $\sigma^2 = 5$, only three of the numbers are free (there are three degrees of freedom). Once three members of the set are known, the other two are inevitable, given the mean and variance (Table 3.3).

In the same way, in any calculation about a set of numbers, the number of degrees of freedom is limited by the number of properties of the set (the number of constraints on the set) you must already know in order to complete the calculation.

In the calculation of the sample estimate (s^2) of a population's variance (σ^2), there are n numbers in the set (the sample), of which the sample mean ($\bar{X}$) must be known. Thus, there are $(n - 1)$ degrees of freedom.

Table 3.3. Demonstration of the operational meaning of degrees of freedom

1. There is a set of $n = 5$ numbers, with a total of 20.

Suppose you discover that the first 4 numbers are: 1, −17, 8, −4. To make the total = 20, the fifth number is not "free". It must be 12. If, instead, the first four numbers in the set were: 1, −17, 12, 8, to make the total = 20, the fifth number must be −4.

2. A set of $n = 5$ numbers has $\mu = 5$ and $\sigma^2 = 10$.

If you are given three members of the set (say 3, 4 and 7), the last two are not free – they are known from the mean and variance. They are calculable from the following equations (where X_1 and X_2 are the two unknown members of the set):

$\mu = (3 + 4 + 7 + X_1 + X_2)/5 = 5$
therefore $X_1 + X_2 = 11$
$X_2 = 11 - X_1$
$\sigma^2 = ((3 - \mu)^2 + (4 - \mu)^2 + (7 - \mu)^2 + (X_1 - \mu)^2 + (X_2 - \mu)^2)/5 = 10$
therefore $(X_1 - 5)^2 + (X_2 - 5)^2 = 41$

Substituting X_2:
$(X_1 - 5)^2 + (6 - X_1)^2 = 41$
therefore $2X_1^2 - 22X_1 + 20 = 0$
$X_1 = 1$ or 10
therefore $X_2 = 10$ or 1
The last two members of the set are 1 and 10.
$\mu = (1 + 3 + 4 + 7 + 10)/5 = 5$
$\sigma^2 = ((1 - 5)^2 + (3 - 5)^2 + (4 - 5)^2 + (7 - 5)^2 + (10 - 5)^2)/5 = 10$

3.11 Representative sampling and accuracy of samples

Note that the requirement to take a representative sample is not the same as the need to take a *random* sample. A random sample is one in which every member of the whole population has an equal and independent chance of being in the sample. Random samples will, *on average*, be representative. Random samples are not, however, necessarily representative in any one particular case (Figure 3.3). This is one reason why random samples are common and why procedures to acquire them are so important. Furthermore, random sampling is generally considered to be more likely to lead to independence of the data within a sample – an important topic covered in more detail in Chapter 7. The estimate of any parameter calculated from a random sample will be unbiased – which means that the average expected value of the estimates from numerous samples is exactly the parameter being estimated.

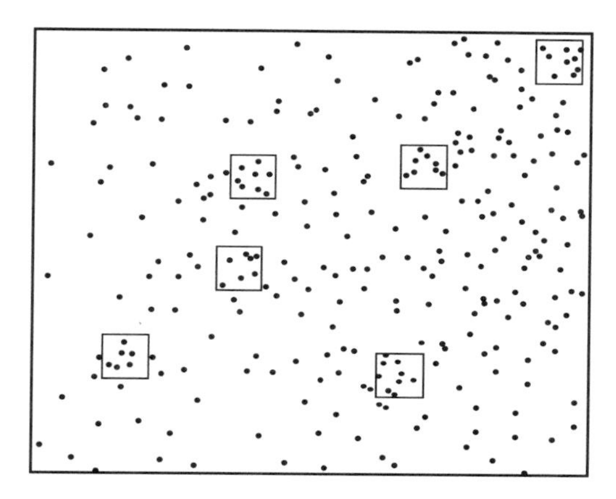

Figure 3.3. Random sampling with quadrats, placed in a field by randomly chosen co-ordinates. By chance, the sample is unrepresentative because most of the quadrats have landed in patches of dense plants.

Why are simple random samples not always ideal? Consider the case illustrated in Figure 3.3. To estimate the density of animals, a sample of quadrats was randomly placed across a field. It is possible that this set of quadrats might provide an accurate representation of the actual frequency distribution of densities per quadrat across the whole field (i.e. across the entire population of quadrats). In fact, however, this sample provides an overestimate of the mean density and an underestimate of the variance (Figure 3.3). This has happened because, by chance, the small but randomly chosen number of quadrats all happened to land in patches of large density. If we knew that this had occurred, those quadrats would not be used (because we would know that they were not representative). Even though we do not know what is the pattern of dispersion of the animals, or whether there is, in fact, a problem with these quadrats, we would normally assume that the animals are not likely to be uniformly spread over the area. Much empirical ecological work has demonstrated the existence of gradients and patchiness in the distributions of animals and plants. If we had a map of densities across the field, we would reject an, ever-so-random, sample of quadrats which had a spatial arrangement like that in Figure 3.3.

The problem of non-representative samples is that estimates calculated from them are biased. They are inaccurate; they do not estimate correctly the underlying parameters of the distribution being measured. The degree to which a sample is biased is the degree to which it over- or underestimates the true value of the parameters being investigated. It is necessary to do whatever can be done to ensure that samples are representative.

So, why is there so much emphasis on random sampling? The reason is that there are many sources of bias, including conscious and unconscious biases introduced by the observer or sampler. Random samples are not subject to some of these biases. As an example of conscious bias, consider what happens on a cold, rainy day during an undergraduate students' field excursion in an ecological course. The task is to count and identify the flowering plants in quadrats (1 m × 1 m) on a wind-swept hillside. Each student is to count and identify the plants in one quadrat placed in the area. A student throws the quadrat, apparently at random, across the field, walks to it and discovers it is covered by hundreds of small and differently coloured plants of numerous species. Many students will be very tempted to kick the quadrat a few metres to a relatively sparse patch of plants of few species. The rewards are less work and a quicker departure from inclement weather. The sample obtained by such conscious bias is, of course, one with fewer plants on average than occur in the population of quadrats. Placing quadrats in random positions *and then sampling those quadrats* ensures that the placement of the sampled quadrats cannot be influenced by the plants in them. The positions of the quadrats are determined by random numbers and not by properties of the spots where the quadrats land.

An example of unconscious (i.e. not deliberate) bias is selecting a sample of five fish from a tank containing 100 fish (the population being sampled to estimate mean rate of swimming under some defined laboratory conditions). It is possible (it may even be very likely) that the speed of swimming is inversely related to catchability in a small dip-net (i.e. the faster fish are harder to catch). If the fish are sampled by chasing and catching them one after another with a small dip-net, their average speed is going to underestimate that of the whole population, i.e. you have the five slowest fish! Here, the bias in sampling would lead to inaccurate (under) estimates of the mean speed. If, however, the fish were individually tagged with a number, or were all caught and placed in individually numbered containers, an unbiased sample could easily be obtained. Random numbers could be used to pick the individual fish (or containers), thus ensuring that the individuals sampled were chosen with no regard to their rate of swimming – which could then be measured objectively.

Another example, which is more complicated and demonstrates some of the subtlety of biases that can creep into sampling programmes, was described in detail by Pielou (1969). It concerns attempts to estimate the mean diameter of trees (say, the diameter at 1 m above ground). Suppose

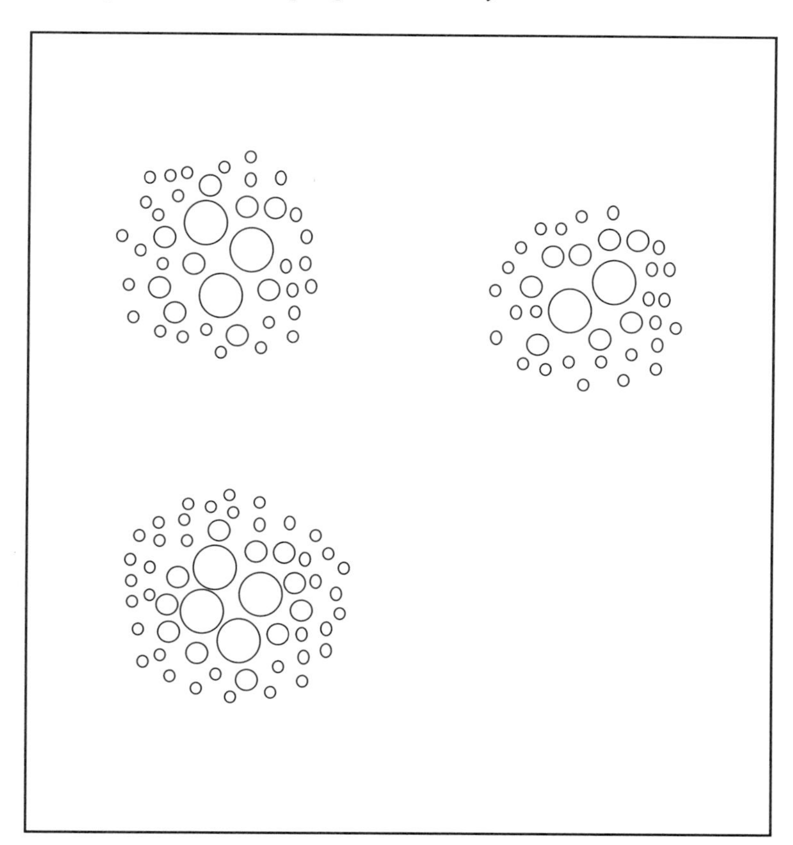

Figure 3.4. An aerial view of trees to demonstrate non-representative sampling to measure their diameters. If a sample of trees is chosen by picking the nearest tree to each of a set of randomly chosen points, the trees are not representatively sampled. Most of them will be the smaller trees on the outside edges of clumps.

that an objective sample of trees is to be measured by choosing the nearest tree to each of several randomly placed points in the area to be examined (Figure 3.4). If the trees are in clumps (as is often the case) and the clumps are formed by vegetative reproduction or by seeds that are dropped near to a parental tree, then smaller and thinner (because younger) trees will tend to be on the outside of clumps. Now, because clumps occupy a relatively small proportion of the total area sampled (as in Figure 3.4), randomly placed points are far more likely to be outside than inside clumps. Thus, the nearest trees to the random points are generally the smaller ones. Large trees, in the centres of clumps, have very little chance of being in a sample. They have much less chance of being included in a sample than should be the case, even allowing for the much greater abundance of the smaller trees. So sampling by this method started out in a conscious attempt to be objective – but finished up being biased in favour of smaller trees (thereby underestimating the mean diameter). In this case, a safe method of getting a random sample would be to number all the trees in the area to be investigated. Then random numbers could be used to pick the sample of trees to measure.

Using random numbers to pick samples is very easy if the whole population of individual entities to be sampled is available to you. You number the individuals ($1 \ldots N$) and then pick the ones you want or need in your sample ($1 \ldots n$) by n different choices of random numbers in the range 1 to N, i.e. discarding numbers that have already been chosen (otherwise those individuals have a greater chance of appearing in the sample). Most calculators and all computers have the capacity to generate pseudo-random numbers (approximately random numbers in the interval 0 to 1). It is a simple matter to transform these to digits in the range 1 to N, as required.

If the population to be sampled is very large, it may prove impracticable to number all the individuals. So, for example, sampling fish in a fish-market containing thousands of specimens or the number of florets on stalks of plants scattered over a large field may prove problematic. There are far too many individuals to number them all. Some form of 'two-stage' sampling is the answer. The fish could be divided into arbitrarily sized groups (of say 50 fish). Then n groups would be chosen at random and one fish randomly chosen out of the 50 fish in each sampled group. An unbiased sample will result if every fish in a group has an equal chance of being picked and every group of 50 fish has an equal chance of being picked.

For the plants, a sample could probably best be chosen by picking n randomly placed small quadrats in the field. In each quadrat, the plants can be numbered and an individual chosen at random. This will be an appropriate procedure to generate an unbiased sample, provided that every quadrat in the field is equally likely to be chosen and that every plant in any quadrat has an equal chance of being chosen. Both conditions are probably fulfilled by using random procedures to pick quadrats and then individuals. Plants will not have an equal chance of being chosen if some quadrats have very large and others have very small numbers of plants (the former are less likely to be picked in their quadrat than are the latter). So, again, it may be better to contemplate grouping quadrats or plants to ensure that similar numbers (and therefore probabilities) prevail in each quadrat or group of quadrats examined.

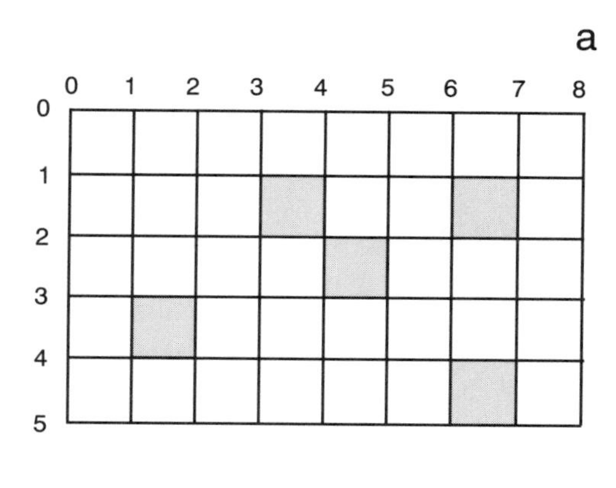

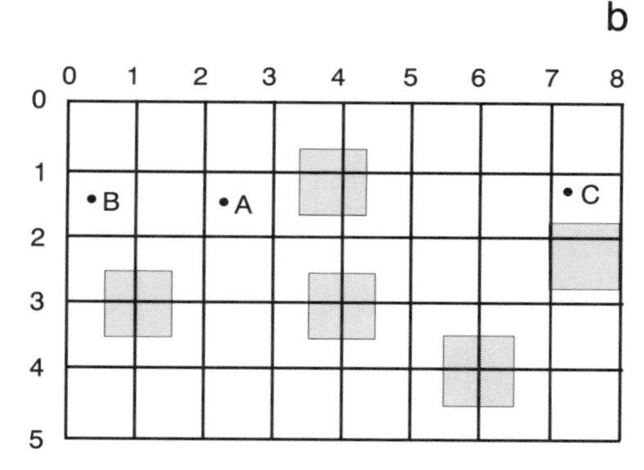

Figure 3.5. Random (representative) sampling of a specified study-site of sides 8 m × 5 m. Quadrats (1 m × 1 m) are placed by co-ordinates measured from the top left-hand corner. (a) Five (shaded) quadrats of a total of 40 are chosen by random co-ordinates at 1 m spacing. (b) Quadrats are sited at 0.1 m intervals. Points A, B and C are discussed in the text.

These are relatively straightforward procedures. There are, however, traps in attempting to find representative samples using random numbers. Consider trying to pick a sample of quadrats in a garden so that the density of plants per unit area can be estimated for the garden as a whole. There are two commonly used procedures. In one, the population of quadrats is sampled by choosing *n* quadrats at random. The simplest method would be to pick a random number to represent a position along one side of the study area and another random number to pick a position along a side at right angles. In Figure 3.5a, a sample of quadrats would be picked by choosing a random number in the range 1 to 8 for the long axis of the area and a random number in the range 1 to 5 for the short axis. These co-ordinates define a particular quadrat in the study area. n such quadrats would be chosen.

The alternative method consists of taking a random distance along each of the two sides of the area and using this as the position for a particular corner of a quadrat to be sampled. This does not lead to an unbiased sample, as illustrated in Figure 3.5b. If the quadrats are 1 m × 1 m and, for illustration, the left-hand top corners of quadrats can be placed at 0.1 m intervals from the top left-hand corner (i.e. random numbers between 0 and 80 and between 0 and 50 are used for the long and short sides of the area, respectively), then position A has more chance of being included in a sampled quadrat than is the case for points B and C. Point A (at 2.15 m from the left-hand edge of the area) can be included in any quadrat starting at 1.2, 1.3, 1.4 . . . 2.0, 2.1 m from the left-hand edge of the area, i.e. 10 intervals of 0.1 m can be chosen on the long axis. In contrast, point B, at 0.45 m from the left-hand edge can be included only in quadrats starting at five intervals (0, 0.1, 0.2, 0.3, 0.4). Point C, at 7.45 m from the left-hand edge can be only in six quadrats starting at six intervals, 6.5, 6.6, 6.7, 6.8, 6.9, 7.0, because a quadrat starting at 7.1 from the left-hand edge will be partially outside the area to be studied and would have to be discarded.

There is no solution to this. A border of width of one quadrat could be excluded from the sampling and quadrats chosen from the remaining area as before. This moves only the area that is underrepresented in quadrats. Only the previous method provides an equal probability of sampling every part of the area. Obviously, great care must be taken to ensure that an apparently random sample is not inadvertently biased.

Wherever there is sufficient information to do better, simple random samples are not the most appropriate samples. What is required is representative sampling and any information available that will increase the likelihood of getting representative samples should be used to get them. Random sampling is appropriate where there are no alternatives or where biases are otherwise difficult to avoid.

3.12 Other useful parameters

3.12.1 Skewness

It is traditional in most fields of biology and ecology to be concerned with estimates of populations' means and variances. There are, however, other parameters of frequency distributions that have great potential for

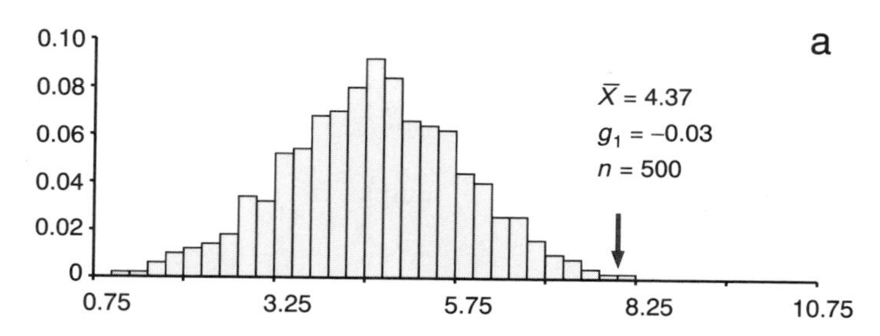

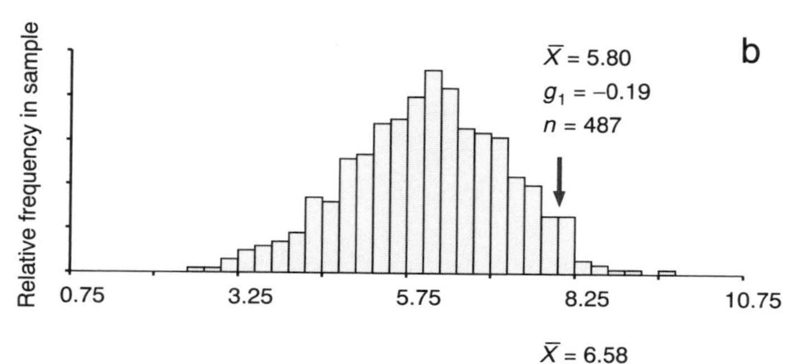

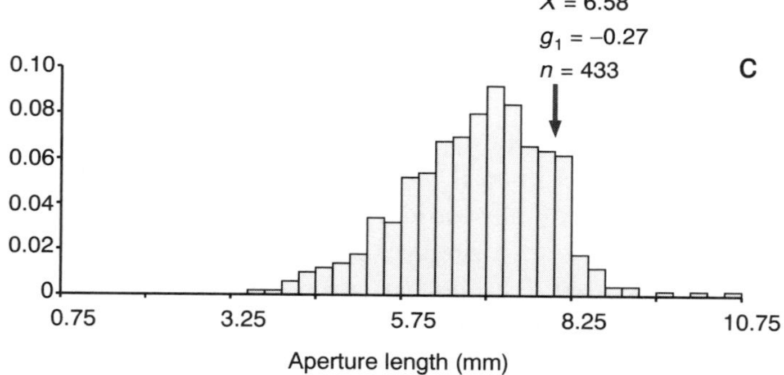

Figure 3.6. Skewness in a population of barnacles as they grow and are consumed by a predator. (a), (b) and (c) are the sizes (as lengths of apertures) at three successive times of sampling, in a sample initially of 500 barnacles. The sample mean ($\bar{X}$), skewness (g_1) and the number of surviving barnacles are shown for each time. The arrow indicates the minimal size of barnacle eaten by the predators.

providing useful information (and sometimes for testing hypotheses) about ecological processes.

For example, the skewness of a frequency distribution is a measure of its asymmetry around its central location (as illustrated in Figure 3.6).

Snedecor & Cochran (1989) described tests for magnitude of skewness in samples from populations compared with that expected for normal distributions. The tests use the statistic g_1 (the coefficient of skewness):

$$g_1 = \frac{\text{Estimate of skewness}}{\text{Estimate of variance}^{3/2}} = \frac{m_3}{(\sqrt{m_2})^3} \tag{3.10}$$

m_3 is an estimate of skewness in a sample of size n of variates $(X_1, X_2 \ldots X_i \ldots X_n)$:

$$m_3 = \sum_{i=1}^{n} (X_i - \bar{X})^3 / n \tag{3.11}$$

m_2 is an estimate or index of variance, dividing squared deviations of the sampled values from their mean by n instead of $(n - 1)$, which cancels the n in Equation 3.11, so that g_1 is not affected by size of sample:

$$m_2 = \sum_{i=1}^{n} (X_i - \bar{X})^2 / n = (n - 1)s^2 / n \tag{3.12}$$

The coefficient of skewness is scaled (by division by $m_2^{3/2}$) so that it is dimensionless. Snedecor & Cochran (1989) provided tables for tests of magnitudes of skewness.

Here is an ecological example of the potential value of studying skewness of a population. Imagine a population of small individuals that become subject to predation when they grow to a certain size. Connell (1970, 1975), among others, has discussed the selection by predatory whelks (snails) of barnacles once the barnacles reach a minimal size. Assume, for the purposes of illustration, that growth of barnacles is fairly similar for all animals of a similar size (it almost certainly is not in the real world) and that juvenile barnacles smaller than the sizes consumed by whelks are normally distributed (a shape of frequency distribution very commonly encountered in biology; see Chapter 4). Now, as the barnacles grow, some of those that first reach the sizes vulnerable to predation will be killed by whelks (Figure 3.6b). Mortality of slightly larger barnacles will be faster than that of smaller ones until all barnacles have grown large enough to be eaten.

Under these circumstances, the frequency distribution of sizes of barnacles will become asymmetrical (there will be fewer individuals in those larger, above average, size classes). The skewness of the population will therefore increase (Figure 3.6c). At first, skewness will become negative because there is an excess of barnacles smaller than the mean and there are more negative deviations of individual sizes from the mean $(X_i - \bar{X})$; see Equation 3.11 above. When the barnacles grow larger and more of the population becomes vulnerable to predation, the mean will become smaller than the mode and skewness will become positive.

3.12.2 Kurtosis

Another 'shape' parameter of frequency distributions is the kurtosis parameter. It determines the 'peakiness' of a distribution near its central mode. Thus, if a population is markedly *leptokurtotic*, it has a very peaked, overcentralized distribution. If it is *platykurtotic*, there are relatively few values near the mean compared, say, with a normal distribution.

Kurtosis would be a useful attribute to examine in ecological studies where, for example, different processes affect survival of relatively small compared to relatively large individuals in an age- or size-cohort. For example, consider, yet again, a cohort of barnacles that are normally distributed in sizes, with a mean size in the range consumed by predators. Then a group of predators arrive in the habitat. The predators (whelks) choose to eat animals over a minimal size (as in Figure 3.7), but cannot consume individuals larger than the size they can handle (for discussions of handling time by whelks, see Connell, 1970; Moran, 1985; Fairweather & Underwood, 1983). As the whelks consume their prey, there is a decrease in the number of barnacles of average sizes, but the largest and smallest barnacles are unaffected. Kurtosis declines and the shape of the distribution of sizes is flattened in its centre (Figure 3.7b and c compared with Figure 3.7a).

Again, Snedecor & Cochran (1989) provided statistical tables to test whether kurtosis is different from what would be expected in a normal distribution (see once again the definition and some uses of a normal distribution in Chapter 4). They used the test statistic g_2, a scaled coefficient of kurtosis that is dimensionless. In a normal distribution, the ratio of kurtosis to variance squared $(\sigma^2)^2$ is 3. Thus, in g_2, 3 is subtracted from the scaled coefficient, so that the estimate is zero for a normal distribution. Positive values indicate increased peakiness towards the centre of the distributions (leptokurtosis). Values less than zero indicate platykurtosis (flatter distributions)

$$g_2 = \frac{\text{Estimate of kurtosis}}{(\text{Estimate of variance})^2} - 3 = \frac{m_4}{m_2^2} - 3 \qquad (3.13)$$

m_4 is an estimate of kurtosis in a sample of n values of X_i:

$$m_4 = \sum_{i=1}^{n}(X_i - \bar{X})^4/n \qquad (3.14)$$

and m_2 is the estimate of variance given previously $((n-1)s^2/n)$.

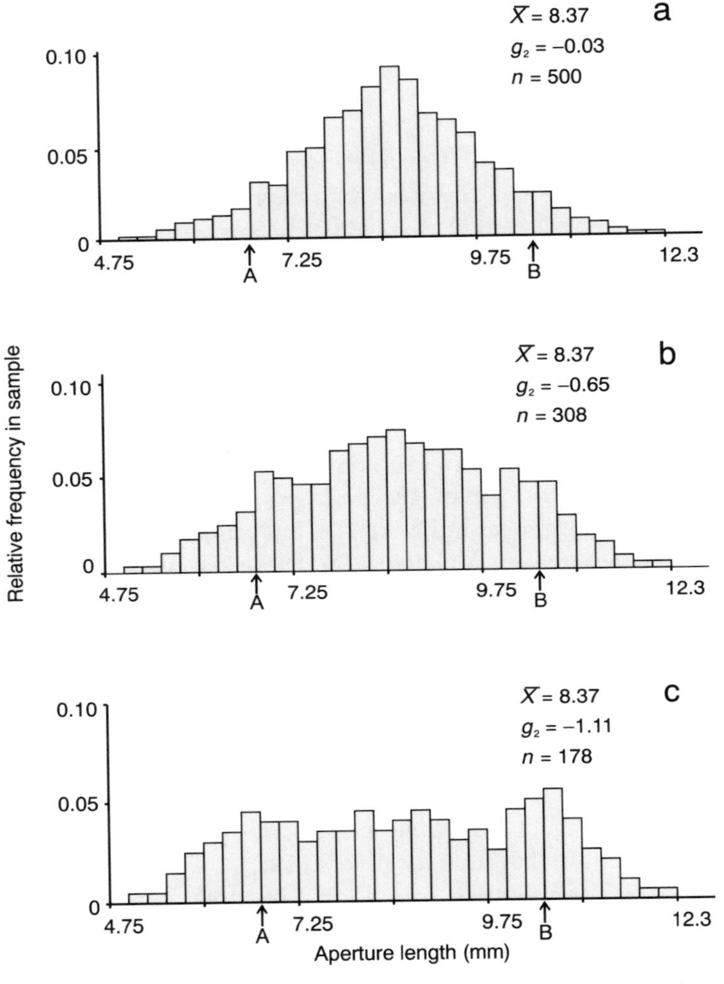

Figure 3.7. Kurtosis in a population of barnacles as they grow and are consumed by a predator. (a), (b) and (c) are the sizes (as lengths of apertures) at three successive times of sampling, in a sample initially of 500 barnacles. The sample mean ($\bar{X}$), kurtosis (g_2) and the number of surviving barnacles are shown for each time. The minimal and maximal sizes of barnacle eaten by whelks are shown as A and B, respectively.

The tables in Snedecor & Cochran (1989) cover only samples larger than 200. For smaller samples, there are useful statistical tables in Geary (1936).

In ecological studies of such processes as predation, competition and effects of physical stresses, determinations of kurtosis through time would provide indications of such things as differential mortality at different sizes. For example, relatively small individuals might be more subject to mortality due to physical factors. Relatively large individuals may be more deleteriously affected by shortages of food. Kurtosis should increase in magnitude if these two processes operate to reduce survival of relatively large and relatively small individuals.

At the same time, examination of skewness would be a method for determining the relative importance of two processes that operate at different ends of a frequency distribution. In the example discussed, skewness will increase if, for example, mortality due to physical stresses acting on small individuals is faster than mortality due to competition affecting larger animals.

Uses of these higher-order parameters are rarely encountered in the ecological literature partly because estimates of skewness and kurtosis require large samples (but partly because ecologists are so mean-minded!).

4

Statistical tests of null hypotheses

4.1 Why a statistical test?

From the earlier account of the logic of experimentation, it is not clear why we might need statistical tests. There could well be null hypotheses that require very straightforward, invariant observations to cause their rejection. For example, the null hypothesis 'There are no squirrels in this forest' can be rejected if someone sees a squirrel. The observations, model and hypothesis leading to this being the relevant null hypothesis are, however, hard to imagine. Most teachers of biology have long since learned not to make such categorically definitive statements. In my experience, stating that there are no crabs in a rock-pool leads almost immediately and universally to the unmistakable scuttling of (at least) one crab across the pool!

Our observations are, far more commonly, not ones that lead us to null hypotheses of such absolutes as 'no entities present'. Usually, we are concerned with quantitative measurements. As explained above, these are rarely taken on such small, manageable populations that we can measure every unit in the population about which we are hypothesizing. We must therefore deal with samples. These can provide only statistical estimates of the required parameters of the population. They are subject to sampling error and are variable and inexact.

Consequently, any statement about them or conclusions drawn from them are themselves under the influence of the uncertainty inherent in the measurements. Dealing with this uncertainty involves trying to calculate how uncertain our conclusions are. Thus, we must estimate the probabilities associated with the relationships between our samples and the populations from which they are taken. We must estimate how likely it is, under the circumstances required by our hypotheses and null hypotheses, that we would obtain a sample with the mean, variance, frequency distribution, etc., that we got.

We are therefore dealing with probabilistic statements, rather than absolute conclusions. These are based on statistical estimates, rather than known values of parameters, of specified populations. As a result, we need statistical tests to help us to decide whether to reject or retain our null hypotheses. We have no simple, black-and-white, absolute data to make up our minds for us.

4.2 An example using coins

One of the simplest ways to illustrate the components and logic of a statistical test is to use a very simple example. A normal (so-called 'fair') coin, when tossed, should be equally likely to come down heads or tails (on one side or the other; on the 'obverse' or 'reverse' to use the technical jargon). Discounting the rare occasions when it lands on its edge (and therefore does not come down on either side), this is a reasonable expectation. Suppose you wish to examine this proposition – that a fair coin has a 50 : 50 (1 : 1) chance of landing on either side. For example, coins can be doctored to make them more likely to come down on one side than on the other, allowing their owner to win money on bets (purely, of course, to add to research grants). You might be betting on a coin. Forget why or that you don't gamble and just accept the example. Much statistical theory is developed for games of chance (i.e. games played in bars for money). This may reveal a great deal about the working habits of statisticians.

Suppose, before you bet, someone suggests that your opponent is cheating, by using a doctored coin. The observer believes this because he or she has seen the coin come down very often on one side – more than the observer thinks is likely as a result of chance. You therefore need to test the coin.

Obviously, tossing the coin once will not tell you anything. It must come down one way or the other, whether it is fair or doctored. Therefore, it will. But, what if you toss it twice? There are now four possible sequences of outcomes (Heads, Heads; or Heads, Tails; or Tails, Heads; or Tails, Tails). For a fair coin, the chances of two Heads or two Tails in a row are each 0.25 ($\frac{1}{4}$); the chances of one Head and one Tail are 0.5 ($\frac{1}{2}$). A doctored coin should be more likely to come down Heads then Heads or Tails then Tails depending on which direction it has been fiddled. Of course, with only two tosses of the coin, it is not easy to see how you could tell the difference between a doctored coin, which comes down twice the same and a fair coin which comes down twice the same with a

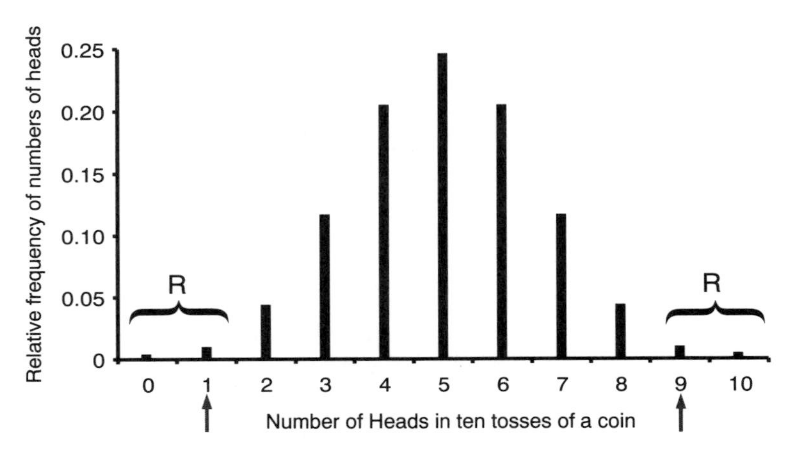

Figure 4.1. Frequency distribution of numbers of Heads in ten tosses of a 'fair' coin (probability of Head = probability of Tail = 0.50). The region of rejection is indicated by R and the critical values are indicated by arrows.

probability of 0.5. What happens if you toss it three times? There are now eight possible outcomes (HHH, HHT, HTH, HTT, THH, THT, TTH, TTT, where H indicates Head and T indicates Tail). The chances of getting 0 or 1 or 2 or 3 Heads out of three tosses are 0.125 (1 in 8), 0.375 (3 in 8), 0.375 (3 in 8) and 0.125 (1 in 8), respectively. The probability of coming down all Heads or all Tails is now 0.25 (the sum of the chances of coming down 3 Heads and 0 Heads). Thus, a fair coin still has a 1 in 4 chance of coming down the same way three times and it may be impossible to distinguish this in any meaningful way from a doctored coin.

The same calculations can be done for any chosen number of times of tossing a coin. The case used as an example here is that of tossing the coin 10 times (it does not matter why 10 as opposed to a larger number). If you work out all the possible outcomes of sequences of Heads and Tails for a 'fair' coin, you could calculate the chances of getting 0, 1, 2, 3 . . . 9 or 10 Heads (or Tails, which is the inverse). The result is given in Figure 4.1. This is, of course, a frequency distribution of possible number of Heads in ten tosses of a fair (unbiased, true) coin. The number of Heads is the variable. The frequencies are over a theoretical population of infinite number of times a coin is tossed ten times (i.e. assuming that all possible sequences of Heads and Tails are equally likely and there is no error due to only a small number of tosses being considered). If we toss the coin ten times, we must get one of the possible sequences and therefore one of the possible results (0, 1, 2 . . . 9, 10 Heads in ten tosses).

So, we have a theoretical distribution constructed assuming that the coin is a fair one.

What have we achieved so far? The logical structure of the preceding is as follows and summarized in Figure 4.2, in relation to the scheme in Figure 2.1. The *observation* is that the coin appears to come down on one side more often than is likely from chance. The *model* to explain this is that the coin has been made biased or unfair. The *hypothesis* proposed is that, if the coin is biased (i.e. if the model is true) and it is tossed ten times, it will come down on one side more often than would be predicted for a normal, fair coin. The *null hypothesis* is straightforward. It presumes that the hypothesis is wrong. The coin is fair and should come down on one side or the other exactly as dictated by chance for a normal, fair coin. We have constructed (Figure 4.1) the outcomes of the null hypothesis. So, we could do the *experiment* of tossing the coin ten times to test the null hypothesis (that it is a fair coin) versus the hypothesis (it isn't).

Suppose the result, in terms of number of times the coin came down on the Heads side, was five. This seems entirely consistent with the coin being fair, for which five is the most probable result (in Figure 4.1). For a fair coin, this result is expected to occur with probability 0.246, or about $\frac{1}{4}$ of such experiments. We would retain the null hypothesis that it is a fair coin because we have no evidence contrary to it. We have not falsified it. A similar conclusion would result if the coin came down Heads four or six times. These are quite likely results if the coin is fair.

Suppose, instead, we got the result that the coin came down Heads in all ten tosses. What might we conclude? It could be that the coin is fair and we simply have this result. This is possible, but only occurs with a probability of 0.001 (see Figure 4.1). The chance of getting the same side of the coin in ten tosses is 0.002 (the chance of all Heads plus the chance of getting all Tails, which is zero Heads). Such a result occurs by chance about once in every 500 series of 10 tosses of a fair coin. We could accept the result as being unlikely, but a possible outcome with a fair coin. We would then retain the null hypothesis.

Alternatively, we could consider that such a result is extremely unlikely to have occurred by chance (as the 1 in 500 cases suggested by theory). Rather, we could argue that it represents an extremely likely outcome if the coin has been altered to make it more likely to come down Heads more often than Tails. If we use this argument, we must reach the decision that the null hypothesis should be rejected, because we have a result that suggests it is wrong. It is more reasonable to assume that our experiment

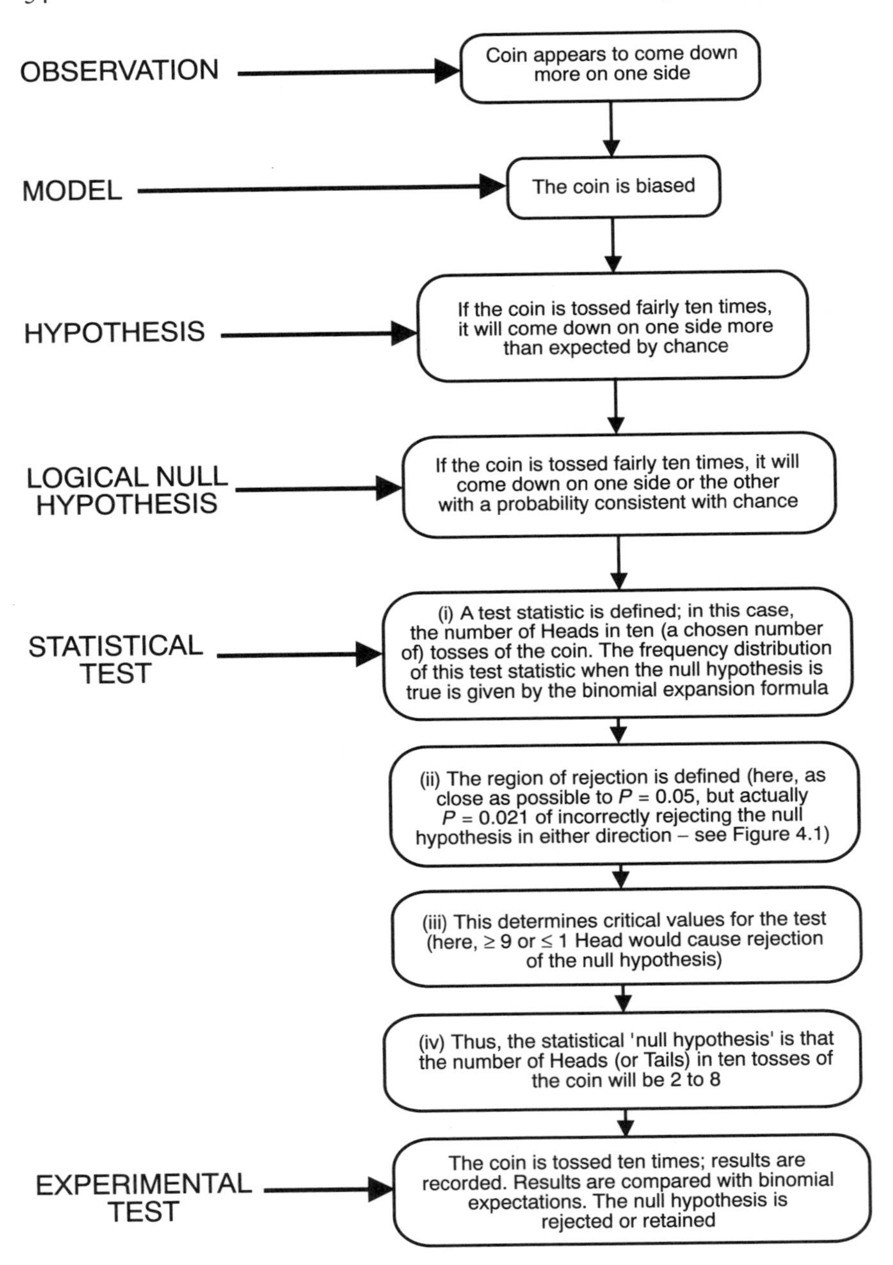

Figure 4.2. The logical framework for an experimental test of the likelihood of bias of a coin.

produced the likely outcome of tossing a biased coin than that it produced the very improbable result from tossing a fair one. We should reject the null hypothesis.

What if the coin comes down Heads in only one of the ten trials? The probability of a fair coin coming down the same way in at least nine tosses is 0.022. This is made up of the probabilities of getting nine Heads, nine Tails, ten Heads or ten Tails, i.e. all the chances of getting nine results the same. Again, this is an unlikely result, occurring about once in every 50 such experiments with fair coins. We should, again, reject the null hypothesis.

If, in contrast, we got the result that the coin came down Heads eight times out of ten tosses, the result is quite likely with a fair coin. Such a result (at least eight tosses the same) occurs with probability 0.11, or about one in nine occasions when a fair coin is tossed ten times.

Conventional wisdom considers (arbitrarily) that events are likely if they occur more than about 1 in 20 times (i.e. with probability > 0.05).

The outcome of this experiment, under this convention, must then be to reject the null hypothesis and conclude that it is a biased coin (i.e. to support the hypothesis) if nine or ten Heads or nine or ten Tails occur. Any other result (2 to 8 Heads or 8 to 2 Tails) would cause us to retain the null hypothesis that it is a fair coin. Notice, as a final comment on the process, that there must be a statistical, probabilistic decision about whether a coin is fair or not, because the frequency of results of ten tosses of a coin will differ between a fair and a biased coin. There is, however, no absolute difference between the *possible* results of tossing one of the two types of coin.

4.3 The components of a statistical test

This example indicates the steps and components of a statistical test.

4.3.1 Null hypothesis

First, the null hypothesis must be defined in some quantitative manner. This turns a *logical* null hypothesis into a *statistical* null hypothesis – that defines a frequency distribution that can be sampled to examine the likely validity of the logical null hypothesis. Here, the logical null hypothesis is that the coin is fair and is therefore equally likely to come down Heads or Tails. The statistical null hypothesis therefore defines the frequency distribution of possible results of tossing the coin (in this case a *binomial* distribution with the probability of Heads = the probability of Tails = 0.50).

4.3.2 Test statistic

Then, a test statistic is chosen. This has two properties. First, it must be measurable in some possible experiment. Second, it must be possible to calculate in advance of the experiment the frequency distribution of the test statistic if the null hypothesis is true. Here, the test statistic is the number of Heads (or Tails) expected in ten tosses of the coin. Its frequency distribution, given the probability of Heads or Tails being 0.50 was calculable (Figure 4.1).

4.3.3 Region of rejection and critical value

Now a decision must be made about how unlikely an event must be (if the null hypothesis is true) so that we would consider the event more likely to represent an alternative hypothesis. Above, the conventional, but arbitrary, probability of 0.05 (1 in 20) was used. Let us keep this as a 'rule' for the moment; it is discussed later. This defines the region of rejection of the null hypothesis. We would reject the null hypothesis if the experiment (ten tosses of the coin) produces a result that has less than a probability of 0.05 of occurring by chance if the null hypothesis were true. So, the region of rejection of the null hypothesis is all parts of the frequency distribution of the test statistic that would cause us to reject the null hypothesis. In this case, the region of rejection consists of the values 0, 1, 9 and 10 of the test statistic (Figure 4.1). These are the outcomes causing us to reject the null hypothesis because they occur with sufficiently small probability when the null hypothesis is true.

The boundaries of the region of rejection are defined by the critical values of the experiment. Here, the critical values are 1 and 9, which define the upper and lower frequencies of Heads (or, for that matter, Tails in this experiment) that would cause rejection of the null hypothesis. A result of the same size or greater than the upper critical value would cause rejection of the null hypothesis. So would a result of the same size or smaller than the lower critical value.

All statistical tests have the same component steps (as summarized in Figure 4.2). Note that the various steps listed under 'Statistical test' in Figure 4.2 must be completed before the experiment is done. Otherwise, there is no way of interpreting the outcome of the experiment. For example, if the region of rejection and therefore the critical values are chosen *after* the experimental result is known, they could be chosen to be outside or inside the observed value of the test statistic. Choosing to place the critical values outside the observed test statistic means the latter does not exceed the former and the null hypothesis will be retained. Putting the critical values inside the observed value means the latter is outside the former and therefore in the region of rejection, causing the null hypothesis to be rejected. By choosing where to put the critical values (by defining the region of rejection) after the experiment, the procedure by which a conclusion is reached from the experiment could be altered to produce whatever conclusion is chosen by the experimenter. This obviously makes no sense in the construction of a tool to aid in making decisions. If the decision were made without use of the procedure (which it would be if it were decided where to place the region of rejection after knowing the outcome of placing it either side of the observed value) then the procedure and the experiment would not have been necessary.

The correct procedure is to identify how a choice would be made to retain or reject the null hypothesis before the data are obtained and then stick to the decision once the data cause it to be made.

4.4 Type I error or rejection of a true null hypothesis

Obviously, any outcome possible when tossing a coin ten times could be the result of chance events for a fair or for a biased coin. So, rejection of the null hypothesis is not a certain or sure conclusion. We might reject the null hypothesis because an unlikely outcome has occurred with a fair coin (e.g. it came down Tails nine times out of ten, by chance). The conclusion to reject the null hypothesis when it is, in fact, true (i.e. the coin is fair) is a mistake. It is known as a Type I error and occurs in such an experiment with the probability chosen as the region of rejection. This is so because the region of rejection was chosen as the ranges of values considered unlikely if the null hypothesis were true. Later (Section 5.11), we will consider Type I error in detail. For now, just accept that the conventional decision about Type I error is to use the probability of 0.05 as an appropriate level of acceptable error. In other words, convention has it that we are prepared to reject null hypotheses with a 1 in 20 chance of doing so incorrectly (i.e. when unlikely data occur from a population for which the null hypothesis is actually true). Why we do not make this risk of a mistake smaller is considered later (Section 5.11).

4.5 Statistical test of a theoretical biological example

Consider another simple, hypothetical example. Suppose that a particular morphological feature (say, the length of beak) of some population of birds in a particular mountain range has been measured. This would, in practice, require catching and measuring all the birds present in the relevant stated period of time, but never mind the impractical aspects of the example. Further, suppose the lengths of beaks form a normal distribution of mean $\mu_B = 11$ mm and variance $\sigma_B^2 = 2.34\,\text{mm}^2$ (where subscript B indicates length of beak). The standard deviation of the population is $\sigma_B = 1.53$ mm. Normal distributions have a particular shape, as shown in Figure 4.3. The frequency distribution of a normal distribution is defined by:

$$f(x) = \frac{1}{\sigma\sqrt{(2\pi)}}\, e^{-(x-\mu)^2/2\sigma^2} \qquad (4.1)$$

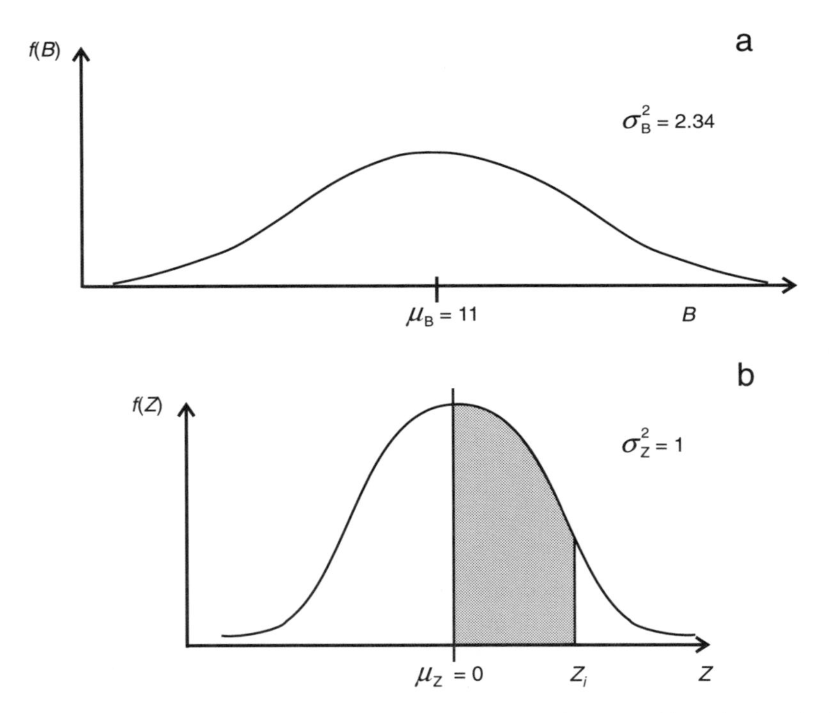

Figure 4.3. Normal distributions. (a) A normal distribution of lengths (mm) of birds' beaks (B_i) with mean $\mu_B = 11$ and variance $\sigma_B^2 = 2.34\,\text{mm}^2$. (b) Transformation to a standard normal distribution of Z values ($Z_i = (B_i - \mu_B)/\sigma_B$) with mean, $\mu_Z = 0$ and variance, $\sigma_Z^2 = 1$. The shaded area in (b) represents the proportion of the entire population of Z values that falls between the mean (0) and the value Z_i.

Most of the distribution is close to the mean and decreasingly few values of the population are a long way away from the mean in either direction.

4.5.1 *Transformation of a normal distribution to the standard normal distribution*

Normal distributions have useful mathematical properties. The most interesting is that all normal distributions can be transformed into a particular normal distribution that has mean $\mu = 0$ and variance $\sigma^2 = 1$. This is the standard (or sometimes unit) normal distribution, usually described as the frequency distribution of a variable Z (see Figure 4.3). If there is a normal frequency distribution of some variable, say X, with mean, μ_B and variance, σ_B^2, the formula to transform any value of X to the corresponding value of Z in the standard normal distribution is:

$$Z_i = (X_i - \mu_B)/\sigma_B \tag{4.2}$$

as in Figure 4.3. The standard normal distribution has been tabulated (see e.g. Fisher & Yates, 1953; Pearson & Hartley, 1970; Snedecor & Cochran, 1989). The area under the standard normal curve is also tabulated so that this area from the mean (zero) to a particular value (Z_i) can be ascertained from tables. An example is given in Figure 4.3.

Because the standard normal distribution is so well known, it can be used to identify intervals containing a defined proportion of the distribution. For example, in Figure 4.4, the shaded area of the curve contains

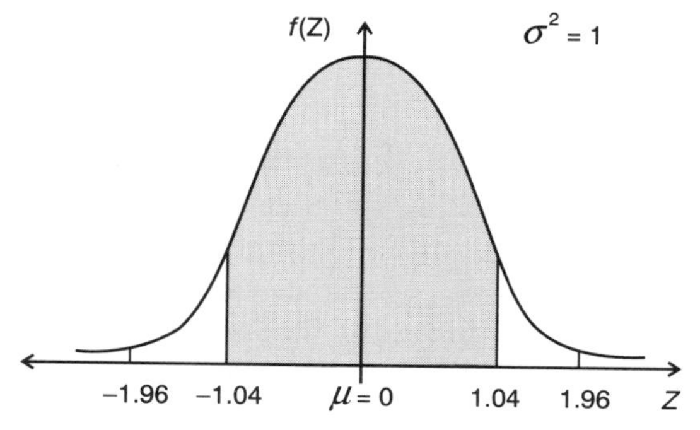

Figure 4.4. Standard normal distribution ($\mu = 0$, $\sigma^2 = 1$) showing 70% confidence interval (confidence limits are $Z = -1.04$ to $Z = 1.04$). The shaded area contains 70% of the entire distribution. 95% of the distribution are in the 95% confidence interval defined by upper and lower limits, $Z = 1.96$ and $Z = -1.96$, respectively.

70% of the distribution (which all lie between $Z = -1.04$ and $Z = 1.04$ in the standard normal distribution). In general, limits (Z_{upper} and Z_{lower}) contain any stated proportion of the distribution. Commonly, limits of $Z = \pm 1.96$ are used, because these contain (or include) 95% of the distribution; these are known as the 95% confidence limits. Thus, in a standard normal distribution there is a 95% chance that, for any value of Z:

$$-1.96 \leq Z \leq 1.96 \quad (4.3)$$

And, generally, for any chosen proportion A, there are limits Z_{upper} and Z_{lower} that contain A. The probability is A that any value of Z is:

$$Z_{lower_A} \leq Z \leq Z_{upper_A} \quad (4.4)$$

The transformation used to convert any normal distribution to the standard normal distribution is:

$$Z_i = (X_i - \mu)/\sigma \quad (4.5)$$

where X_i comes from a normal distribution with mean μ and variance σ^2. The transformation can be used in reverse:

$$X_i = \mu \pm \sigma Z_i \quad (4.6)$$

This formula transforms any chosen upper and lower boundary from the standard normal distribution to the equivalent boundaries in any normal distribution for which the mean and variance are known. If Z_{upper} and Z_{lower} include proportion A of a standard normal distribution (as in Figure 4.5), then X_{upper} and X_{lower} include proportion A of a normal distribution of X values, with mean μ_X and variance σ_X^2 because:

$$X_{upper} = \mu_X + \sigma \times Z_{upper} \quad (4.7)$$

$$X_{lower} = \mu_X - \sigma \times Z_{lower} \quad (4.8)$$

(noting that X_{lower} is negative).

Now, suppose someone tells you that he or she has caught a bird in the mountains and measured its beak and considers it too large to come from the population you have previously described. So, the observation is a very large-beaked bird, but you don't yet ask for its size, so that you can determine the appropriate region of rejection for the statistical test you will do, without being influenced by the outcome of the measurement. The model to explain this observation is that it comes from some population with larger beaks than those on the mountains you have investigated. The length of the beak will therefore be predicted (the hypothesis) to be larger than is consistent with the previously measured distribution. The

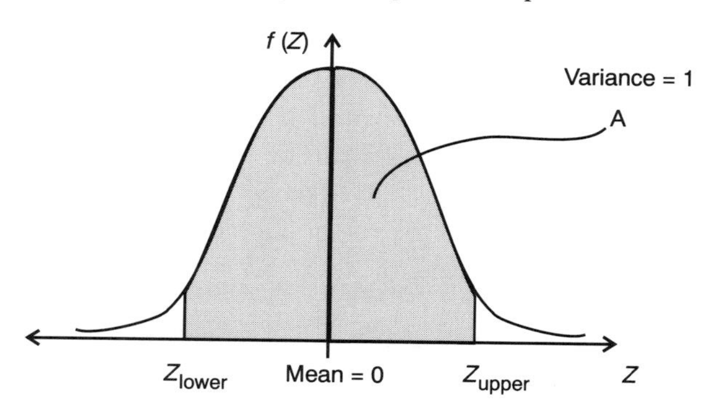

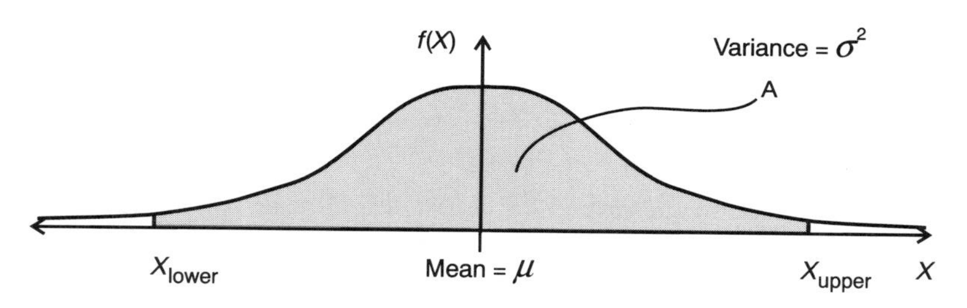

Figure 4.5. Transformation of confidence limits from a standard normal to any normal distribution. (a) Shaded area A represents a proportion of a standard normal distribution of Z values, with mean 0 and variance 1. Z_{upper} and Z_{lower} contain A. (b) Shaded area A is also contained between X_{upper} and X_{lower} in a normal distribution of X values with mean μ and variance σ^2 ($X = \mu + \sigma Z$; see text).

logical null hypothesis is all alternatives – the bird's beak is of a length consistent with the known distribution of lengths of beaks or is, in fact, smaller than would be consistent with the population. Rejection of this null hypothesis would support the hypothesis.

The statistical null hypothesis is the frequency distribution of lengths of beaks previously measured on the mountain (as in Figure 4.3a). The test statistic in this case is simply the length of the bird's beak. In this totally unrealistic example, we know the distribution of lengths of beaks for the null hypothesis (i.e. for the population previously observed on the mountain; see p. 58). The region of rejection can only be in the right-hand (large) end of the distribution – small beaks are all consistent with the stated null hypothesis. Let us keep the same convention as used before and be willing to reject null hypotheses when they are true (i.e. to make a Type I error) with probability of 0.05. So, the region of rejection is defined as the part of the normal distribution of beak-lengths that contains the largest 5% of beaks, as illustrated in Figure 4.3a. There is only one critical value, the size that defines the region of rejection. This must be found from the standard normal distribution using the transformation described in Equation 4.5 above.

Consultation of the appropriate table (e.g. Pearson & Hartley, 1958; Snedecor & Cochran, 1989) indicates that $Z_{\text{upper}} = 1.645$ cuts off 0.05 (or 5%) of the right-hand side of the standard normal distribution. Therefore:

$$\text{Probability}\,(Z \geq 1.645) = 0.05 \tag{4.9}$$

Using the previous transformation to convert this to the relevant value for the distribution of lengths of birds' beaks (B values in Figure 4.3a):

$$\begin{aligned} B_{\text{upper}} &= \mu_{\text{B}} + \sigma_{\text{B}} \times Z_{\text{upper}} \\ B_{\text{upper}} &= 11 + (1.53 \times 1.645) = 13.52 \end{aligned} \tag{4.10}$$

The critical value is therefore 13.52 mm.

Using the procedure described earlier, a single beak can be tested to determine how likely or unlikely it is to come from the stated distribution. Large beaks will fall in the region of rejection (i.e. are greater than the critical value) and cause rejection of the null hypothesis. It turns out that the length of the interesting beak is 15.1 mm. We should therefore reject this null hypothesis (with probability of Type I error chosen as 0.05) in favour of the alternative that the measured bird is from a population with larger beaks. The observed beak is said to be *significantly* large, i.e. larger than the defined critical value. We have used this simple, mensurative experiment to determine that it is unlikely for the observed bird to come from the defined population.

4.6 One- and two-tailed null hypotheses

The previous, simplified and contrived examples illustrate an important point about the nature of hypotheses and null hypotheses. In the first case, your informant did not state in which direction the bias of the coin was supposed to be (i.e. an excessive number of Heads or an excessive number of Tails). Of course, the informant must have known this, but did not tell you and you did not ask. As a result, the hypothesis, null

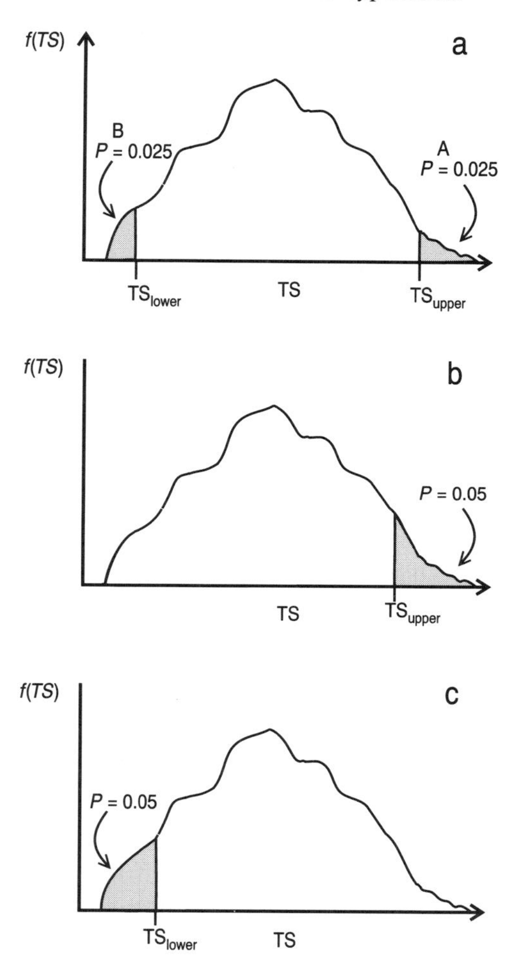

Figure 4.6. Distinction between one- and two-tailed null hypotheses and their tests. Data are frequency distributions of a test statistic (TS) plotted when the relevant null hypothesis is true. The region of rejection for a probability of Type I error of 0.05 is shown (a) for a two-tailed null hypothesis. Region A indicates values of TS that would suggest an alternative hypothesis prevails that generates large values of TS. B indicates values of TS that would suggest an alternative hypothesis which generates small values of TS. Each region causes a probability of Type I error of 0.025. (b) The one-tailed (shaded) region of rejection when the only alternatives to the null hypothesis must generate large values of TS. (c) The situation where all alternatives to the null hypothesis generate small values of TS. The one-tailed region of rejection is the left-hand shaded part of the distribution.

hypothesis and region of rejection were phrased in terms of either possibility. This allows the null hypothesis to be rejected by an observed value of the test statistic (the number of Heads in ten tosses of the coin) that is large (the coin is considered to be biased in favour of Heads) or small (the coin is considered to be biased in favour of Tails), compared with what is usually expected from a fair coin. There are therefore two 'tails' or regions of the null hypothetical distribution of the test statistic that can cause rejection of the null hypothesis. Such an hypothesis is called 'two-tailed'. The region of rejection and the probability of Type I error are divided into two components, one in each direction of possible contradiction of the null hypothesis.

In contrast, in the second case, the null hypothesis could be rejected only by observing a large beak. A small one could not contradict the hypothesis that the observed bird was larger than was usually found in the known population. The region of rejection was therefore defined solely in terms of one end (the larger sizes) of the test statistic. The entire probability of Type I error must be in this single end of the null distribution of the test statistic. Hence, such hypotheses and tests are called 'one-tailed'.

The differences between these two types of test are illustrated in Figure 4.6. Note that the distinction is entirely determined by the logical structure of the hypothesis and therefore entirely a property of the logical processes used by the experimenter. As a general consideration, one-tailed tests are a result of either more information being available in the original observations or the models proposed to account for them being more complex than is the case for two-tailed situations. For example, the coin-tossing example would have been one-tailed rather than two-tailed if the given observation had been that there was an excess of Tails compared with the observer's expectations. This is a more detailed piece of information than originally supplied and would lead to a more precise model – the coin is doctored in favour of Tails rather than simply 'the coin is doctored'.

The only crucial thing about one-tailed versus two-tailed tests is that the region of rejection must be correctly identified so that there is no confusion about how to interpret the observed value of the test statistic when the experiment is over. The probability of Type I error must be associated with the appropriate possible values of the test statistic (Figure 4.6). This is considered for some specific situations in later parts of the book.

5

Statistical tests on samples

The previous discussion of statistical tests used only very (and over-) simplified examples involving hypotheses about single entities – for example, a single bird's beak. Such examples are not useful except to illustrate the point and purpose of the procedures. Most ecological examples, even the simplest ones, concern the analysis of statements about populations – either real biological populations or populations of variables. They therefore require statistical tests on samples from populations. The theory and procedures for these sorts of tests are considered here.

5.1 Repeated sampling

Before discussing an actual example, first consider what would happen if you repeatedly sampled a population (i.e. a frequency distribution). Imagine a population of animals for which you wish to estimate the mean length. You take a sample of n individuals and measure their lengths. The sample provides an estimate of the population's mean and variance (μ and σ^2); these estimates are $\bar{X}_1$ and s_1^2, respectively. Now another sample is taken, also of size n. It provides another estimate of μ and σ^2, this time $\bar{X}_2$ and s_2^2.

The two values of $\bar{X}$ are not going to be identical, unless the two samples happen, by chance, to contain exactly the same individuals, or the individuals sampled happen, by chance, to produce exactly the same estimates. Each sample $\bar{X}$ estimates μ, but with some error because each sample does not have exactly the same frequency distribution as that of the whole population. Samples are unlikely to be exactly representative of the entire frequency distribution of the population of measurements.

It is very important to consider the frequency distribution of sample means of a population. For many real problems in biology, this will allow estimates to be made of the closeness of a sample mean to the real,

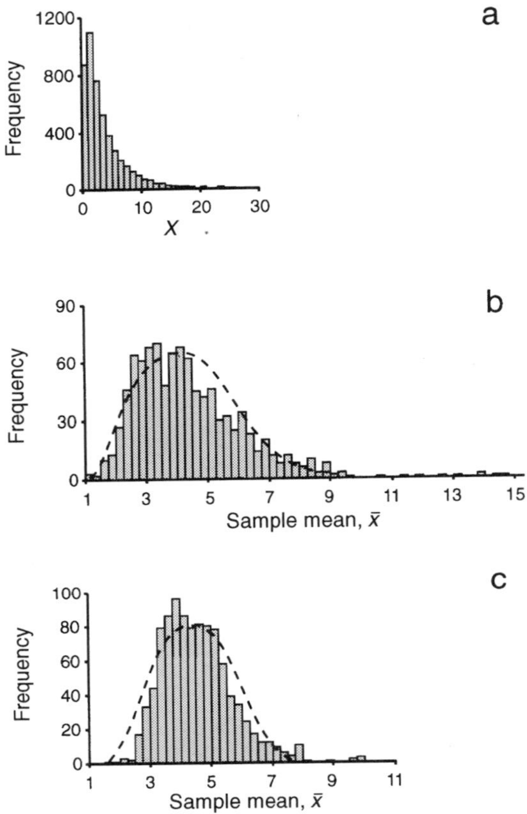

Figure 5.1. Frequency distributions of independent sample means from a skewed distribution (a) of a variable X, with mean $\mu = 4.20$ and variance, $\sigma^2 = 27.4$. (b), (c) Means with samples of size $n = 9$ and 25, respectively. In each case, a normal distribution of the same mean and variance is shown. Note that the similarity to a normal distribution is greater and the variance of the distribution of sample means is smaller for $n = 25$. In (b), the mean of all samples is 4.16, with variance 3.32. In (c), the mean and variance are 4.31 and 1.45, respectively.

but unknowable mean of the population (i.e. μ). To approach this topic, imagine sampling a very large population over and over again, each time with a sample size of $n = 5$. After hundreds of samples have been taken, we have hundreds of estimates of the population's mean. These means are variable and are therefore distributed as a frequency distribution.

The frequency distribution of sampled means can be examined, as in Figure 5.1. In this case, the shape is distinctive, as it is for many types of population. The distribution of sample means from a normal distribution is normally distributed (see Section 4.5). In Figure 5.1, the exact normal distribution has been calculated with the same mean and variance as that of the frequency distribution of sample means. Even though the original population from which the samples came was not anything like a normal distribution (Figure 5.1), the sample means are approximately normally distributed.

Why the estimated means from repeated samples should have this sort of shape is fairly simple to understand. Common sense dictates that the distribution of sample means should be centred on (located at) the mean (μ) of the original population. Obviously, if samples were representative, each sample should be a representation of the population's mean. Sample means must therefore be centred on the mean of the population from which they come. Furthermore, representative samples should not often produce sample estimates of the population's mean that are very different from the population's mean. Only fairly unrepresentative samples can produce an estimated mean that is very different from the population's mean. Such samples should occur rarely if sampling is unbiased, representative or random. Thus, there is a decreasing chance of getting a sample that produces means very different from the population's mean, which explains the decline in frequencies in the normal distribution of sample means (Figure 5.1) as you move away from the centre of the distribution.

The normal frequency distribution of sample means has another property. In general, the variance of the frequency distribution of sample means is equal to that of the original population divided by the size of sample (n; Figure 5.2). This is also true for normal distributions that are repeatedly sampled. From the central limit theorem, it is also approximately true that the distribution of sample means is normal even if the samples are taken from other types of frequency distribution, particularly those that are unimodal (only have one hump or peak) and are not very skewed (i.e. are symmetrical).

The degree to which the distribution of sample means conforms to a normal distribution is also dependent on the size of the sample. Generally, larger samples are more likely to be normally distributed (Figure 5.1) and, as above, their means will be distributed with smaller variances. This comes about because large samples are more representative of the original distribution and the sampled means from large samples are therefore much closer to the population's true mean. The variation among the means of several samples must therefore be smaller as sample sizes increase.

The effect of different sizes of samples on frequency distributions of different shapes is shown in Figure 5.2. It is clear that even when the original frequency distribution is nothing like a normal distribution, the

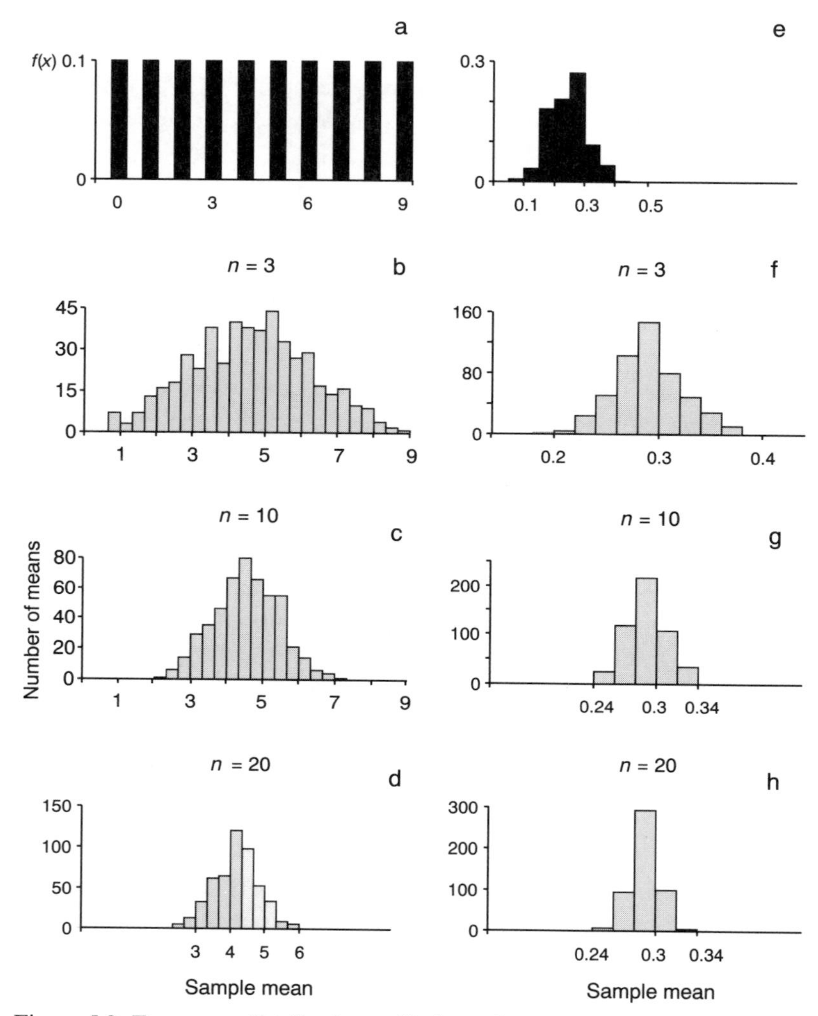

Figure 5.2. Frequency distributions of independent sample means from (a) uniform distribution of digits. (b), (c) and (d) 1000 means with sizes of sample $n = 3$, 10, 20, respectively. Samples from (e), a binomial distribution with mean $\mu = 0.3$ gave 1000 means in (f), (g), (h) with $n = 3$, 10 and 20, respectively.

distribution of sample means is often very like the theory suggests. Also, it is obvious empirically that increasing the size of samples makes the variance of the distribution of sampled means smaller and makes the distribution of sampled means more like a normal distribution. This simulated sampling provides some faith in the use of a theory based on sample means being normally distributed.

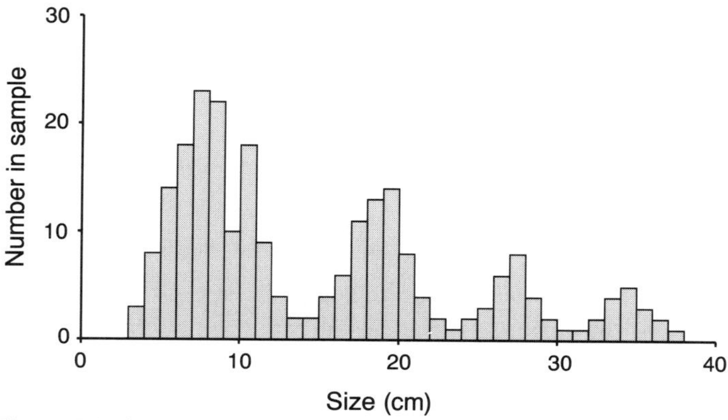

Figure 5.3. A multimodal frequency distribution of sizes of, say, a fish (length measured in centimetres).

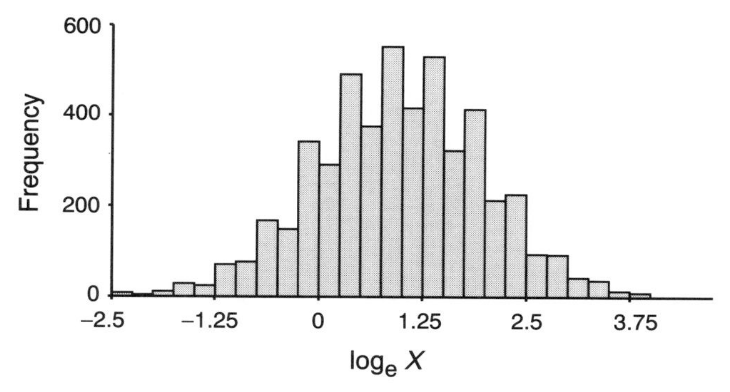

Figure 5.4. Transformation to natural logarithms of data from a very skewed distribution. The data are in Figure 5.1a. The log-transformed data are much more like a normal distribution.

The approximation to a normal distribution of sample means will not be so good when the population being sampled has a frequency distribution that is multimodal or very skewed. In these cases, there are alternative ways to proceed. For multimodal distributions (Figure 5.3), it is often worth considering breaking the population into different components to be measured separately. In the case illustrated in Figure 5.3, this is a sensible thing to do, because there is no compelling reason to oversimplify a study of the sizes of an animal that clearly comes in different categories of size (or age). Obviously, there will be much to be gained from treating each size-class separately, however crudely they are divided.

Where distributions are very markedly skewed (as in Figure 5.1), it is often the case that transforming the data to logarithms will remove the asymmetry and sample means of the log-transformed data will probably be much more normally distributed. This is shown in Figure 5.4; log transformation of the values made the distribution much more normal. The distribution of sample means of log-transformed data also became more like a normal distribution as a result.

5.2 The standard error from the normal distribution of sample means

For any distribution of sample means that can be approximated by a normal distribution, the variance is the variance of the population being sampled divided by the size of sample used (σ^2/n; see Figure 5.2). The standard deviation of a distribution of sample means is the square root of the variance (as with any other distribution, see Section 3.8). For some reason lost in time, this is known as the standard error (or standard error of the mean) and is:

$$\text{Standard error} = \sqrt{(\sigma^2/n)} = \sigma/\sqrt{n} \qquad (5.1)$$

where σ^2 is the variance of a distribution from which samples of size n have been taken. Thus, a single sample from a distribution provides estimates as follow:

Sample mean $= \bar{X}$

Sample variance $= s^2$

Sample standard deviation $= s$

Sample standard error $= s/\sqrt{n}$

The estimate of variance indicates how variable the original data are in the population being sampled. The standard error indicates how variable sampled means would be if numerous samples were taken from the population (see Table 5.2 below).

As an aside (to be skipped if you don't care or don't like algebra), now that a standard error has been introduced, it is possible to revisit the calculation of the sample variance to explain why it is appropriate to use the divisor $(n - 1)$ rather than n, as described in Section 3.9. The explanation is provided in Table 5.1.

5.3 Confidence intervals for a sampled mean

Previously (Chapter 4) it was explained how the standard normal distribution could be used to construct confidence intervals for any chosen

Table 5.1. Why the divisor for a sample estimate of variance is the degrees of freedom

For a population of N values of X_i, with mean μ, the variance is:

$$\sigma^2 = \sum_{i=1}^{N} (X_i - \mu)^2/N$$

In a sample of n values of X_i, with mean $\bar{X}$, the sample variance s'^2 could be calculated as:

$$s'^2 = \sum_{i=1}^{n} (X_i - \bar{X})^2/n$$

$$= \sum_{i=1}^{n} [(X_i - \mu) - (\bar{X} - \mu)]^2/n$$

$$= \frac{\sum_{i=1}^{n}(X_i - \mu)^2}{n} - \frac{2\sum_{i=1}^{n} (X_i - \mu)(\bar{X} - \mu)}{n} + \frac{\sum_{i=1}^{n} (\bar{X} - \mu)^2}{n}$$

By expansion and simplification:

$$s'^2 = \frac{\sum_{i=1}^{n} (X_i - \mu)^2}{n} - (\bar{X} - \mu)^2$$

The first term on the right-hand side clearly estimates σ^2.
If $E[\ ]$ indicates what is expected under an enormous amount of repeated random sampling:

$$E\left[\frac{\sum_{i=1}^{n}(X_i - \mu)^2}{n}\right] = \sigma^2$$

The second term on the right-hand side is a squared deviation of a sampled mean from a population mean and, therefore, a single unbiased estimate of variance in a population of sample means (as in Section 5.2). The variance of a population of sample means is σ^2/n:

therefore $E[(\bar{X} - \mu)^2] = \dfrac{\sigma^2}{n}$

therefore $E[s'^2] = \sigma^2 - \dfrac{\sigma^2}{n} = \dfrac{(n-1)}{n}\sigma^2$

therefore $E\left[\dfrac{n}{(n-1)}s'^2\right] = \sigma^2$

therefore $s^2 = \sum_{i=1}^{n}(X_i - \bar{X})^2/(n-1)$ estimates σ^2 unbiasedly probability. The procedure involved identifying upper and lower boundaries that included or contained some chosen proportion of the standard normal distribution. These values could then be transformed to give the appropriate upper and lower boundaries that included the chosen proportion of any empirical frequency distribution. The same procedure can be used with the normal distribution of numerous sampled means from a frequency distribution, provided that these means are normally distributed (as discussed above in Section 5.1).

As an entirely theoretical example, suppose that there is, as used before (Section 4.5), a population of birds with average beak length $\mu_B = 11$ mm and variance $\sigma^2_B = 2.34\,\text{mm}^2$. According to the theory about sampling a normal distribution (see above, Section 5.1), the distribution of all possible sample means from such a distribution, if samples are of size $n = 10$ is a normal distribution with the same mean ($\mu_{\bar{B}} = 11$ mm; $\bar{B}$ indicates sample means of $n = 10$ values of B, i.e. 10 lengths of beaks of the birds). The variance of the normal distribution of sampled means is $\sigma^2_{\bar{B}}$ and, from the relationship between the population's variance and the variance of sampled means of a population given in Section 5.1:

$$\sigma^2_{\bar{B}} = \frac{\sigma^2_B}{n} = \frac{2.34}{10} = 0.234 \tag{5.2}$$

The standard deviation of the population of sampled means, i.e. the standard error (see above) is the square root of the variance:

$$\sigma_{\bar{B}} = \sqrt{\sigma^2_{\bar{B}}} = 0.484 \tag{5.3}$$

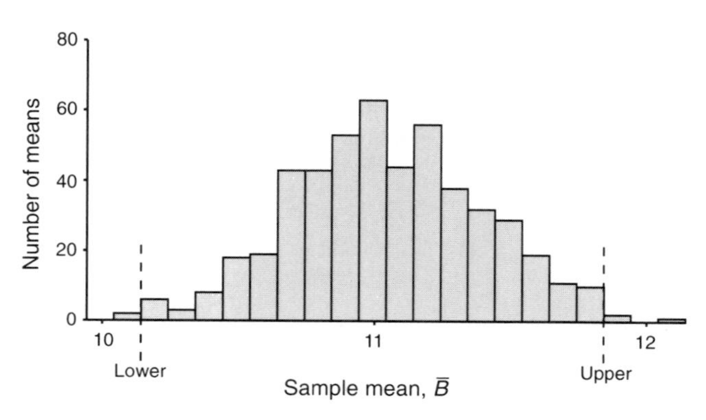

Figure 5.5. Frequency distribution of 500 randomly sampled means ($\bar{B}$; $n = 10$ in each sample) from a normal distribution of lengths of birds' beaks, with mean $\mu = 11$ mm and variance $\sigma^2 = 2.34\,\text{mm}^2$. Lower and upper 95% confidence limits are 10.15 mm and 11.85 mm, respectively, as shown by the dashed lines. See text for further details.

The 95% confidence limits can be calculated from the appropriate values in the standard normal distribution (i.e. $Z_{\text{upper}} = 1.96$, $Z_{\text{lower}} = -1.96$; see Section 4.5.1; these values include 95% of a standard normal distribution). These are transformed as before (Section 4.5) to give the appropriate values of the variable of interest here (i.e. $\bar{B}$; see Figure 5.5). The transformation uses the standard deviation of the distribution, which in this case is the standard deviation of the distribution of means. Thus, the 95% confidence limits for the distribution of sample means from the distribution of lengths of birds' beaks are:

$$\bar{B}_{\text{upper}} = \mu_{\bar{\text{B}}} + Z_{\text{upper}} \times \sigma_{\bar{\text{B}}} = 11 + 1.96 \times 0.484 = 11.948 \tag{5.4}$$

$$\bar{B}_{\text{lower}} = \mu_{\bar{\text{B}}} - Z_{\text{lower}} \times \sigma_{\bar{\text{B}}} = 11 - 1.96 \times 0.484 = 10.052 \tag{5.5}$$

5.4 Precision of a sample estimate of the mean

The preceding, entirely theoretical exercise turns out to be extremely useful. We can take the constructed 95% confidence interval and use it to estimate how precisely a given sample mean estimates the unknown mean of the population. This requires a simple trick – an inversion of the previous formula (Equation 5.5), as follows. The 95% confidence interval just constructed specifies that 95% of the entire population of possible sample means from the distribution of birds' beak lengths will fall in the stated range. Thus, for any sample estimate, $\bar{B}$, there is a 95% chance that $\bar{B}$ falls within the range from 10.05 to 11.95, i.e. within the calculated confidence interval for 95% of the entire population of sampled means. Thus, there is a probability of 0.95 that, for any sample mean $\bar{B}$:

$$\mu_{\text{B}} - \sigma_{\bar{\text{B}}} \times 1.96 \leq \bar{B} \leq \mu_{\text{B}} + \sigma_{\bar{\text{B}}} \times 1.96 \tag{5.6}$$

or:

$$\mu_{\text{B}} - 1.96(\text{SE}) \leq \bar{B} \leq \mu_{\text{B}} + 1.96(\text{SE}) \tag{5.7}$$

where SE is the standard error of the distribution.

The practically useful inversion of this formula is as follows:

$$\bar{B} - 1.96(\text{SE}) \leq \mu_{\text{B}} \leq \bar{B} + 1.96(\text{SE}) \tag{5.8}$$

for 95% of all possible samples ($\bar{B}$ values). This region is the 95% confidence interval for μ_B. This is a very powerful statement about the unknown mean of the population. If we were to know the standard error of the population (i.e. $\sigma_{\text{B}}/\sqrt{n}$ or $\sigma_{\bar{\text{B}}}$), for a single sample of a population (i.e. one sample estimate, $\bar{B}$) we could estimate a range that has a 95%

Table 5.2. Estimates of variance used to describe results of sampling

Statistical estimate	Symbol	Purpose
Sample variance	s^2	Describes your estimate of the population's parameter σ^2
Sample standard deviation	s	Most useful description of the variability among replicate units in the sample, i.e. variation among the things counted or measured. Is in the same scale and units as the sample mean
Sample standard error	$s/\sqrt{n}$	Useful description of the variability to be expected among sample means, if the population were sampled many times with the same size of sample. Is in the same scale and units as the sample mean
Confidence interval	$\bar{X} \pm t_p \times s/\sqrt{n}$	Indicates (with a stated probability $(1 - p)$) the range in which the unknown mean (μ) of the population should occur. For a given size of sample and probability, specifies the precision of sampling

chance of containing the true mean. If this range were small, we would consider our single sample to be providing a *precise* estimate of the unknown mean (Table 5.2), because there is a very good chance that the true mean lies in a small range around our sample estimate. In contrast, if the calculated confidence limits were large, our estimate is not precise – the unknown mean could be anywhere in a large range of values.

5.5 A contrived example of use of the confidence interval of sampled means

To illustrate the procedure, consider an entirely silly example. We shall eventually reach the real world, but the framework for dealing with it is much simpler to understand if we use unreal examples! Suppose that you are minding your own business in a bar, trying to drink your way through tomorrow's assignment for the experimental design course (or, perhaps, doing novel statistical theoretical work – see Section 4.2). Up lurches a drunk, waving a collection of brightly painted, battered beer-cans, which he claims come from Mars. He informs you that you can verify this because the cans are larger on average than the ones found on earth. In other words, he has observed that the cans are larger than expected from earth (or so he thinks) and he has proposed the model that this occurs because the cans come from Mars, where beer-cans (for some reason) are larger. The relevant hypothesis is that the mean size of the cans will be larger than expected for a sample of cans from earth. The null hypothesis is that the sample of cans he is holding are the average size of those on earth or are, in fact, smaller (i.e. don't forget all alternatives to the hypothesis). To test this null hypothesis, you must do some research about beer-cans on earth. It is, however, possible to ascertain (at least for the purposes of this example!) that the mean volume of terrestrial beer-cans is 300 ml, with a variance of 3000 ml^2 (i.e. $\mu = 300$, $\sigma^2 = 3000$) and that the frequency distribution of beer-cans on earth is approximately normal. These facts can be found in the *Journal of the Beer-Can* or some similar data-base. The owner of the purportedly martian cans has five of them and the average volume of those five cans is 360 ml (i.e. $\bar{X} = 360$ ml and $n = 5$). Using the logic of the example of birds' beaks in Section 4.5 and the theory developed in Section 5.4, we can construct the 95% confidence interval for the population of possible sample means to determine that there is a 95% chance that any sample of beer-cans from earth will have a mean volume in the range:

$$\mu \pm 1.96\sigma/\sqrt{n} = 300 \pm 1.96\sqrt{(3000/5)} = 300 \pm 48 \qquad (5.9)$$

In other words, 95% of all possible samples of beer-cans (with $n = 5$) on earth would have sample means in the range 252 to 348 ml. It is therefore extremely unlikely that the sample in the possession of your acquaintance has come from earth! The cans have a mean of 360 and are too large to be considered a sample of those on earth.

Of course, for the purpose of the example, we are assuming that the sample he has is representative of the population he got them from. (More realistically, of course, we probably imagine that he has simply taken some of the largest cans he can find on earth, but don't spoil the example. In fact, we shall treat it as referees of scientific papers must – we assume he is telling the truth!)

We consider the sample mean of cans to be improbably (although still possibly) large to have come from earth. We reject the null hypothesis. In the absence of other information ('How did they get here?' comes immediately to mind, followed rapidly by 'Of what are they made?', 'How did this drunk get hold of them?' and, given the state of scientific research funding everywhere these days, 'What are they worth?' and, above all, 'How can I get hold of them?') we must reject the null hypothesis in favour of the alternative – they might indeed come from Mars.

5.6 Student's *t*-distribution

The preceding theory about confidence intervals, as outlined in Section 5.4, Equations 5.6 and 5.7, is, unfortunately, useless in its present form for any practical purpose because it requires knowledge of the variance, σ^2. To know the variance of a population requires measurement of the entire population being studied. Then, of course, the mean would be known and there can be no imprecision or uncertainty. For a while, this theory could not progress. There was no solution available to the problem that the variance itself could only be estimated. This situation was altered when Student (1908) (a pseudonym for Gossett, see below) calculated the distribution of the confidence interval from a sample to take into account the uncertainty of the estimate of s^2.

In summary, he determined the frequency distribution of a number (called t) to use instead of 1.96 (or any other value from the standard normal distribution) when the variance is not known, but is estimated. These distributions use the frequency distribution of sampled estimates of variance and are calculated for different sizes of sample, usually expressed as degrees of freedom ($n - 1$). This is appropriate because, as discussed earlier, when sample sizes are large, an estimate will, on average, be closer to the population's parameter (i.e. there is less uncertainty in the use of s^2 compared to σ^2). He claimed that the solution to the problem was simple enough for a student to have solved it and signed his paper in the name of A. Student. The distributions are therefore known as Student's t-distributions. An example is shown in Figure 5.6; the calculation of t is demonstrated later (Section 5.9).

Suppose we have a sample of 15 lengths of lizards from a population in some valley, during summer, giving an estimated mean of $\bar{X} = 18.4$ mm and an estimated variance, $s^2 = 29.6\,\text{mm}^2$. With $n = 15$, we consult Student's t-table with 14 degrees of freedom (see Section 3.10) to find the tabled value that includes 95% of the distribution of the t-table. This is the same procedure done previously to find $Z = \pm 1.96$ in the unit normal distribution (see Section 4.5). Now we use this value (say, $t_{0.05}$) to replace the value from the standard normal distribution and we

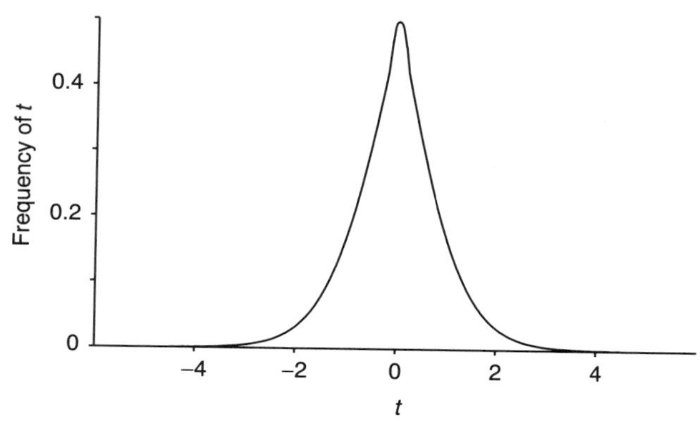

Figure 5.6. Frequency distribution of Student's t-statistic for samples of size $n = 15$, i.e. 14 degrees of freedom. The use of the distribution is described in Sections 5.6, 5.7 and Chapter 6.

use the estimated variance from the sample to determine that, for any mean, $\bar{X}$, there is a 95% chance that:

$$\mu - t_{0.05}\surd(s^2/n) \leq \bar{X} \leq \mu + t_{0.05}\surd(s^2/n) \quad (5.10)$$

Compare this with Equation 5.7, earlier. From this, using the previous manipulation (Equation 5.8) we have a 95% chance that:

$$\bar{X} - t_{0.05}\surd(s^2/n) \leq \mu \leq \bar{X} + t_{0.05}\surd(s^2/n) \quad (5.11)$$

In our particular case, we consult a table of the t-distribution with 14 degrees of freedom to discover that $t_{0.05} = \pm 2.14$. Thus, for our population, $\bar{X} = 18.4\,\text{mm}$, $s^2 = 29.6\,\text{mm}^2$, $t_{0.05} = \pm 2.14$ and there is a 95% chance that:

$$\begin{aligned} 18.4 - 2.14\surd(29.6/15) \leq \mu \leq 18.4 + 2.14\surd(29.6/15) \\ \text{or } 15.39 \leq \mu \leq 21.41 \end{aligned} \quad (5.12)$$

The sample suggests that the mean length of the population of lizards is very likely to be somewhere between these two limits, except for the circumstance where the sample is not representative of the population.

5.7 Increasing precision of sampling

This result leads to some interesting consequences. First, as shown below, it leads directly to statistical tests of quantitative hypotheses. Second, it allows a good description of precision of the sampling. This topic is covered in greater detail in an ecological context in the review by

Andrew & Mapstone (1987). The smaller the confidence interval, the smaller the range in which there is a good (95%) chance that the unknown mean may be found. Thus, precision is the inverse of the size of the confidence interval.

Precision of sampling is obviously a function of three things.

5.7.1 The chosen probability used to construct the confidence interval

Precision decreases with increased probability used to construct the confidence interval. In the case discussed, 95% was used. If you need to be more certain about the possible range of the population's mean, you need to increase the proportion of all possible sample means that are included in your confidence interval. Thus, you could be 99% or 99.5% sure about the possible value of the population's mean, but the value of t will be correspondingly larger. Hence, you can be more sure (99% or 99.5% compared with 95%), but the range of values is greater. In this case, with 14 degrees of freedom, the value for 99% from the t-table is $t_{0.01} = 2.98$. Using Equation 5.11, these give 99% confidence limits of 14.21 and 22.59. Increased certainty is coupled with decreased precision. The probability is entirely the choice of the experimenter, but is conventionally 0.95. See below (Section 5.13) for a discussion of how to make a better choice.

5.7.2 The sample size (n)

For any given distribution, precision of an estimate of the mean will increase with increasing size of sample. This is common sense; as more of a population is observed (or measured), there are fewer possibilities that the sample will not represent the entire population. The sampled mean must therefore become closer to the parameter being estimated. This is illustrated in Figure 5.7a, where a population of birds' beak lengths (as used before in Section 4.5) has been repeatedly sampled with different sizes of samples. These are plotted against the size of sample, to show that they rapidly converge on μ as n increases. The range of sampled means at each size of sample is what is measured by the confidence interval. Thus, precision increases with n as confidence interval decreases.

With increasing size of sample (n), the confidence interval declines for a second reason – the value of t for any chosen probability also decreases. $t_{0.05}$ decreases to reflect the greater precision of s^2 as an estimate of σ^2 when a greater representation of the entire population is available in the

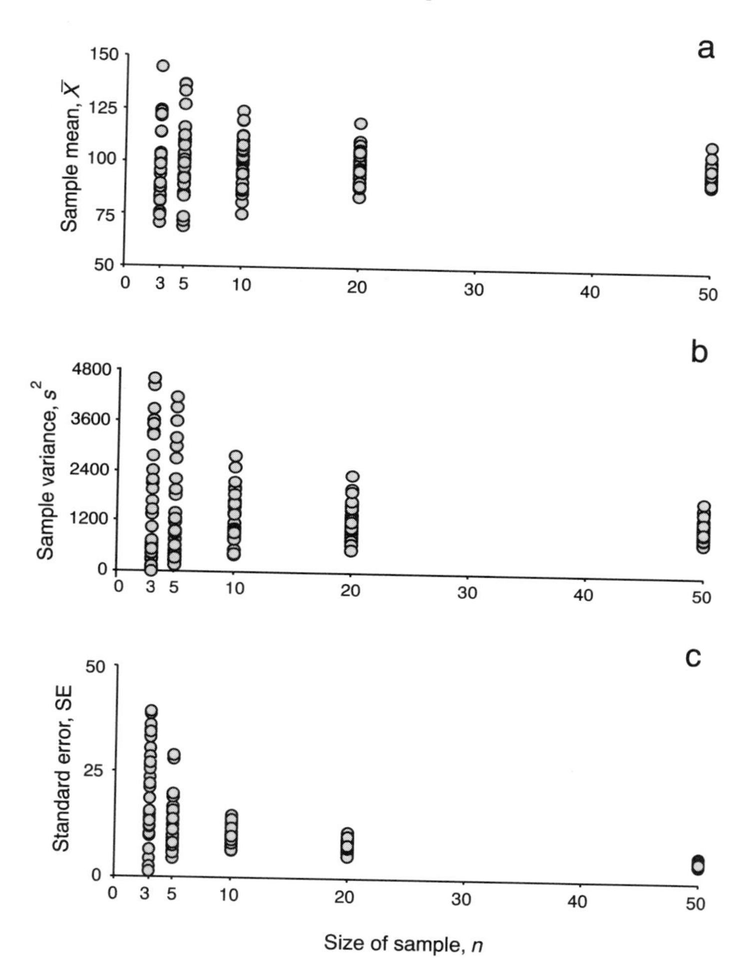

Figure 5.7. Effect of size of sample on various statistical estimates from a population with $\mu = 100$, $\sigma^2 = 1200$. For each size of sample ($n = 3, 5, 10, 20, 50$), 30 samples were taken. (a) The sample mean ($\bar{X}$) is not influenced, on average, by size of sample, but the scatter of sampled values decreases as n is larger. (b) The same increase in precision occurs with the sampled variance, s^2, which is also not affected, on average, by a change of size of sample. (c) The standard error decreases with increasing size of sample.

sample. Theoretically, when the entire population is sampled (when n equals infinity), $t_{0.05}$ becomes 1.96 as in the standard normal distribution. This is the asymptotically smallest value – when σ^2 is known, not estimated. As n increases, s^2 is more likely to be close to the true value, σ^2. This is illustrated in Figure 5.7b, again by repeatedly sampling the population of birds, with different sizes of sample. The scatter of s^2 around σ^2 decreases with increasing size of sample. The t-distribution was invented to cope with this scatter.

In passing, it is worth noting that variance (σ^2) is a parameter and therefore a property of the distribution being sampled. Despite commonly expressed misconceptions to the contrary, estimates of variance (i.e. s^2) are not smaller if samples are larger. In Figure 5.7b, there is no trend in magnitude of s^2 with increasing size of sample. It always averages σ^2. What is changing with increasing n is the decreasing range of values of s^2 – i.e. the precision of s^2 is increasing. Thus, the standard error decreases with increasing size of sample (Figure 5.7c). The size of sample is obviously under the control of the experimenter, although logistic constraints, lack of finance for the research and, sometimes, inertia by the investigator usually determine how many replicates are taken.

5.7.3 The variance of the population (σ^2)

Obviously, for a very variable population (with σ^2 relatively large), a single sample will not provide as precise an estimate of the mean as in a less variable population. In Equation 5.11, this is obvious because increasing σ^2 (and therefore s^2) would increase the confidence interval. It is also obvious from the point of view of common sense. A random sample of some chosen size (chosen n) will be less representative of a more variable population than of a population which has limited variability around its mean.

This is shown in Figure 5.8, where numerous, but independent samples of size $n = 20$ have been taken from several normal populations with the same mean ($\mu = 100$), but with different variances. The scatter of sampled means obeys the central limit theorem because each population sampled is normally distributed. So, the sampled means for each distribution are normally distributed with mean $\mu = 100$ and variance $\sigma^2/20$ (see Section 5.1).

The intrinsic variance of a population being sampled is a parameter of the population. It is not under the control of the experimenter. Once a defined population is to be studied, σ^2 is set because it is a property of the population. The precision of sampling can then be controlled by choice of probability used to construct the confidence interval and the size of sample taken. The former is often set by convention (but see below, Section 5.13). The latter is often set by logistic or other constraints. Whether it matters to have precise samples depends entirely on what use will be made of the data. This topic is intimately associated with the alternative hypotheses being examined (see Section 5.13). Hence, there is no simple answer to

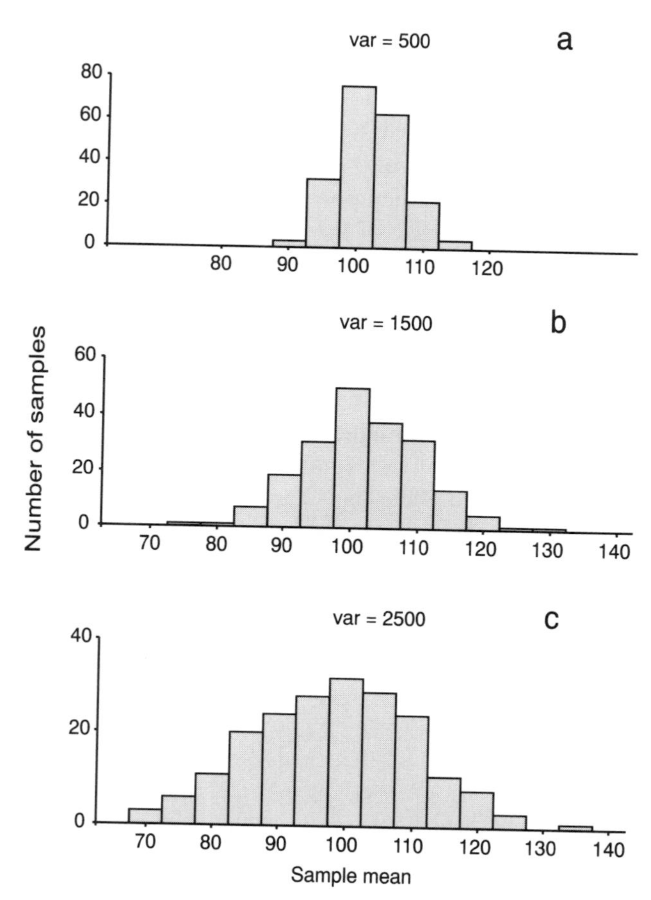

Figure 5.8. Frequency distributions of 200 randomly sampled means (all with $n = 20$) from a normal distribution with $\mu = 100$. The variance of the sampled means increases with increasing variance of the distribution sampled, which was (a) $\sigma^2 = 500$, (b) $\sigma^2 = 1500$, (c) $\sigma^2 = 2500$.

the oft-repeated question 'How many replicates shall I take in my samples?' The answer is not separable from considering the consequences of having too many (at cost to the project) or too few (at the expense of precision and power). This is discussed below (Section 5.13).

5.8 Description of sampling

The preceding account of a measure of precision allows a coherent description of the results of sampling a population to estimate its mean. There must be careful definition of the hypotheses being addressed (Section 2.5), so that it is possible to determine what is the appropriate population to be sampled. Then there must be careful definition of the population sampled, to ensure that it conforms to that specified in the hypotheses (Section 3.4). Finally, there must be a complete account of how the sample is taken, so that its validity as a representation of the appropriate population can be assured. Whenever it is possible, the size of the sample should be chosen to provide appropriate precision (see later).

After the sample has been taken, the results should be described by reporting the sample estimate of the mean (i.e. $\bar{X}$). Then sufficient information should be provided to allow readers or users of the data to determine the precision of sampling. To achieve this, you must report the size of sample (n) and some estimate of variance (s^2), standard deviation (s), standard error ($s/\sqrt{n}$) or confidence interval, at some stated probability (for details, see Sections 5.2 and 5.3). Any of these estimates of variation would be appropriate because each could be used to generate the others. Which you use really depends on what emphasis you wish to give your data (see Table 5.2). None of these measures provides a useful statistical 'test' of whether means differ, as sometimes assumed or claimed from graphical plots of sampling using confidence intervals. This point is illustrated when tests of hypotheses involving more than one population are discussed in Chapter 6.

5.9 Student's *t*-test for a mensurative hypothesis

Earlier (Section 2.10) it was pointed out that measurements are rarely, if ever, taken purely for their own sake. Usually, the sampling is to estimate a mean to determine whether it is in agreement with some hypothetical value, predicted from some model of the world. It is therefore important to identify how to test such hypotheses using a sample estimate of the mean measurement. This is presented as a simpler version of the same test using confidence limits as described for the test of martian beer-cans in Section 5.5.

This time, however, we will use the example of eggs per nest considered briefly in Section 2.10, except that we will examine the weight of the eggs. Previous research on the birds suggests an average weight of eggs per nest of 15 g, that being the average recorded elsewhere. The null hypothesis being examined for southern Europe is as follows:

$$H_0: \quad \mu = 15 \tag{5.13}$$

If the origins of this, given the starting observations and model in Section 2.10 are confusing, or you have already seen a problem with the underlying logic of this case, just wait until the quantitative aspects of it have been considered. The logic and its difficulties are discussed below (Section 5.10). The test statistic to be used is the t-statistic. Its calculation will be done by transforming the equation for construction of confidence limits, using the following argument. Assume that the population of nests is such that samples of the mean weight of eggs per nest would be normally distributed (note: the sampled means, *not* the original distribution of weights of eggs per nest). We know from theory that, under these circumstances, for 95% of all samples (using 2-tailed t values):

$$\bar{X} - t_{0.05}\sqrt{(s^2/n)} \leq \mu \leq \bar{X} + t_{0.05}\sqrt{(s^2/n)} \tag{5.14}$$

We can rearrange this, so that for 95% of all possible samples:

$$\bar{X} - \mu \leq t_{0.05}\sqrt{(s^2/n)} \quad \text{and} \quad -t_{0.05}\sqrt{(s^2/n)} \leq \bar{X} - \mu \tag{5.15}$$

Therefore:

$$\frac{\bar{X} - \mu}{\sqrt{(s^2/n)}} \leq t_{0.05} \quad \text{and} \quad -t_{0.05} \leq \frac{\bar{X} - \mu}{\sqrt{(s^2/n)}}$$

So

$$-t_{0.05} \leq \frac{\bar{X} - \mu}{\sqrt{(s^2/n)}} \leq t_{0.05}$$

This leads naturally to defining a test statistic as:

$$t_{\text{obs}} = (\bar{X} - \mu)/\sqrt{(s^2/n)} \tag{5.16}$$

where μ is the mean specified in the null hypothesis. For 95% of all possible samples from a normal distribution with mean, μ (as stated in H_0):

$$-t_{0.05} < t_{\text{obs}} < t_{0.05} \tag{5.17}$$

Thus, if the observed value, t_{obs}, is larger than $t_{0.05}$ or smaller than $-t_{0.05}$, we should reject the null hypothesis because we have obtained an unlikely value of t_{obs} if the null hypothesis were true. It is more reasonable to conclude that the sample has come from a population with a mean different from the value (μ) proposed in the null hypothesis than to assume that an unusual sample has been obtained from that distribution (see Section 4.5).

In this example, we have a null hypothesis (Equation 5.13). As in Section 4.3, we decide to use the observed value of t as the test statistic.

This is simple to measure in a sample and its distribution is known when the null hypothesis is true, provided that means of samples are normally distributed. In this case, they are probably close to normal because there will be a limited range of weights of eggs. The weight of eggs per nest is also likely to have a unimodal distribution around its mean. For conventional reasons (only), we shall choose a probability of Type I error (rejecting the null hypothesis when it is true) of 0.05 and, for convenience, we shall take a sample of 14 nests. We therefore have critical values for $t_{0.05}$ with 13 degrees of freedom.

To test the null hypothesis proposed for the weight of eggs per nest of the bird in southern Europe, we take a sample of $n = 14$ nests, providing an estimate $\bar{X} = 10.35$ (1 nest had no eggs, 2 had 5 g of eggs, 7 had 10 g, 3 had 15 g and 1 had 20 g weight of eggs) and a sample variance $s^2 = 24.75\ \text{g}^2$. The critical value for the test (i.e. $t_{0.05}$) with 13 degrees of freedom is ± 2.16 (from a table of the t-distribution). Our observed value of t is calculated for the null hypothesis ($\mu = 15$). Therefore:

$$t_{\text{obs}} = \frac{10.35 - 15}{\sqrt{(24.75/14)}} = -3.50 \tag{5.18}$$

and we should reject the null hypothesis because the observed value of the test statistic is in the region of rejection (outside the range defined by the chosen critical value – see Section 4.3.3). We reject the proposed mean of 15 g of eggs per nest in favour of a smaller mean, estimated to be 10.35.

It may be wrong to reject the null hypothesis (H_0) because the sample could have been one of the 5% of possible samples when the mean is 15 g that generate values of t larger than 2.16 or smaller than -2.16. The probability of being wrong is the chosen value of probability of Type I error, i.e. 5%.

We have finally done a realistic statistical test and rejected a null hypothesis. We conclude, ecologically, that the birds in southern Europe have an average clutch-weight smaller than is recorded elsewhere for that species. This would serve as an objective observation leading to research programmes to explain why (i.e. to propose and test models to explain why there are smaller or fewer eggs per nest in southern Europe).

5.10 Goodness-of-fit, mensurative experiments and logic

Observant and perceptive readers will have noticed something wrong with the logic of the previous example. The logical framework used did not fit with that discussed in Chapter 2. The observations and model were explained in Section 2.10. The observations are that this species elsewhere lay 15 g of eggs per nest and there is a persistent population breeding in southern Europe. The model to explain the latter is that the southern European population persists because it reproduces at the same rate as populations elsewhere. Thus, the logical hypothesis is that there will be an average of 15 g of eggs per nest in the population in southern Europe.

This poses a logical dilemma. The logical null hypothesis now includes all other possible weights of eggs per nest – theoretically an infinite number of possibilities. There is therefore no way to identify what might serve as a test statistic. There is no way to specify a measurable quantity to serve as a statistic such that its frequency distribution is known for an infinite number of possible alternatives. There is an infinite number of frequency distributions associated with the infinite number of possibilities included in the null hypothesis.

The only statistical solution to this problem is to specify that the test statistical frequency distribution be based on the specified *hypothesis* (i.e. $\mu = 15$), which is unique, rather than by use of a logical null hypothesis (for details, see Underwood, 1990). Therefore, the statistical null hypothesis is equal to the logical hypothesis. When we rejected the null hypothesis in the example in Section 5.5, we were using good falsificationist logic, but we were then also rejecting the hypothesis and the model from which it was derived (as in Figure 5.9). This is completely the opposite conclusion to that used and advanced earlier (contrast with Figure 2.1). There is, however, no logical interpretative problem as a result of this procedure.

The problem arises when we *fail* to reject the null hypothesis in a goodness-of-fit test (or mensurative experiment). We make a predictive hypothesis ($\mu = 15$) and gather data to test it. Suppose, for example, the 14 nests had contained 5 g more of eggs each when we sampled them (i.e. one nest had 5 g of eggs, two had 10, seven had 15, three had 20 and one had 25 g of eggs). The mean is now $\bar{X} = 15.35$, with the same sampled estimate of variance, $s^2 = 24.75$. Therefore,

$$t = \frac{15.35 - 15}{\sqrt{(24.75/14)}} = 0.26 \qquad (5.19)$$

which is not a significant departure from what one would expect if the sample really did come from a population with a mean of 15 g. We would retain the null hypothesis and, therefore, the hypothesis and model (see Figure 5.9).

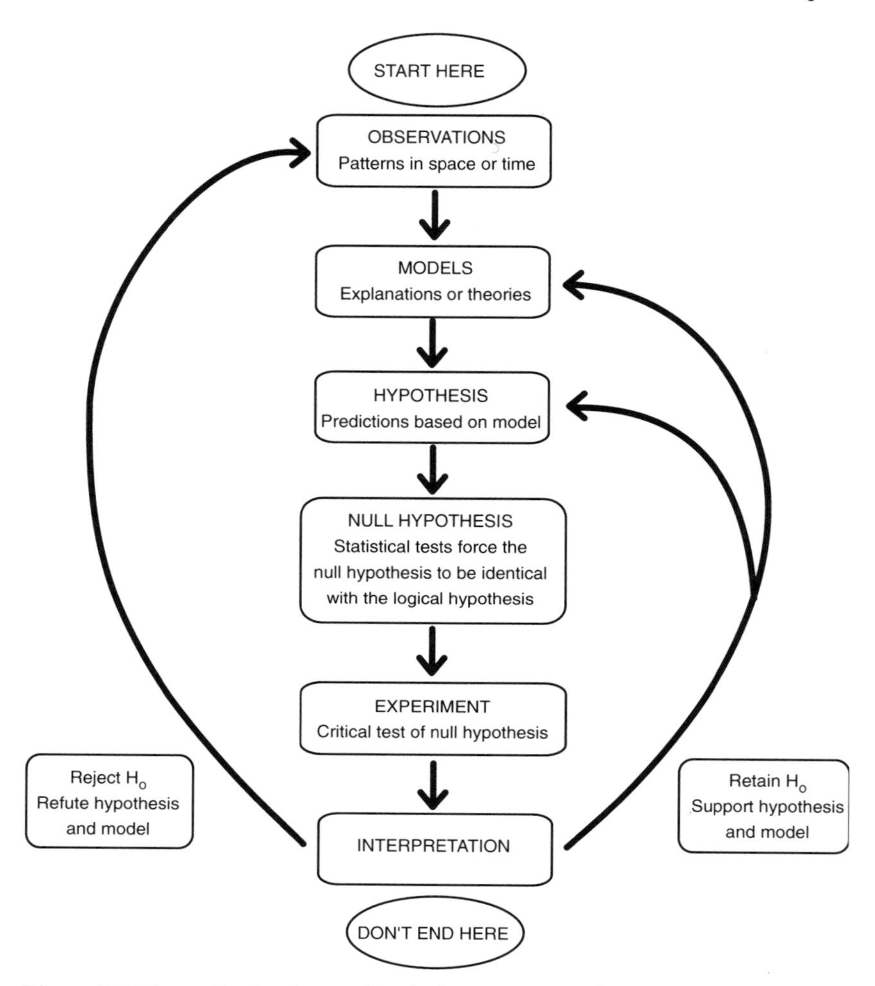

Figure 5.9. Generalized scheme of logical components of a research programme when the statistical tests require that the statistical null hypothesis be identical to the logical hypothesis. Relationships are discussed in the text and described in detail by Underwood (1990). Note the opposite conclusions and pathways from the corresponding, logically more sound procedure in Figure 2.1.

The conclusion to support the model is not based on any falsificationist procedure – nothing has been shown to be wrong. Drawing conclusions from such corroboratory data is a logical fallacy – an error. It has long been known to be an error and is called 'affirming the consequent' (e.g. Lemmon, 1971; Hocutt, 1979; see the detailed discussion in Underwood, 1990). Thus, demonstrations that data produce estimates in accordance with (that fit well) the hypothesis (the prediction made) are not useful in any logical framework. Data that do not fit provide a rejection of the underlying model.

There is a further problem with such mensurative experiments in that there is no theoretical or empirical basis from which to attempt to calculate the power of the statistical test used (Underwood, 1990). This topic is discussed below (Section 5.12).

Unfortunately, there is no solution to the logical problems: the illogicalities of corroborating hypotheses when no opposite null hypothesis has been constructed. In some cases, it is possible to use a different hypothesis derived from the underlying model. Some examples were discussed by Underwood (1990). For the supposedly simple example of birds' nests considered here, there is no obvious way to by-pass the problem (which is why in Section 5.5, I carefully ensured that the null hypothesis was rejected rather than retained!). All this demonstrates is that apparently simple cases are *not* simple. The supposedly easy bits of a biological research programme – the description and measurements of observations at the beginning – are often very complex.

The crucial message from this discussion is that a great deal of careful thought must go into the procedures used and their logical framework, even when the task seems straightforward and simple.

5.11 Type I and Type II errors in relation to a null hypothesis

Consider a slightly more complex, but similar example of goodness-of-fit. It is known that the average weight of adult house-mice in Europe is 23 g. You have found an isolated population of these mice on an island off the coast of Spain and you wish to spend many more of your holidays there. You consider, therefore, a research project, probably under the headings of biodiversity and conservation, on the island and the mice. You therefore consider, for reasons discussed above, that this new population will conform to the pattern in the rest of Europe. You therefore propose the null hypothesis that the mean weight of the mice will be 23 g.

You do, however, also realize that some variation in mean weight probably does occur without much importance from place to place and population to population. You therefore decide that any difference of about 10% or smaller in mean weight is not 'important'. It is the sort of thing that is expected among populations of mice and would not serve to indicate any special characteristic of the mean size of the population you wish to study. You now sample the mice with a representative sample of

Table 5.3. Type I and Type II errors in interpretations of results of experiments

Outcome of statistical test is to:	Null hypothesis (H_0) is (unknown to us): True	False
Reject H_0	Type I error occurs with probability α chosen by experimenter	Correct conclusion: false H_0 is rejected
Retain H_0	Correct conclusion: true H_0 is retained	Type II error occurs with probability β sometimes chosen by experimenter

$n = 20$ individuals (an arbitrarily chosen number). The sample estimate of the mean was $\bar{X} = 27.4$ and of the variance $s^2 = 176.8$. As a result:

$$t_{obs} = \frac{27.4 - 23}{\sqrt{(176.8/20)}} = 1.48 \tag{5.20}$$

which is not significant ($t_{0.05} = 2.09$ with 19 degrees of freedom). We therefore retain the null hypothesis, even though the estimated mean is more than 10% larger than the value predicted and therefore has already been determined to be 'important' to us.

This example introduces the other statistical error, that of failing to reject the null hypothesis when it is, in fact, false and a specified alternative hypothesis is true. Failure to reject it occurs because the sample data provide no statistically significant evidence to reject it.

Consider Table 5.3, which shows the only four possible outcomes of a statistical test. If the null hypothesis is true and is retained as a result of the statistical test or is false and is rejected, the correct inference has been made. If the null hypothesis is true, but is rejected, the experimenter has, unwittingly, made a mistake. The probability of making such a mistake is, however, entirely under the control of the experimenter. It is chosen, in advance of the statistical test by the choice of critical values and the region of rejection (Sections 4.3.3 and 4.4). So far, this has been arbitrarily determined to be a probability of 0.05, 1 in 20. This probability of Type I error is commonly known as the probability α.

Rejection of a true null hypothesis occurs whenever a sample estimate of the parameters of the population generates a test statistic in the region of rejection. We keep α small so that such samples are unlikely.

There is, however, the opposite error, which occurs when the null hypothesis should be rejected because it is wrong. The sample, however, generates a test statistic that is consistent with the null hypothesis. How this comes about is illustrated in Figure 5.10. The first distribution (a) is the normal distribution of weights of mice, centred on the mean specified by the null hypothesis ($\mu = 23$). The second distribution is the distribution of weights (with the same variance as for the previous distribution) with a mean of 25.3, some 10% larger than the one specified in the null hypothesis (Figure 5.10b). Below these (Figure 5.10c) is the null distribution of t_{obs}; this is the distribution of values of t_{obs} that would be obtained from repeated independent sampling (with samples of size $n = 20$) of the distribution of weights when the null hypothesis is true and the mean is 23. Finally, there is the distribution of values of t_{obs} that would be obtained from repeated sampling (again with $n = 20$) of the alternative distribution, with mean weight 25.3.

Note that the two distributions of t overlap. Thus, some samples obtained from either the null or the alternative distribution of weights will generate values of t_{obs} that could come from either of the two frequency distributions of t.

The region of rejection was specified in advance by $\alpha = 0.05$ (i.e. the probability of Type I error is 0.05), as shown by the shaded area in Figure 5.10c. A value of t_{obs} of, say, 1.5 would therefore not cause rejection of the null hypothesis. If, on the one hand, the null hypothesis were true, this value of t_{obs} would have come from the null distribution and the conclusion would be correct. On the other hand, if the null hypothesis were false, this value of t_{obs} would have come from the alternative distribution and the conclusion would be incorrect – a Type II error.

In this example, because we have specified the smallest alternative to the null hypothesis that we think should cause us to reject it, if it is false (i.e. a mean weight 10% larger or smaller than that in the null hypothesis), we could calculate the probability of Type II error. It is the shaded region in Figure 5.10d – the set of values of t_{obs} in the alternative distribution of t that are smaller than the critical value (which is 2.09, the value of $t_{0.05}$ with 19 degrees of freedom from a table of the distribution of t). Any of these values would cause us to retain the null hypothesis. When the alternative hypothesis is true, the probability of getting, from the alternative distribution of t, a value of t_{obs} that is smaller than the critical value ($t_{obs} = 2.09$, see above) is the probability of Type II error. This probability (the chance of retaining the null hypothesis when the alternative is correct) is known as β. So, the probability of incorrectly retaining the null hypothesis when a particular alternative is true is β. The power of the test to reject the null hypothesis when the specified alternative is

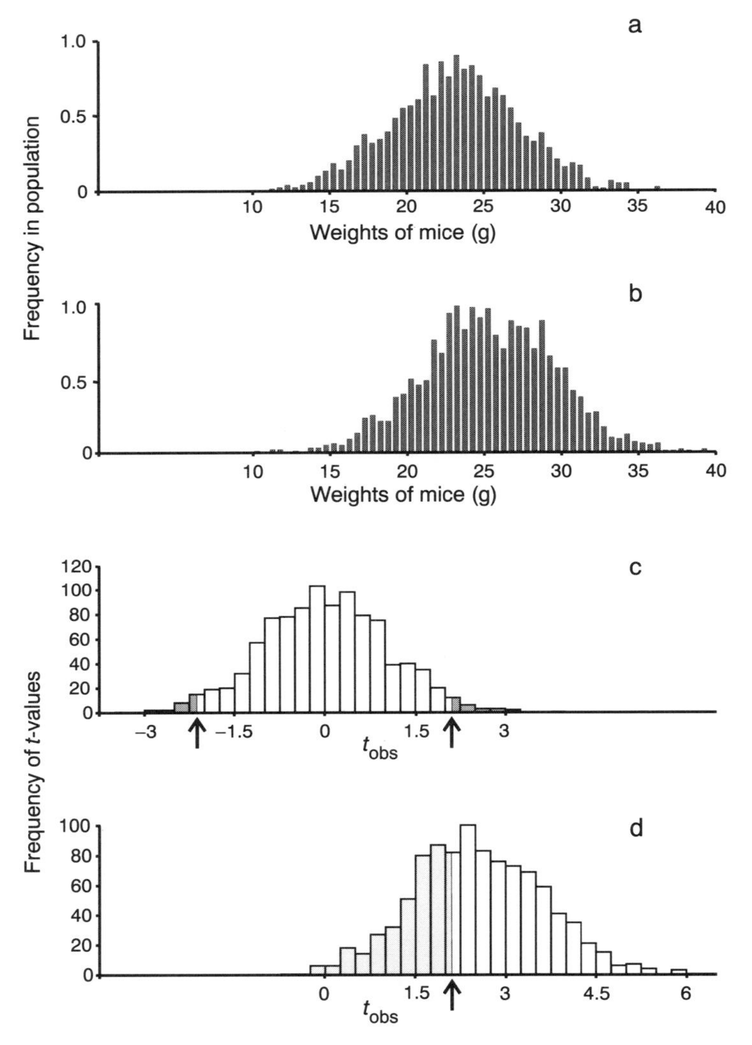

Figure 5.10. Demonstration of Type I and Type II errors in t-tests. The mean weight of mice is sampled to determine whether they weigh, as claimed, 23 g (i.e. the null hypothesis is H_0: $\mu = 23$). The weights are normally distributed. In (a) and (b), the distributions of weights of mice are shown (a) when the null hypothesis is true and $\mu = 23$ and (b) when H_0 is false and the mean is larger, $\mu = 25.3$. In (c) are observed values of t_{obs} from numerous samples of population (a), with $n = 20$. The probability of incorrectly rejecting the null hypothesis (α, the probability of Type I error) is chosen to be 0.05. The arrow indicates the critical value of t at $P = 0.05$, $t_{crit} = 2.09$, with 19 degrees of freedom. The shaded areas ($t > 2.09$ and $t < -2.09$) show this probability. In (d) are the observed values of t_{obs} from numerous samples of population (b). The probability of incorrectly retaining the null hypothesis (β, the probability of Type II error) is shown as the shaded area, i.e. all values in distribution (d) that are smaller than 2.09, the upper critical value for distribution (c).

true is therefore the probability of correctly rejecting the null in favour of the alternative. Power for a specified alternative hypothesis is therefore defined as:

$$\begin{aligned} \text{Power} &= 1 - \text{Probability of Type II error} \\ &= 1 - \beta \end{aligned} \tag{5.21}$$

5.12 Determining the power of a simple statistical test

The power of a test is an important element in the design of good experiments. Obviously, there is little point in doing an experiment that has only a small chance of detecting a departure from some null hypothesis, unless you want to spend your entire career failing to corroborate any of your theories. Biologists, however, have been obsessed with conventional and arbitrary views about the probability of Type I error, so that we use $P = 0.05$ as though it were written on a tablet of stone by some god of statistical theory. It is convention, but it affects the power of the experiments we do. It also leads to the notion that rejecting null hypotheses is useful because it leads to publications of results (and fame, fortune, tenure). It has not led to rational consideration of how to ensure that the experiments we do are sufficiently powerful that, when we fail to reject the null hypothesis, the proper inference is that the null is correct. In other words, if experiments are not powerful or are of unknown power, there is a constant temptation to assume that when a null hypothesis is retained (the test statistic is not in the region of rejection), we should consider that the experiment has failed. Retention of null hypotheses is a very important component of the logic of experimentation (see Figures 2.1 and 5.9), yet could really only be useful if the experiments were powerful and the probability of Type II error (retention of an incorrect null hypothesis) unlikely.

Here, we shall consider the components of hypotheses and experimental designs that influence the power of the experiment. Power is defined in terms of the probability of Type II error. Accordingly, it is influenced by the critical value set for the null hypothesis, i.e. the probability of Type I error. It is also influenced by the size of the experiment, i.e. the sample size, n in the case of a simple mensurative experiment of the sort considered so far. It is also influenced by the intrinsic variability in the population being sampled, where the variance is used as part of the calculation of the test statistic, such as the t distribution. Finally, the power of a test is influenced by what is known as the '*effect size*', i.e. how large a difference there is between the null hypothesis and the specified alternative hypothesis. Each of these is considered in turn.

5.12.1 Probability of Type I error

Examination of Figure 5.10 immediately reveals why and how the chosen level of probability of Type I error influences the power of an experiment. The choice of α sets the criterion of rejection in the null distribution of the test statistic. In Figure 5.10c, the null distribution of the *t*-statistic is its frequency distribution if the null hypothesis is true. Setting α to a different probability moves the boundary that is used to decide whether to reject or to retain the null hypothesis. Consequently, it moves the boundary that determines whether or not we make a mistake in retaining the null hypothesis when the specified alternative is true. As a result, decreasing α increases β as shown in Figure 5.10d. If α is made smaller, the critical value moves to the right and there is an increase in the proportion of the alternative distribution of t that is smaller than it. If you choose a probability of Type I error of 0.01, or 0.001, to try to avoid rejecting null hypotheses by accident, you make it very difficult for the test to detect the alternative when it occurs. The chance of erroneously retaining the null hypothesis is increased.

In many experimental situations, it would be sensible to hedge your bets by having the probability of making either of the two kinds of mistake (Table 5.3) equal. In other words, if you are going to make a mistake, it would seem sensible to try to be equally likely to make the mistake in either direction – falsely rejecting or falsely retaining the null hypothesis as opposed to the specified alternative hypothesis.

In some situations, this is not the correct thing to do. For example, the precautionary principle as it applies to testing of drugs or to environmental monitoring would suggest very strongly that you should be prepared to make Type I errors much more often than Type II errors. In other words, α should be quite large relative to β. The argument for this is that, when a new drug is to be released onto the human market, the usual hypothesis is that it has no untoward side-effects. These are usually defined in some quantitatively estimable form. The statistical null hypothesis (which, in this case, is the same as the logical hypothesis) will be that there is no increased incidence of whatever the defined side-effect is for patients taking the drug (the experimental population) compared with controls taking a placebo (the control population). If there is no effect of the drug, but the statistical test suggests there is one, a Type I error occurs and the drug cannot be released for sale. This would normally be followed by much more research into the nature and causes of the unpredicted side-effects. If done well, this should eventually lead to the discovery that the previous conclusion was a mistake. In contrast, if there are side-effects, but the experimental procedures to detect them are not powerful enough, a Type II error occurs and the statistical test will suggest there are no side-effects. The drug is then sold with potentially disastrous consequences before the mistake is identified by increased sickness in the population. Thus, it is more important to make a mistake preventing the release of the drug than to make a mistake causing increased illness.

In the case of environmental impact, it is often predicted as the null hypothesis that there will be no effect of some development or discharge (for examples, see Fairweather, 1991). If the sampling and experimental work done to detect the potential impact is inadequate because it is of insufficient power to detect quite reasonable alternatives (i.e. quite realistic impacts), then the development or discharge continues even though there is an impact. Instead the sampling could be done with a larger probability of Type I error. If a Type I error is made so that an apparent, but unreal, impact is detected in the sampling and experimental work, presumably the development or discharge is re-assessed and may be delayed until the problem is fixed. The research to fix the problem will potentially also identify that a mistake has been made in the outcome of the experiment, but no damage has been done to the environment.

Thus, in both cases, the probability of Type II error should be kept small, even at the expense of the probability of Type I error. For any given alternative to the null hypothesis, the two probabilities are intimately and reciprocally related. How to manage the appropriate determination of these probabilities must wait for later discussion.

5.12.2 Size of experiment (n)

The power of an experiment is increased by having larger samples. This accords with common sense in that the more work you do on the experiment, the more likely it should be that the experiment is useful in rejecting an incorrect null hypothesis. In the case of a simple mensurative experiment of the sort discussed here, the issue is entirely the size, n, of the single sample taken. If you increase n, the standard error of the sample decreases (Figure 5.7). Correspondingly, the magnitude of t increases (Equation 5.16), because the standard error is a divisor in the calculation of t. At the same time, the degrees of freedom for the test (i.e. $n - 1$)

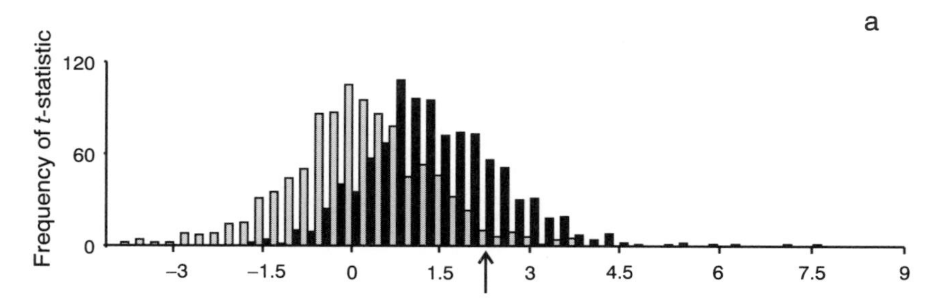

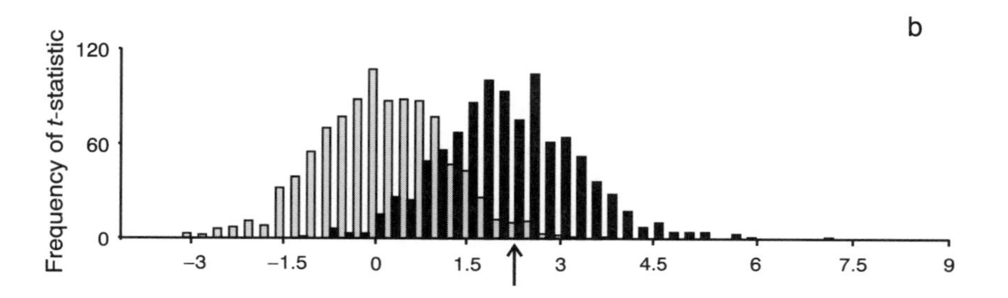

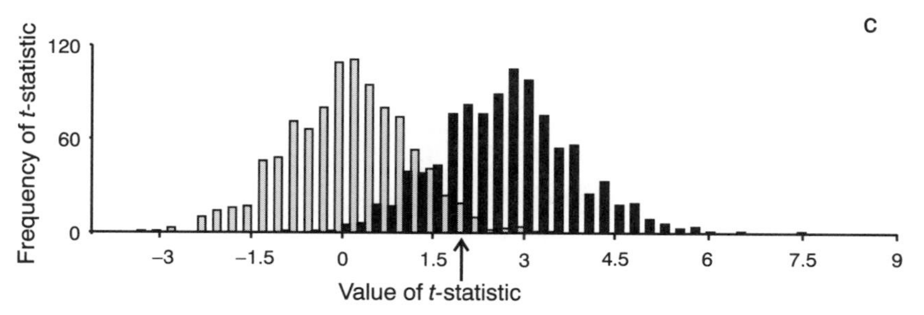

Figure 5.11. Effect on power of a t-test of increasing sizes of samples (n). In each pair of sets of data are results from 1000 samples of a normal distribution. The data are t-values for each sample calculated to test the null hypothesis that $\mu = 100$. Shaded bars represent samples from a normal distribution when the null hypothesis was true ($\mu = 100$). Filled bars represent samples when the null hypothesis was false ($\mu = 110$ in all cases). The variance of sampled distributions was $\sigma^2 = 800$ in all cases. Sizes of sample were $n = 10, 30, 50$ for (a), (b), (c), respectively. The arrows indicate the upper critical value of t (at probability of Type I error = 0.05) for two-tailed tests. These values are 2.262, 2.045, 2.009, for (a), (b), (c), respectively. Shaded bars to the right of the arrow indicate α, the probability of Type I error when H_0 is true. Filled bars to the left of the arrow indicate β, the probability of Type II error when H_A is true (see also Table 5.4).

Table 5.4. Power of t-test of a mensurative null hypothesis (H_0: $\mu = 100$) under different conditions. H_A specifies the value of μ for the alternative hypothesis; σ^2 specifies the variance of the normal distributions sampled

Probability of Type I error = 0.05

			One-tailed test		Two-tailed test		
n	H_A	σ^2	Critical value of t	Power	Critical value of t	Power	Refer to Fig.
(a) Different sizes of sample							
10	110	800	1.83	0.277	±2.26	0.175	5.11a
30	110	800	1.70	0.548	±2.04	0.460	5.11b
50	110	800	1.68	0.779	±2.01	0.669	5.11c
(b) Different variances							
40	110	800	1.68	0.705	±2.02	0.598	5.12a
40	110	1000	1.68	0.620	±2.02	0.490	5.12b
40	110	1200	1.68	0.555	±2.02	0.429	5.12c
(c) Different effect sizes							
40	110	800	1.68	0.705	±2.02	0.598	5.13a
40	115	800	1.68	0.943	±2.02	0.906	5.13b
40	120	800	1.68	0.999	±2.02	0.996	5.13c

increase, changing the critical value. The best way to illustrate the effects of increasing n is by a graph of different distributions of t under a specific null and alternative hypothesis, with different sample sizes, as in Figure 5.11 and Table 5.4. Increasing the size of sample increases the power of the test to detect an alternative.

5.12.3 *Variance of the population*

The intrinsic variability of the population being sampled also influences the power of any statistical analysis. A moment's thought suggests that the more variable a population, the less likely it is that sampling will allow a distinction between the null and alternative hypotheses. In the calculation of t, the divisor obviously increases with increasing variance (Equation 5.16). Thus, the value of t will, on average, be smaller with larger variances. This is illustrated in Figure 5.12 and Table 5.4. Where σ^2 is larger, the power of the test is decreased.

This is the only component of sampling and experimentation that affects the power of the experiment and is not a function of the hypothesis (as is the effect size, see below) nor under the control of the experimenter (as are the size of sample and the choice of probability of Type I error).

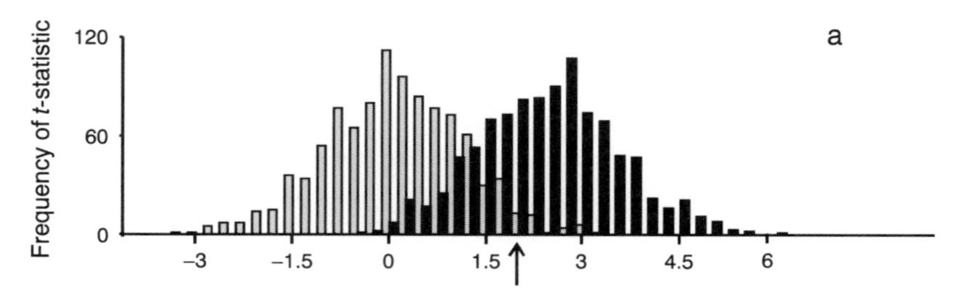

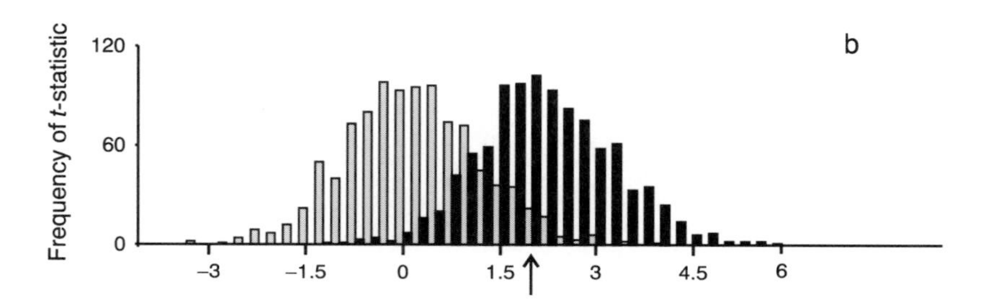

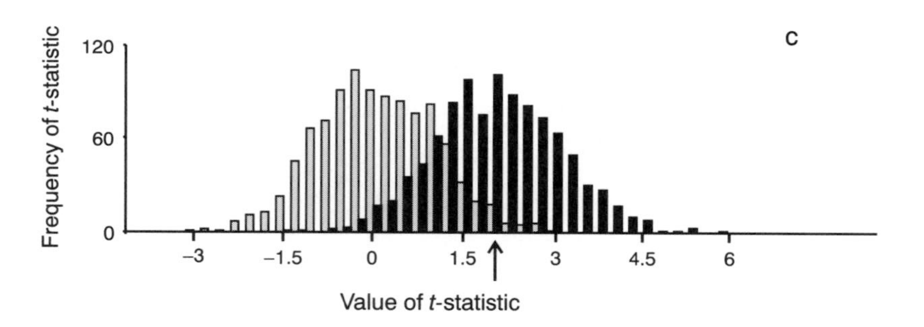

Figure 5.12. Effect on power of a t-test of increasing variance of samples. In each pair of sets of data are results from 1000 random samples of a normal distribution. The data are t-values for each sample calculated to test the null hypothesis that $\mu = 100$. Shaded bars represent samples from a normal distribution when the null hypothesis was true ($\mu = 100$). Filled bars represent samples when the null hypothesis was false ($\mu = 110$ in all cases). Sizes of sample were $n = 40$ in all cases. Variances of the sampled distributions were 800, 1000, 1200 for (a), (b), (c), respectively. The arrows indicate 2.02, the upper critical value (at probability of Type I error = 0.05) of t for two-tailed tests. Shaded bars to the right of the arrow indicate α, the probability of Type I error when H_0 is true. Filled bars to the left of the arrow indicate β, the probability of Type II error when H_A is true (see also Table 5.4).

5.12.4 Effect size

This ungrammatical term refers to the difference between the null hypothesis and the specified alternative hypothesis. Suppose a new outfall is proposed in some river in which there is an estimated mean abundance of a species of polychaete of 100 per m^2. The nutrients to be discharged from the outfall may cause a change in abundance of these animals and all previous knowledge suggests an increase of about 10% (i.e. an increase to 110 per m^2) during the first three months of discharge. The proposers of the outfall claim that the rate of dilution and dispersal of effluents will result in no effect on the animals (no environmental impact). So, their claim is the null hypothesis (i.e. after three months, H_0: $\mu = 100$) and the alternative specifies a difference of 10 animals per m^2 (H_A: $\mu > 110$). The 'effect size' is the minimal difference specified by the alternative hypothesis.

The influence of effect size on the power of an experiment is illustrated by Figure 5.13, where three different scenarios are illustrated. In each, the null hypothesis specifies a mean of 100. The alternative is 110, 115 or 120 in each of the three cases, i.e. effect sizes of 10, 15 or 20. Each situation is modelled with the same variance and size of sample ($n = 40$). As dictated by common sense and the formula for t (Equation 5.16), the larger effect sizes are much more likely to be detected. The power of the experiment increases with increasing effect size (Figure 5.13 and Table 5.4). Obviously, choice of an appropriate effect size is the business of the experimenter, based on the biology of the situation being examined and the nature of the hypotheses being investigated.

5.13 Power and alternative hypotheses

The preceding illustration of influences on the power of an experiment demonstrates one enormously important point about the logic and design of experiments. The alternative hypothesis must be specified before it is possible to determine the power of any planned experiment or to design the experiment to have a pre-determined power.

This fact requires biologists to spend much more time thinking about the alternatives to the null hypotheses. It requires more effort in phrasing hypotheses in quantitative, rather than qualitative terms. The production of better quantified hypotheses will require much more specific and precise models than are often used. Of course, all of this will depend on clearer, more formally described quantitative observations. Models to explain them will need more coherent quantitative components and the hypotheses

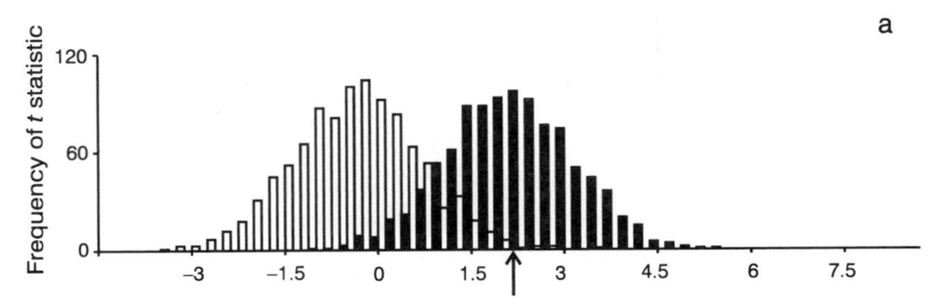

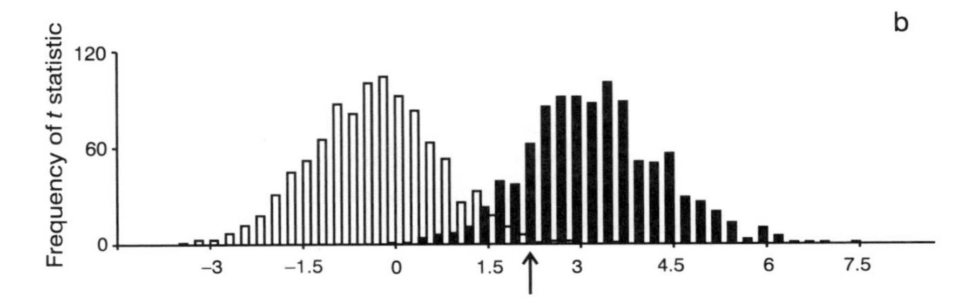

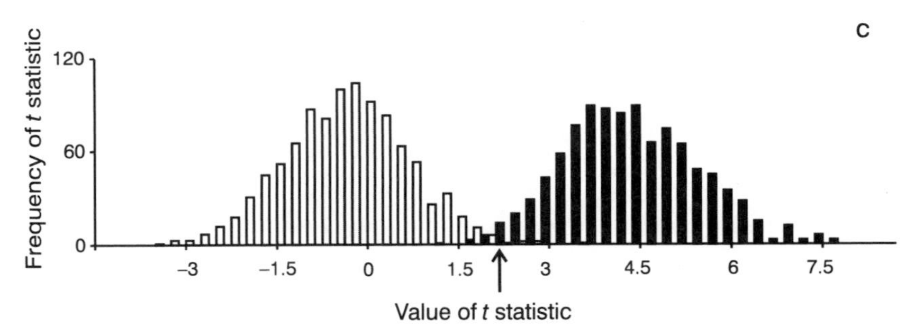

Figure 5.13. Effect on power of a t-test of increasing effect sizes. In each pair of sets of data are results from 1000 samples of a normal distribution. The data are t-values for each sample calculated to test the null hypothesis that $\mu = 100$. Shaded bars represent samples from a normal distribution when the null hypothesis was true ($\mu = 100$). Filled bars represent samples when the null hypothesis was false ($\mu = 110, 115, 120$ for (a), (b), (c), respectively). The variance of sampled distributions was $\sigma^2 = 800$ and $n = 40$ in all cases. The arrows indicate 2.02, the upper critical value of t (at probability of Type I error $= 0.05$) for two-tailed tests. Shaded bars to the right of the arrow indicate α, the probability of Type I error when H_0 is true. Filled bars to the left of the arrow indicate β, the probability of Type II error when H_A is true (see also Table 5.4).

will need to be realistically phrased in numerical terms. Sometimes, a semi-quantitative approach may be helpful (Cohen, 1977), but it would be better to make better predictions.

In various parts of this book, the power of experiments is discussed in detail for different types of statistical procedures. A complete description of procedures is provided by Cohen (1977) for many common statistical tests. Some consequences of retrospective attempts to calculate the power of an experiment are also discussed later (Sections 8.3 and 8.4). An example of calculation of the power of a t-test for a specified alternative hypothesis is described in Section 6.7.

A major challenge for modern biology is to break out of the constraint of obsession with Type I error. This will involve much more thought about biological and ecological processes operating in nature, so that better models and more structured predictions can be made. Through increased attention to the power of experiments, there will be increased primacy of the biology underlying research programmes. Statistical procedures will become much more subservient to the core of the discipline. Finally, better quantification of predictive hypotheses will go a long way towards answering the criticism raised by Peters (1991) in his accurate reflection on the inadequacies of much of modern practical ecology.

6

Simple experiments comparing the means of two populations

The logic and procedures used in the mensurative or goodness-of-fit tests described in the previous chapter can be used in much more useful and less contrived circumstances. The procedures of t-tests underpin a variety of different topics and tests (including, for example, analyses of regressions). Here, the usefulness of t-distributions and their underlying theory are discussed in comparisons of two populations where there are hypotheses about their means. These occur in a great variety of biological experiments and are one of the most widespread forms of experimentation used. The simplest cases concern what are known as paired data.

6.1 Paired comparisons

The classical case of paired comparisons is known as a 'before–after' contrast. This sort of experiment involves knowing something about a population before the application of an experimental treatment and then measuring it again after the treatment to test the hypothesis that the treatment makes some predicted difference. The same elements in the population are measured before and after the experiment.

Other situations involve naturally paired data. For example, the eldest male and eldest female siblings are naturally paired data – they must come from the same family. Another case involves measurements on the right- and left-hand side of an animal's body to test a null hypothesis in which one side of the animal has been experimentally treated in some way. A final situation where paired data can be used to test hypotheses about means is when there are two analytical procedures to be compared, for example a new and a previously standard method for determining the concentration of chlorophyll in leaves. A replicate sample of leaves can each be split into two and one half of each leaf analysed by each method. The divided material provides naturally paired data, each half using a different method.

Consider the observation that survival of adults of some population of desert rodents decreases during the summer. The model proposed to account for this is that there is a chronic shortage of food or other resources during the summer months and the animals therefore lose condition and die. From this, the hypothesis can be proposed that the mean body-weight of the animals should decline before they die and therefore mean body-weight towards the end of summer will be less than at the start. Note that, yet again, there are problems in trying to identify realistic, but simple, examples. The logic of this one is awry in that the hypothesis does not directly address the issues of availability of resources, nor does it take into account growth of the individuals, but let us ignore that for a moment. The null hypothesis is therefore that the mean weight of the animals at the end of the summer will be similar to that at the start, or that it will, in fact, increase. It is a one-tailed hypothesis and therefore a one-tailed null hypothesis.

A representative sample of animals is caught at the start of summer, weighed, tagged and released. During the summer, mortality of the population is monitored to ensure that there is no chance that all the tagged animals will die before the experiment is complete. When mortality in the population starts to increase, as many as possible of the tagged animals are caught again and weighed. For each individual, there are two weights, one before and one after the summer. At the end of summer, the null hypothesis can be tested. The hypothesis is:

$$\mu_{\text{start}} > \mu_{\text{end}} \qquad (6.1)$$

The null hypothesis is:

$$H_0: \quad \mu_{\text{start}} \leq \mu_{\text{end}} \qquad (6.2)$$

but the same individuals are sampled both times, as shown in Table 6.1. Thus, for these paired data, there is an alternative way to consider the situation. For each individual, there should be no decrease in weight during summer unless the hypothesis is true. The individuals will therefore either increase in weight (one possibility consistent with the null hypothesis) or stay the same. When we measure them, they will not be exactly the same, because of differences in when they last ate, or defaecated, or due to errors in the weighing equipment. Such differences from one individual to another should be expected to cancel out, on average. Thus, we have a sample of differences from before to after; one difference

Table 6.1. t-tests on paired "before–after" data

(a) General case of two-tailed hypotheses

H_0: $\mu_{before} = \mu_{after}$

A random sample of n paired data is collected $(X_{11}, X_{21}, X_{1i}, X_{2i} \cdots X_{1n}, X_{2n})$ where $1 \cdots i \cdots n$ is the replicate and 1. and 2. indicate the paired treatments (1 = before, 2 = after).

$d_i = X_{1i} - X_{2i}$

$$\bar{d} = \frac{\sum_{i=1}^{n} d_i}{n} \text{ estimates } \mu_d$$

$$s_d^2 = \frac{\sum_{i=1}^{n}(d_i - \bar{d})^2}{n-1} = \frac{\sum_{i=1}^{n} d_i^2 - \left(\sum_{i=1}^{n} d_i\right)^2 / n}{n-1} \text{ estimates } \sigma_d^2$$

H_0: $\mu_d = 0$

$$t_{obs} = \frac{\bar{d}}{\sqrt{(s_d^2/n)}} = \frac{\bar{d}}{SE_{\bar{d}}} \text{ with } (n-1) \text{ degrees of freedom}$$

(b) An experiment on weights of tagged mice before and after summer. Each of $n = 14$ randomly selected individuals is weighed at the start and again at the end of the experiment. The differences are analysed by a paired t-test (Section 6.1) and the deviations of differences are analysed by a binomial test (Section 6.8.1)

Replicate	Before $X_{before\,i}$	After $X_{after\,i}$	Difference $d_i = X_{before\,i} - X_{after\,i}$	Direction +/−
1	80.6	72.8	7.8	+
2	88.4	35.8	52.6	+
3	90.6	100.4	−9.8	−
4	103.8	65.4	38.4	+
5	84.9	48.1	36.8	+
6	117.4	100.6	16.8	+
7	72.3	47.7	24.6	+
8	96.8	66.4	30.4	+
9	81.7	57.3	24.2	+
10	91.3	98.6	−7.3	−
11	101.5	96.4	5.1	+
12	92.0	101.8	−9.8	−
13	97.2	92.4	4.8	+
14	99.7	89.1	10.6	+

t-test: $\bar{d} = 16.09$ $s_d^2 = 372.14$ $SE_{\bar{d}} = 5.16$ $t_{obs} = 3.12$, 13 df, $P < 0.05$

---

df, degrees of freedom.

for each animal. If there is no change in their weight, on average, during the summer, these differences will have a mean of zero. So, we could propose the following hypothesis and null, based on the differences (*d* values, see Table 6.1). The hypothesis is:

$$\mu_d > 0 \tag{6.3}$$

because we predict that weight will decline; data will be larger before than after summer for the individuals measured. The null hypothesis is:

$$H_0: \quad \mu_d \leq 0 \tag{6.4}$$

We now have a mensurative experiment of the sort discussed in Chapter 5. We propose that there is a population of differences (d_i values) with an hypothetical mean of zero. We have a sample of these (d_i values), from the individuals measured at the start and end of summer. Consequently, under the assumption that the mean of such a population of differences will be normally distributed (see Section 5.9), we could do a *t*-test, using the mean ($\bar{d}$) and the variance (s_d^2) of the sample of differences.

The entire procedure is demonstrated in Table 6.1 and is identical with that followed for the mensurative experiment in Chapter 5. Thus, we have used logical argument to turn a complex situation (two populations) into a simple one (one variable being measured).

In the example, t_{obs} was sufficiently large to cause us to reject the null hypothesis. Note that in this one-tailed case, the probability of Type I error (α) was 0.05, all in the direction of being larger than zero, i.e. positive values of *t*. The appropriate probability in a two-tailed table of the *t*-statistic is that for $t_{0.10}$, because half of this (i.e. $t_{0.05}$) is in each tail (Section 4.6). Be careful which type of table you consult. If you use a two-tailed table of *t*, you need the $t_{0.10}$ value for a one-tailed test. Such a table gives critical values of *t* when 5% of all values occur in each tail if the null hypothesis is true. If, in contrast, you consult a one-tailed table of *t*, you would need to use the $t_{0.05}$ value, because, in a one-tailed tabulation, 5% of all values fall in one tail, as tabulated. The $t_{0.05}$ value in a one-tailed table is the $t_{0.10}$ value in a two-tailed table. Here, the data seem to support the model (except note the comments about confounding below).

All considerations of power, etc., as discussed before are relevant to this case. Much thought should be given before tagging the animals to ensure that sufficient are likely to be recaptured at the end of the summer to have adequate power in the test. Remember that power is only knowable if there is a specified alternative hypothesis. In this case, construction of an alternative that matters is presumably to be based on some knowledge and theory about the relationship between loss of food and other resources and rate of mortality. For example, if it is known that the animals have seasonal fluctuations in body-weight at other times of year, such that they change by at least ± 15 g, then mortality is presumably not brought about until a much larger loss of weight occurs. In such a case, the alternative can be specified by stating that a change of at least 15 g would have to be detectable with adequate power, if the null hypothesis is false. From this and some knowledge of the variance of differences from before to after summer in the weight of the animals, decisions can be made about the choice of probability of Type I error and the size of sample.

6.2 Confounding and lack of controls

There is a major logical flaw in the use of paired comparisons of the sort just described. They will unambiguously provide a valid test of the statistical null hypothesis that there is no difference from before to after in the mean value of the variable being measured. This does not, on its own, support the logical hypothesis. The problem arises because several other influences are almost always present and acting on the experimental units being sampled. Therefore, several different processes may cause the change from before to after and only one of these is the process about which the hypothesis was proposed. These other influences are known as *confounding* influences and the experiment is confounded. There is no way to attribute *causality* (i.e. the process operating according to the model and hypothesis) to the observed difference from before to after. This is an example of a procedure where there is no *statistical* problem – the analysis is not problematic; there is no difference. The problem is for the logical interpretation of the results.

In the previous example, a confounding influence could have been an increase in weight due to growth of the animals during the early part of the summer. An important decrease in weight, leading to death, may have occurred towards the end of summer as specified in the model and hypothesis. It would, however, have been invisible because of the earlier increase due to growth unless it was much larger than the effect of growth. At the end of the experiment, weights would be similar to those at the start, causing retention of the null hypothesis. Yet, the animals could have been losing weight and dying towards the end of summer, exactly as specified in the model and hypothesis.

Even if the animals were not growing, they were still getting older. Loss of weight at that time in their lives may have been a function of ageing and nothing to do with a shortage of resources. Thus, a significant rejection of the null hypothesis may have been due to a completely different process from that specified.

Confounding works both ways and there is no safe conclusion from a confounded experiment. So, if the null hypothesis were retained at the end of the experiment, it is not logically valid to assume that the model and hypothesis are wrong. The lack of significant difference from before to after summer could have been due to a loss in weight, but from some average weight that had increased before the decline. The model might have been correct. Alternatively, a significant difference from before to after summer may have been due to a loss of weight because of some process of increased age (i.e. an opposite confounding influence). No sensible conclusion about the model could be reached from such evidence.

There are many examples of these types of experiment. All are potentially confounded, unless they are not designed to detect the outcome of specific processes. For example, if a comparison is to be made between the efficiency or accuracy of two processes of measurement, a paired test will be adequate. Suppose there are two standard methods (A and B) for estimating the calorific content of items of prey in a study of foraging. Each has been recommended for different purposes, but one (A) is more expensive, although potentially more accurate than the other. To ascertain which to use in your study, you wish to compare them. The observations are that the methods sometimes give different results. The model is that method A is more accurate (i.e. gives results closer to the true calorific value of material) and therefore sometimes gives different results from method B. The hypothesis (two-tailed) is that the mean calorific value from A will differ from that using method B. The null hypothesis specifies no difference.

Material from each of a random sample of items of prey is split into two, one part being measured by each of the two methods. These paired data can then be used to test the null hypothesis. If there were any confounding influences, they are irrelevant to the interpretation of the experiment. You want to know if the methods differ. If not, you can choose the cheaper one. Why they differ or do not differ is not part of the model, nor of the hypothesis. Therefore, it is not illogical to interpret the result as indicating a difference or no difference.

To hammer home the point (because confounding will recur), consider a really silly experiment. Despite its silliness, such cases can be found somewhere in the literature. It has been observed that the heart rate of mice often changes and a model has been proposed that this is due to changes in blood pressure. A drug that alters blood pressure (by constricting arterial walls) is used to examine this model. The hypothesis is that injection of the drug will alter the heart rate. It is two-tailed because (seeing as we haven't read the literature!) we do not know whether there should be a positive or negative effect of increased blood pressure on the heart rate.

We decide to do a paired experiment. Mice are used a great deal in experiments – they represent a sort of furry test-tube on legs. Drugs are also used a lot in experiments (a drug can be defined as a chemical that, when injected into an animal, produces a scientific paper). Very carefully, we select a random sample of mice and measure their heart rates (thus handling them, disturbing them, etc.). We then inject them with the drug (thus handling them, disturbing them, sticking a needle in and injecting a potentially significant amount of fluid into their relatively small bodies, etc.) and measure their heart rate again. The difference from before to after is calculated for each mouse and found to be significant.

There is no validity to the conclusion that changing blood pressure affects heart rate. Any of the numerous confounding influences (handling, disturbance, trauma, change of volume of body fluid) could have caused the change in heart rate. Any of these influences could equally well have changed it in the opposite direction to that caused by the drug (if the drug does affect heart rate), leading to a non-significant value of t and retention of the null hypothesis. No logically valid conclusion can be reached from such a confounded experiment.

The appropriate controls are needed to eliminate the potential influences of the confounding processes. This has been well known since its analysis by J. S. Mill (1865), yet much of it, as discussed in later, more complex experiments, seems to have been forgotten by some modern biologists.

6.3 Unpaired experiments

The simplest form of controlled experiment is the standard type with two populations being examined. One is the treated or experimental population, the other the control. Such an experiment cannot be derived for the mensurative study of the proposed shortage of resources discussed in the earlier example (Section 6.1). We can, however, consider the case of the mice and the injection of a drug, to test the proposition that the drug influences heart rate because it acts on blood pressure. Instead of setting this up as a paired experiment, we could have two initial samples of animals (see below for the logic of this procedure). One of these is handled, injected with the drug and then, after whatever is the appropriate interval for the drug to work, has the heart rate recorded. This is the experimental group. The other is handled in the same way and injected with a control amount of fluid. This involves the same volume of whatever is the solvent for the drug, so that the mice have the same change of volume of fluids in their bodies as in the treated samples. These are the control sample. Now, the only difference between the mean heart rates of the two samples should be due to any effect of the drug if the null hypothesis is wrong and the drug does have the predicted effect.

We now have an hypothesis that the mean heart rate of an experimental population should differ from that of the control group and the null hypothesis is that the two populations are the same:

$$\text{H}_0\text{:} \quad \mu_{\text{Experimental}} = \mu_{\text{Control}} \tag{6.5}$$

This can obviously be re-arranged as:

$$\text{H}_0\text{:} \quad \mu_{\text{Experimental}} - \mu_{\text{Control}} = 0 \tag{6.6}$$

and can be considered as the same sort of null hypothesis as was used for the mensurative experiment (Section 4.5) and the paired t-test (Section 6.1). To test this using the previous procedure involves calculating the mean heart rate of each sample and then comparing their difference using the formula:

$$t_{\text{obs}} = \frac{[(\bar{X}_{\text{Experimental}} - \bar{X}_{\text{Control}}) - \text{null hypothetical mean}]}{\text{standard error}} \tag{6.7}$$

and we must calculate the standard error of the difference between the two sample means. Note also that this is a general formula. In our case, the mean specified in the null hypothesis is zero and the equation consists merely of the differences between the two sample means divided by the appropriate standard error. This is very similar to the formulae used before. The relationship between this version and the previous ones for mensurative and paired situations is that, here, the single sample mean being analysed is made up of the difference between two independent sample means.

6.4 Standard error of the difference between two means

The standard error of the difference between two means can be calculated easily from theory. If the difference is normally distributed, the calculated value can be used in a t-test. Thus, if we assume that repeated samples were taken of each population and the difference between the mean of one and the mean of the other repeatedly calculated, the differences would have a normal distribution. This is the usual assumption underlying the use of a t-test.

In addition, the two populations being sampled should have the same variance (although there are formulae for populations that do not have the same variances). Here, because it is important for the logical structure of the test, I propose to consider only cases where there is equality of variances between the two populations. Finally, as will recur elsewhere, the two populations should be sampled independently of one another.

6.4.1 Independence of samples

Independence is a complex matter and is required for much of statistical analysis, because probability theory, as used in the derivation of statistical formulae, depends on data being considered as independent events taken from distributions of independent entities. What is meant by independence is that the value of one replicate or one sample mean is not affected by, nor related to, the values of other replicates or sample means. A much more detailed discussion of this concept, with examples, is given in Section 7.14, where its importance in many forms of analysis is explained more fully. That discussion also gives several ecological examples of different forms of non-independence. Here, just accept that two things are independently sampled if the probability of getting one of them is not affected by the probability of getting the other.

As a simple case to consider, think about tossing a coin three times in succession. The probability of it coming down Heads is 0.5 for each of the three tosses. It is not influenced by what happens to the previous or subsequent tosses. The probabilities of getting three Heads, two Heads or one Head are 1/8, 3/8, 3/8, respectively. The probability of getting three Tails is also 1/8 (see Section 4.2 for calculation of these probabilities). Now, suppose instead, that the circumstances, for whatever reason to do with the rules of some game, create non-independence in the second and third tosses of the coin. In other words, the outcome of the second and third tosses is dependent on what happens in the first one. The rule of the game might, for example, state that, if the first toss results in a Head, the remaining two cannot both be Tails. If the remaining two come down Tails, that throw is discounted and the coin is tossed twice again, until a result that is not both Tails is obtained. Now, a quick calculation of all possibilities will reveal that the probability of getting three heads is 1/7 as is the probability of getting three Tails. The probability of two Heads is 3/7 and that for one Head in the three tosses is 2/7. This example has no importance except to reveal that non-independence must always change the probabilities associated with statistical calculations. Another way of viewing this is to say that independent observations are not *correlated.*

The calculation of the variance of a difference between the means of two samples when they are independent and the sizes of sample (n values) are the same is straightforward (see e.g. Snedecor & Cochran, 1989).

If the variances are the same, the common variance can be estimated from samples of each of the two populations, by first calculating the combined estimate of sample variance, s^2, as:

$$s^2 = \frac{s_1^2(n_1 - 1) + s_2^2(n_2 - 1)}{(n_1 - 1) + (n_2 - 1)} \tag{6.8}$$

where s_1^2 and s_2^2 are the two sampled variances. Then:

$$\text{Estimated standard error}_{\bar{X}_1 - \bar{X}_2} = \sqrt{\left[\frac{s^2(n_1 + n_2)}{n_1 \times n_2}\right]} \tag{6.9}$$

6.4.2 Homogeneity of variances

Equation 6.9 for the standard error of the difference between two sampled means is only useable in a t-test when the two populations have the same variance (i.e. $\sigma_1^2 = \sigma_2^2$). Otherwise, the calculated value of t_{obs} is not distributed as the calculated values of the distribution of the t-statistic. When one of the variances is larger than the other, the probability of Type I error is not as calculated in the t-distribution, but is larger. This phenomenon is explained, in a different context, in Section 7.16. Thus, use of the formula for t to test the null hypothesis results in excessive Type I error and therefore null hypotheses are rejected more often than they should be.

This requirement of the t-test leads to the need for a preliminary test on the variances before the t-test is done. If the variances are considered sufficiently similar as a result of such a test, the t-test can then be used, as before, to test the stated null hypothesis about means. A test for homogeneity of variances is quite simple. The background to it can very easily be explained, by illustrating the origin of the test statistic used.

Consider this entirely empirically. Two samples, of sizes n_1 and n_2, are taken from a population. Each provides an estimate of the population's variance (σ^2), as s_1^2 and s_2^2, with $(n_1 - 1)$ and $(n_2 - 1)$ degrees of freedom,

Table 6.2. Comparison of weights of livers of rodents from before to after summer. Note that weighing a mouse's liver before summer takes the rodent out of the experiment (you can't put its liver back). Thus, data cannot be paired

(a) General case of two-tailed hypothesis

H_0: $\mu_{\text{before}} = \mu_{\text{after}}$

A sample of n_1 individuals is collected before $(X_{11} \cdots X_{1i} \cdots X_{1n_1})$ and a separate sample of n_2 individuals after $(X_{21} \cdots X_{2i} \cdots X_{2n_2})$ summer. For each sample calculate:

$$\bar{X}_1 = \frac{\sum_{i=1}^{n_1} X_{1i}}{n_1} \qquad \bar{X}_2 = \frac{\sum_{i=1}^{n_2} X_{2i}}{n_2}$$

$$s_1^2 = \frac{\sum_{i=1}^{n_1} (X_{1i} - \bar{X}_1)^2}{(n_1 - 1)} \qquad s_2^2 = \frac{\sum_{i=1}^{n_2} (X_{2i} - \bar{X}_2)^2}{(n_2 - 1)}$$

Compare the two sample variances using an F-test:

$$F = \frac{\text{larger of } s_1^2 \text{ and } s_2^2}{\text{smaller of } s_1^2 \text{ and } s_2^2}$$

with $(n_1 - 1)$ and $(n_2 - 1)$ or $(n_2 - 1)$ and $(n_1 - 1)$ degrees of freedom, according to which s^2 was larger.
If not significant, proceed to t-test:

$$s^2 = \frac{s_1^2(n_1 - 1) + s_2^2(n_2 - 1)}{(n_1 + n_2 - 2)}$$

$$t_{\text{obs}} = \frac{\bar{X}_1 - \bar{X}_2}{\sqrt{\left(\frac{s^2(n_1 + n_2)}{n_1 n_2}\right)}}$$

with $(n_1 + n_2 - 2)$ degrees of freedom.

---

respectively. The larger of these is then divided by the smaller to give a test statistic called F. Then, another similar pair of samples is taken and the process repeated and so on until thousands of such ratios have been produced. These values of F form the null frequency distribution for a test of the null hypothesis that two sampled variances are equal. Every F in this sampling scheme has come from a situation where the two sampled variances each represent the same (and therefore equal) population variance. Thus, F has a distribution that represents the effects only of sampling error in the samples used to estimate σ^2 when the null hypothesis is true.

Table 6.2 (*cont*)

(b) An experiment on weights of livers of rodents before and after summer. The livers are weighed for $n_1 = 10$ randomly sampled individuals before and $n_2 = 11$ randomly sampled individuals after summer

Replicate	Before (X_{1i})	After (X_{2i})
1	8.7	11.9
2	15.0	6.0
3	13.1	11.2
4	16.7	8.0
5	14.9	6.9
6	14.7	5.2
7	13.4	8.7
8	10.9	6.2
9	8.3	12.6
10	14.8	6.1
11	—	11.8

$\bar{X}_1 = 13.05 \quad s_1^2 = 8.06$

$\bar{X}_2 = 8.60 \quad s_2^2 = 7.75 \quad F = \frac{8.06}{7.75} = 1.04$

$s^2 = 7.90$

$$t = \frac{4.45}{\sqrt{\frac{7.90(21)}{110}}} = \frac{4.45}{1.23} = 3.62$$

It turns out that this ratio has a frequency distribution that can be tabulated provided that the samples are independently drawn from their populations and that the populations are normally distributed. Clearly, this F-ratio (invented by Fisher (1928) and named F after him; the price of fame as the most inventive statistician ever is to have an F-word named after you!) can be used to test the null hypothesis that two populations have the same variance.

The procedure is very simple. Choose a level of probability of Type I error, to define a region of rejection for the test. The test statistic is F, the ratio of the larger sampled variance to the smaller sampled variance. This measure is one-tailed (it can take on values of only 1 or more), but the test is two-tailed, because either population and therefore sample can have the larger variance. This observed value of F is then compared to a table of the frequency distribution of F with the relevant numbers of degrees of freedom. This is illustrated in Table 6.2, using data similar to those analysed earlier as a t-test in Section 6.1. Note that the relevant probability of Type I error is for a two-tailed F-ratio. Later (Chapter 7) there will be more use of the F-ratio, but the null hypotheses will all be one-tailed. If the test causes you to reject the null hypothesis of equality of variances, the use of the t-test is not valid.

The sequence of events to compare the means of two experimental populations is first to test the assumption that the two populations have equal variances. Then, if that assumption is supported by the available data, to do the t-test. If the two populations do not have the same variances, there are several alternative procedures. The most important one in the context of this book is to consider why. The most important use of the unpaired t-test is to contrast two populations that are presumed to differ in their means because of some applied treatment or because of some innate difference between them. There is usually no reason to expect that the variances of the two populations should be different.

Where the data come from controlled experimentation, the units sampled in each population were selected to put in the experiment. This is described below, but is based on the idea that the two groups should be similar at the start of the experiment and therefore should be taken as samples from the same population. Under these circumstances, the two samples should start the experiment with the same variance (i.e. they each represent the population from which they came). Thus, unless it is anticipated that the experimental treatment will create some difference between the variances of the two samples, they should finish up the same at the end of the experiment. Where they do not, there is already evidence of the effects of the treatment and therefore the null hypothesis about means may not be any longer of interest – the samples have changed as a result of the experiment and it is presumably of interest to know how and why.

In other circumstances, the mean and variance of populations are related, as discussed in Section 7.18. Under these conditions, an alteration of the mean of the experimental or control group relative to the other sample will also result in an alteration of the variance. Here again, the discovery of unequal variances at the end of the experiment is an indication that the experimental treatment has some effect.

In both of these cases, much thought about the effect of experimental treatments on the variances of the two populations being examined is going to be necessary. Doing some of this thinking before the experiment would be much more useful.

For example, consider an experimental analysis of the effects of disturbance on the diversity of species of plants in patches of woodland.

The hypothesis has been proposed (based on some model for the effects of disturbance on structure of patchy assemblages – see Pickett & White, 1985) that patches disturbed by digging will have a greater mean number of species than untouched control plots. The null hypothesis is that there will be no difference or there will be more species on average in undisturbed plots. To test this, experimental plots are disturbed. Another series of randomly chosen plots is left as controls.

Before the experiment starts, however, this is not all that we can predict. We might have enough knowledge of the system to predict what may happen to the variances. For example, we may be convinced that variance in numbers of species in plots increases with time, because newly disturbed plots have not yet had a long history of different colonizations and local extinctions. Thus, one would predict that the variance of the experimental plots should be smaller than that of the controls. Alternatively (it is, after all, ecology – there are always alternative models around), you may think that the effect of disturbance is to allow variable colonization by different species in a patchy manner. Newly disturbed plots are, in effect, more variable than older plots, because the variety of plants has not yet had time to colonize all the plots. In this case, you would predict that the experimental plots should be more variable in terms of the number of species they contain than are the control plots.

Under either model, you are predicting a difference in the variance and the null hypothesis that the variances of the two groups of plots are the same should itself be tested as a legitimate part of the logic of the experiment.

It would be very helpful to the advance of ecological investigations if variances were examined and considered, instead of just focussing on the means of the distributions. A considerable amount of biology and natural history is involved in determining the variance of a population. Our collective understanding of these processes is not yet very sophisticated; we have relatively little ecological theory to explain the variances from time to time and place to place in what we see.

Nevertheless, there are many circumstances where a difference between the two variances is only a nuisance and has no redeeming features. It serves solely to prevent the orderly completion of a valid t-test or its non-parametric alternatives (see Section 6.8). When this occurs, you should consider the underlying nature of the data and try to identify a transformation of the data to another scale in which the variances are better behaved. This topic is left for later and is discussed in full in Chapter 7.

Also, in Section 7.16, it is pointed out that the effect of heterogeneity of variances is to increase the probability of Type I error relative to that which you chose and which you think is operating in your use of the table of the t-statistic. When the variances differ, if you use a t-test, you are more likely to reject a true null hypothesis than the probability (say, 0.05) you chose before the experiment was done. This is demonstrated in Chapter 7. What matters here is that, if all else fails and you cannot fix the problem of the heterogeneous variances, you could try the t-test anyway. If it is not significant, that result is probably valid because you cannot have made an excessive Type I error (i.e. you cannot incorrectly reject a true null hypothesis) if you do not reject the null hypothesis at all.

Finally, there are alternative procedures for testing null hypotheses about the equality of two means or two variances using non-parametric tests (see Siegel, 1953; Hollander & Wolfe, 1973; Conover, 1980). They may be, but are unlikely to be, suitable alternatives if variances are heterogeneous. They also have assumptions underlying their use, so be careful that you do not abandon one scheme because it assumes normality of distributions and equality of variances, but then fail to notice the assumptions of your chosen procedure (see Section 6.8 for a brief discussion of non-parametric procedures and their assumptions).

Snedecor & Cochran (1989), among others, described a modified, approximate form of the t-test that could be used when the variances were not the same. I believe it would be better not to by-pass the knowledge that may be gained from the discovery that the variances of two experimental populations are no longer the same after an experiment. Therefore, I suggest that you always examine the variances in your experiments and sampling (as recommended in Section 7.17). I therefore recommend against using a standard error for the t-test that is calculated from samples with different variances. The modified form of the t-test should not then be done, because it uses such a standard error.

6.5 Allocation of sample units to treatments

The logical basis for using the unpaired, independent comparison of two samples needs to be considered carefully. The principle on which everything is based is that two independently drawn, representative samples of a population do not differ in their frequency distribution, except for sampling error. The two samples are not, of course, identical (except under the most unlikely circumstances of each having identical elements, i.e. all replicates in one sample are of magnitudes identical with those in the other sample). There should, however, be no significant difference. Of course, there is a chosen probability of α that the means of the two samples will differ statistically due to Type I error.

In some experiments, it may be possible to test the null hypothesis of no difference between the two samples at the start of the experiment, before the treatments are applied. If the procedures and methods of measurement of the experiment are non-destructive and cause no lasting disturbance to the replicates in the samples, a t-test (or alternative) could be applied at the start.

Obviously, this is not possible in cases where the variable being measured is the weight of the animals' gonads or any other measure that causes what physiologists often coyly call the 'sacrifice' of the animals. Nor is it possible in behavioural studies where the measurement involves handling the animals, or prior learning or any other procedure that will influence subsequent behaviour. In many ecological studies, the measurements require some disturbance to plots or study-sites, so initial measurements are often undesirable. Such initial interference would obviously affect the animals, plots and replicates in ways that may well alter their behaviour or other processes operating during the experiment.

These problems require very careful thought about how replicates are chosen and how they are allocated to the two treatments (experimental and control). Each sample must be chosen to represent the population being examined and the replicates must be independent of one another (see Sections 6.4.1 and 7.14) and allocated independently to the treatments.

Here is a not entirely fictitious example of inappropriate sampling, which destroys the logical basis of an experiment. Suppose it has been observed repeatedly that a small, easily bred fish reduces its speed of swimming when in water contaminated with certain toxins. The fish can therefore be used in a bioassay to detect the presence of the toxins. The observation is that fish vary from place to place in their speed of swimming. The model (previously examined in numerous experiments) is that the variation is due to the absence or presence of different concentrations of the toxins. There are other observations to suggest that a particular body of water may be contaminated.

Putting these together leads to the hypothesis that, if the water is contaminated, fish in it will swim more slowly than fish in similar water known not to be toxic (i.e. controls). The one-tailed null hypothesis is that there will be no difference or the controls will be slower than the experimental fish. To test the null hypothesis, 12 fish from a laboratory population are to be used in each of the two samples. The fish should be equally representative of the population in the laboratory.

Suppose, however, the research assistant or technical officer on this project chooses fish in the following, convenient way. Each fish is caught in a dip-net in the large tank which holds the population. The first 12 fish are put into a container to be used as the experimental sample; the last 12 are used as controls. Obviously, the problem here is that the first 12 fish will, on average, almost certainly be slower-swimming than the next 12. That is how they came to be caught first! Thus, the two samples already differ in average speed of swimming.

The allocation of replicates to the two treatments must be properly independent and equally representative. In this case, three better procedures could be used. The fish could be caught exactly as before, but, as each is caught, you toss a coin or choose by chance into which treatment to put it. This is convenient and quick. Alternatively, the 24 fish can be caught and put into individually numbered containers. Twelve different random numbers in the range 1 to 24 can then be generated to choose one of the samples, the remaining fish forming the other one.

These methods are appropriate, but they do alter the population being examined. Only the 24 slowest fish in the entire laboratory population are being caught. Thus, the fish being examined have a smaller mean and variance in speed than is the case for the population. This would matter if the bioassay is done often so that later experiments have faster fish. The capacity to detect a reduced rate of swimming (i.e. the power of the experiment) is probably greater when the fish are fast than when they are already as slow as they come. If so, the experiments done later with any starting population of fish will be more sensitive than earlier ones. This does not seem an appropriate protocol to use as a standard bioassay because the starting conditions vary in some systematic, but uncorrectable manner.

In such cases, rather than proceeding as suggested, it would be much better to tag all the fish in the population, either physically or by keeping them in separate containers, although both procedures are logistic nightmares. Then two sets of fish could be chosen at random from the entire population. Any alternative, more efficient, procedure that will achieve this should be used.

Sometimes the needs of the experimental treatment cause serious problems for the definition of the population being sampled. For example, Cubit (1984) needed to keep grazers out of experimental plots and used copper-based paints to mark out the plots. Grazing snails would not cross the paint. In order to avoid problems of the copper killing plants inside the plots, he used small, raised areas of his study-site. When the edges were painted, any leaching of potential toxins would be downhill, away from the plots. Now, the population of plots sampled for his experimental treatment consists only of small, raised patches of the world. The control, unpainted (grazed) plots must be the same microhabitat, otherwise they are not comparable at the start, nor during, the experiment. See Hurlbert (1984) for more discussion of this point.

Such procedural and methodological requirements have now reduced the population of plots for which the hypothesis is tested. The test and any conclusions now apply only to the specified microhabitat and not to the entire range of places in the study-site. This was not a problem for Cubit's (1984) study, because he interpreted his results only for such places. There are, however, cases where a specified or constrained subset of microhabitats is completely inappropriate because neither the original observations nor the model and hypothesis are defined in terms of such specific places.

Consider the situation where the densities of lizards in some habitat vary through time and can be explained by the model that predatory birds cause sufficient mortality to reduce abundances during the warm seasons of the year. An appropriate hypothesis is that, if birds are prevented from foraging, survival will be greatly enhanced and therefore numbers at the end of the warm season will be much larger compared with control areas where birds are allowed to have continued access. The null hypothesis is that final numbers in experimental areas will be equal to or less than those in controls. Scarecrows, i.e. flapping pieces of coloured plastic stuck on poles, will be used to keep the birds away.

It turns out, however, as anyone who has ever been camping knows, that the poles can be embedded in the ground only where there are no stones. Experimental and control plots are therefore chosen by the possibility of sticking poles in the ground – i.e. areas with few stones. This redefines the population of areas being tested. It is no longer an hypothesis about areas where there are lizards. It is now about areas with lizards and few stones.

The lizards, unfortunately, tend to hide under stones, thus avoiding their predators. The parts of the habitat where there are few stones are therefore the parts where they are subjected to the most predation. The experiment should therefore cause rejection of the null hypothesis and support of the model about predators. This will unfortunately be a valid inference only if the proportion of the habitats without stones, or the proportion of the entire population of lizards in areas without stones, is large enough to have been most of what was originally observed about the densities of lizards.

Otherwise, by doing the experiment in chosen, convenient areas of the habitat, the influence of predation will have been overestimated and the model about predation incorrectly corroborated. The logical structure of the research programme would be destroyed by lack of thought about the consistency of the observations, model, hypothesis and experimental units.

Obviously, considerable care must be taken to ensure that the samples are from the population defined by the original observations and model. Otherwise, the experiment is pointless – it cannot test the stated hypotheses. Then, more care must be taken to ensure that the replicates in each sample are allocated equally representatively to the two treatments.

6.6 Interpretation of a simple ecological experiment

Consider the observations in Figure 6.1a. Where predators are active, there are fewer of a particular species of prey; where there are fewer predators, prey are abundant. The pattern seems to have reality, in that the negative correlation between numbers of prey and predators is significant when tested against the null hypothesis of no correlation (Spearman's (1904) rank correlation coefficient $r_S = -0.69$, 10 degrees of freedom, $P < 0.05$).

One model that can explain this trend is that the predators have caused the trend by eating more of the prey where they, the predators, are numerous. There are, of course, several other models. For example, the predators and prey may have different requirements for some aspect of habitat, so that they tend to be relatively more abundant in different parts of the studied area. To determine whether the first of these models is realistic, several hypotheses might be proposed.

A common one is that removal of predators from the areas where they are numerous should lead, after some specified time, to greater numbers of prey compared with controls where predators are not removed. Under some circumstances, this is not a particularly compelling hypothesis. The observed pattern may be the end of the process, so that no further change due to predation will now occur. The predators may, for example, be moving to new prey in areas where predators are sparse. In such a case,

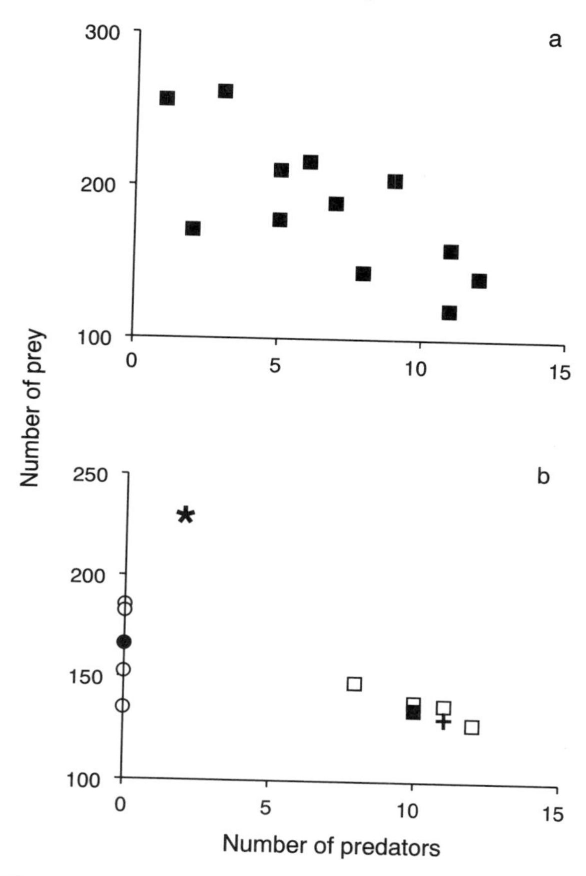

Figure 6.1. An experiment to test an hypothesis about the effects of predators. In (a) are the observations – a negative correlation between numbers of prey and numbers of predators per unit area. In (b) is the outcome of an experimental removal of predators. Open circles are experimental replicates (predators removed), with mean number of prey shown as the filled circle. Open squares are controls (predators present) with mean number of prey shown as the filled square. The asterisk and the cross represent the mean numbers of prey at large and near-zero numbers of predators, respectively, as observed originally (i.e. in (a)).

there would be no subsequent effect of predation where the predators are currently abundant and the hypothesis about predation would be tested at an inappropriate time.

An alternative is that after the next recruitment, or period of birth of the prey, removal of predators would lead to increased numbers of prey compared with control areas where predators are untouched. This may

Table 6.3. Results of experimental removal of predators with $n = 4$ replicates (see text for details and Figure 6.1 for data). Data are numbers of prey alive at the end of the experiment and the average number of predators recorded in each plot during the experiment

	Experimental (predators excluded)		Control (predators allowed in)	
Replicate	No. prey	No. predators	No. prey	No. predators
1	185	0	138	10
2	182	0	137	11
3	135	0	148	8
4	153	0	128	12
	$\bar{X}_E = 163.8$		$\bar{X}_C = 137.8$	
	$s_E^2 = 575.6$		$s_C^2 = 66.9$	

$$F = \frac{575.6}{66.9} = 8.60 \quad \text{3 and 3 degrees of freedom, } P > 0.05$$

$$s^2 = \frac{(575.6) \times 3 + (66.9) \times 3}{6} = 321.25$$

$$t = \frac{163.8 - 137.8}{\sqrt{\left[\frac{321.25(4+4)}{4 \times 4}\right]}} = \frac{26}{12.67} = 2.05 \quad \text{6 degrees of freedom, } P < 0.05 \text{ (one-tailed)}$$

Subscript E denotes experimental, subscript C control.

have problems if the time-course of possible increases in abundance of prey is unknown.

To keep the discussion simple, let us assume that prey recruit in spring and that removal of predators is an easy chore, requiring no controls for scarecrows or electric fences or whatever form of installation is used to keep them out. Suppose an experiment is done with four independent replicate plots where predators are removed and with four control areas where they are present. The replicates are, of course, independently and randomly scattered around the study-area. Suppose the experiment runs for a few months from before to after spring, so that it ends at the time of year when the observations in Figure 6.1a were originally made. At the end, the numbers of prey in the two treatments produce the data in Table 6.3.

There is a significant difference between the experimental removals and the controls (t-test in Table 6.3). As a result, we should reject the null hypothesis and deduce that predators indeed have an effect on the numbers of prey.

This is, however, not an unambiguous interpretation because it is possible that recruitment of prey differed in plots with or without predators. In other words, the results may be that differences in numbers of prey are caused by recruits or immigrants being deterred from establishing themselves in areas where there are numerous predators. Predators would therefore be implicated as the cause of the observed difference in numbers of prey, but not through killing/consumption of prey as specified in the model. To sort this out would have required the eight plots to be used to have the predators removed until recruitment (or immigration) was finished. Four of them would then be altered to allow access to the predators and the experiment then started. In such an experiment, there is no reason to presume any differences in recruitment or migration before the experiment starts (and, indeed, this could be checked before the predators are allowed in).

We shall, however, proceed with the interpretation that the significant difference supports the model, i.e. it is due to mortality caused by predators. So – are we satisfied with this result? We cannot discount the feeling of satisfaction that some researchers gain when they feel that their data support their favourite theory (Dayton, 1979). At one level, therefore, the answer is 'yes'. We have results that could be used as part of a paper in a supposedly prestigious journal.

In the spirit of trying to arrive at a realistic understanding of how the system being investigated actually functions (which was, presumably, the real point of the study), the result and conclusion are *not* satisfactory.

Note the relationship between the experimental results and the proposed explanation for the original observations. The mean values from the experiment are plotted in Figure 6.1b, with the means of the original data. The control values are plotted at the appropriate positions, giving the average number of predators observed in them during the experiment. The experimental values are plotted where the number of predators is zero. Thus, the difference in mean number of prey between experimental and control plots is indicated in relation to the original pattern.

Note that the mean number of prey in the controls matches the numbers originally observed. The mean number of prey where there are no predators is, however, smaller than the numbers observed in nature. Predation on its own does *not* explain the original observations. The model is wrong, or inadequate, even though predation is a demonstrably significant process influencing the numbers of prey in the area studied.

One explanation for what has happened is that the mortality of prey not due to predators (i.e. that due to weather, accidents, competitive

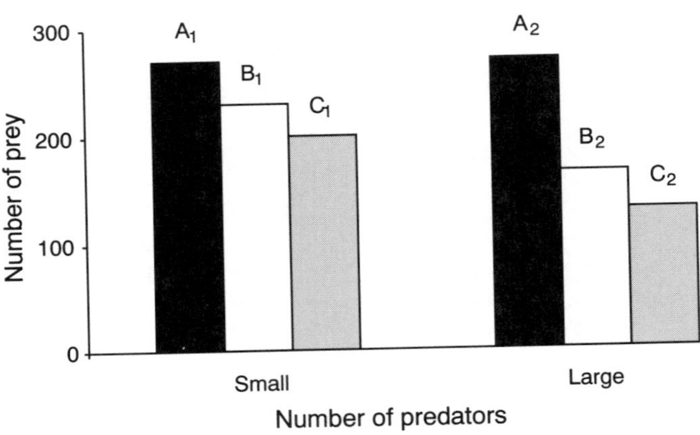

Figure 6.2. Results of an hypothetical experimental removal of predators to examine explanations for observations of numbers of prey as in Figure 6.1a. Prey recruit in equal numbers into all areas, regardless of whether the number of predators per unit area is large (A_2) or small (A_1). Prey then die because of weather and predators. Unknown to the experimenter, a greater proportion of prey would die of non-predatory causes in areas where predators are numerous. Thus, $B_1 > B_2$ and $B_1/A_1 > B_2/A_2$. C_1 and C_2 are numbers of prey surviving where small and large numbers of predators are active. In the experiment, removal of large numbers of predators results in an increase in numbers of prey from C_2 to B_2, which is the effect of predation. It does not increase survival to the density shown by C_1 – the original observations in data such as in Figure 6.1a. Predation matters, but does not explain the original observations.

interactions, diseases, etc.) is not the same in areas where there are many predators compared with where there are few. Thus, predators are most numerous in parts of the habitat where non-predatory mortality of prey is relatively large. Under these circumstances, the experimental plots are only in the parts of the range where there are naturally large numbers of predators and mortality due to other processes is larger than in areas where there are naturally few predators. Thus, the numbers of prey predicted by the model do not occur. This is illustrated in Figure 6.2 to show how another process will influence the magnitude of outcome of an experimental removal of a predator or any other simple comparison.

Here, the experimental result is significant, but the hypothesis should not be considered corroborated. The results do not support the model that the original correlation observed is due to predation. It is apparently partially due to climatic variation. This would, presumably, have been discovered by originally proposing more complex hypotheses about the outcome of removing predators in different parts of the habitat, i.e. from areas with different natural densities. Alternatively, more complex hypotheses could be proposed about the numbers of prey that should survive when numbers of predators are manipulated in several parts of the habitat to mimic the numbers found in other places. This would require experimental manipulations to remove some or half or all predators in various places to create a gradient of numbers of predators in different parts of the habitat (on the need for such experiments, see also Underwood & Denley, 1984).

A useful approach to thinking about your experiments is to use all the available information in order to make the hypothesis much more quantitatively predictive. Therefore, you should, wherever possible, determine in advance of the experiment how much difference the model should predict to occur between the control and experimental treatments. Examination of the data in Figure 6.1a demonstrates that if predators are responsible for the observed correlation, their removal should result in numbers of prey equivalent to those originally observed where there were no predators. If, as before, predators are removed from the areas where they are most dense, the mean difference in numbers of prey should be an increase from about 130 to about 230 (as shown in Figure 6.1b). Thus, the predicted effect size as an alternative to the null hypothesis is about 100. We now have:

$$\begin{aligned} H_0&: \quad \mu_{\text{Predators removed}} = \mu_{\text{Predators not removed}} \\ H_A&: \quad \mu_{\text{Predators removed}} = \mu_{\text{Predators not removed}} + 100 \end{aligned} \qquad (6.10)$$

If, as here, the outcome of the experiment is only a difference of 26 prey individuals per sample unit (Table 6.3), it is not consistent with the alternative hypothesis. The effect size is not known absolutely, but was estimated from the original data and therefore has variance. If we ignore this and treat it as a constant, we could test the observed difference in the experiment against that predicted in the alternative hypothesis. This is done in Table 6.4, where it is shown that the observed experimental difference in means (26 individuals per unit area) is significantly smaller than that predicted in the alternative hypothesis based on the original observations (100 individuals per unit area). So, although there is a significant effect of removing predators (the null hypothesis is rejected), the alternative proposed is not achieved. This demonstrates the point made in the previous section.

Note that the test done in Table 6.4 against the alternative difference of 100 individuals was not a realistic test, because it ignored the variability in the available estimate of 100, calculated from a few data points on a graph. The next section describes how to deal with this.

Table 6.4. Results of experimental removal of predators as in Table 6.3. The t-test is done for the effect of predation expected from the original observations, i.e. that the effect of predation is a difference in density of 100 individuals per replicate (see text for details)

$$H_0: \quad \mu_{\text{Predators removed}} = \mu_{\text{Predators not removed}} + 100$$
$$H_0: \quad \mu_{\text{Predators removed}} - \mu_{\text{Predators not removed}} - 100 = 0$$
$$t = \frac{(163.8 - 137.8) - 100}{12.67} = \frac{-74}{12.67} = -5.84$$

6 degrees of freedom, $P < 0.005$ (two-tailed)

6.7 Power of an experimental comparison of two populations

The example illustrated in the previous section will be used to demonstrate the calculation of power of an experiment involving two samples. Remember, power is the capacity of an experiment to reject a null hypothesis, measured as the probability that the null hypothesis will be rejected if a specified alternative hypothesis is true (see Section 5.12). The power of the previous experiment was obviously more than sufficient to detect the difference that actually eventuated when the experiment was done. If it turns out that there is a greater difference between the null hypothesis and the real world (i.e. an even greater departure from the null hypothesis occurs), the experiment will have more power to reject the null hypothesis than is the case for the defined alternative. So, we usually calculate power for the smallest 'effect size', i.e. the smallest difference between the null and alternative hypotheses. Then, power for greater differences will be even larger.

Here is a second, but similar case, involving predation as an explanation for an observed negative correlation between numbers of prey and numbers of predators. If, as proposed in the model, predation *per se* explains this pattern, the removal of predators where they are dense (D in Figure 6.3), should cause an increase in numbers of prey from about 200 (at +P in Figure 6.3) to about 300 (as in −P in Figure 6.3). This is a predicted effect size of 100.

The original observations also provide an estimate of the variance among replicates that we might expect to measure during the experiment, provided that the experimental units are of a size similar to that of the units sampled in the original data. The experimental plots must also be scattered over the same spatial extent as the original observations so that all other

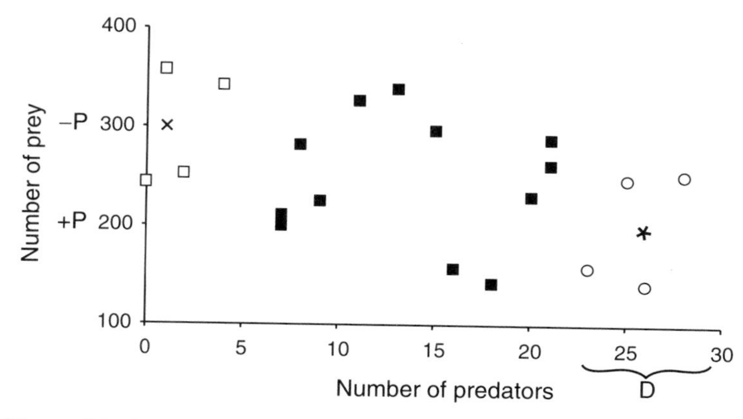

Figure 6.3. Calculation of power in a simple ecological experiment to test the hypothesis that removal of predators will increase numbers of surviving prey to match those observed where predators are sparse. Twenty data points (numbers of predators and prey) show the originally observed correlation. D indicates the density of predators that would be left intact in controls. Removal areas will have no predators. Empty squares indicate data points used to estimate the mean number of prey where there are no predators (−P; $n = 4$). Empty circles indicate data points used to estimate the mean number of prey where there are many predators (+P; $n = 4$).

environmental influences on numbers of prey operate in the same way during the experiment as during the previous period of observation. For example, if the original observations were sampled in quadrats about 2 m apart, the variance among quadrats is a result of the processes operating at that scale. Other processes may influence numbers of prey per quadrat or variability in numbers of prey per quadrat at, say, 10 m distances. The variance among quadrats at that scale may be much larger or smaller. If the experimental plots were at this larger distance apart, the variance estimated from the original observations (at the smaller spatial scale) is unlikely to be a realistic estimate of what will happen during the experiment. In this case, the sample variance among data where predators are numerous (as indicated in Figure 6.3) is 3453, with three degrees of freedom. In addition, it is also worth while to note that the sample variance among observations where there are no or few predators is 3519, also with three degrees of freedom, which is satisfactorily similar to that calculated where predators are numerous. Thus, there is no reason to anticipate heterogeneous variances between the two experimental treatments if predation is (as in the model) the cause of the observed trend.

Now it is possible to determine the sample size necessary in an experiment designed to detect the chosen effect size as significant (using α, the

Table 6.5. Calculation of power of a t-test to detect the effect of predation. The experiment will be to remove predators from n replicate areas, keeping a further n as untouched controls. It is proposed that predators remove at least about 100 individuals of prey per unit area (see Figure 6.3). Other details are in the text

Average variance expected (from data in Figure 6.3) $= \frac{3453 + 3519}{2} = 3486$

Effect size $= 100$

Hypothesis: $\mu_E - \mu_C > 100$
H_0 $\mu_E - \mu_C \leq 100$

To use Cohen's (1977) tables, you need to calculate:

$$d = \frac{\text{Effect size}}{\text{Standard deviation}} = \frac{100}{(3486)^{0.5}} = 1.69$$

Consideration of the procedures in Cohen (1977) gives:

	Power =	0.70	0.75	0.80	0.85	0.90	0.95
	Probability of Type II error (β) =	0.30	0.25	0.20	0.15	0.10	0.05
Probability of Type I error (α) = 0.05	Sample size (n) =	4	4	5	6	7	8
Probability of Type I error (α) = 0.10	Sample size (n) =	3	3	4	4	5	6

probability of Type I error as the conventional 0.05) using a t-test. The calculations are shown in Table 6.5, using various powers (0.70 to 0.95). To have an equal probability of Type I and Type II error (which seems an appropriate goal for this experiment so that a mistake is equally likely in each direction) requires power of 0.95 (i.e. $\beta = 0.05$) and therefore $n = 8$ replicates in each of the two treatments. This may or may not be reasonable given the budget, personnel, logistic difficulties, etc. If the experiment is done with $n = 8$, it has the power (i.e. the probability of rejecting the null hypothesis in favour of the specified alternative) of 0.95. Any larger effect – any greater difference in mean numbers of prey due to removal of predators – would be even more likely to be detected. Such an outcome would cause a problem for interpretation in relation to the proposed model because predation should be causing the observed pattern – not a larger difference than was observed.

If the budget or space available does not allow eight replicates per treatment, I would change α to keep it and β equal for the number of replicates that can be used. In this example, if there can only be five replicates (a total of ten plots), to have α and β equal requires them to be set at 0.10 (see Table 6.5). Now, there is still an equal chance of rejecting the null hypothesis incorrectly, or failing to reject it when the alternative is true. The chance of making either one of these mistakes is 0.10, one in ten. If the budget were even more constrained and only a total of eight plots can be used (i.e. $n = 4$ in each treatment), the power of the experiment becomes ridiculously small (at $\alpha = 0.10$, $\beta = 0.20$) so that there is a one in five chance of failing to detect the effect of predators if it occurs, but a one in ten chance of claiming to find one when there is no effect as large as that hypothesized.

These risks of being wrong are, to me, unacceptably large. It is, however, usually argued that if the hypothesis is to be tested at all, this is the best we are going to be able to do, given the constraints. I doubt it is worth doing at all. I would prefer to attempt to justify more funding. If I were unsuccessful, I would, of course, as everyone does in ecology(!), accept the views of my peers that the project is not worth funding adequately (as required for n to be 8) and therefore not thought to be worth doing. It is not realistic to plough ahead with the experiment if it is already known not to be powerful enough to identify the specified alternative, nor likely to enable you to reject or retain the null hypothesis without large chances of error.

As a much more likely and better alternative, if money or space to do the experiment is a limitation for the number of replicates, it might be possible to do it with $n = 4$ and to consider doing it again in some later year, using independent replicates, or in some other place where the observed trend in numbers of prey and predators also applies. Methods for combining different experiments are discussed in Section 14.4. Designs involving experimental treatments replicated in several places or at several times are discussed in Chapter 10. An example of a small experiment repeated many times to determine the outcome is given by Underwood & Chapman (1992).

One advantage of using numerous independent experiments is that the experimental results are likely to have great generality (the model may be able to explain more or be more consistent in its usefulness) because results will have been obtained for several situations not just one (Connell & Sousa, 1983). Of course, the observations may not recur much through time or space, which would certainly give you some reason to pause in the pursuit of an experimental evaluation of your chosen explanation. The explanation would need to be able to explain the lack of correlations in those times and places where they did not recur. Until such a model has been proposed, the experiment is worthless. When such a model is proposed, it is inevitable that a much larger and more complex experiment is going to be needed.

Of course, like other branches of science, ecology would benefit greatly from much more repeated experimentation. Consistent results from two or more independent experiments are much more convincing than any amount of replication and power in a single experiment at one site and time. There is an all-prevailing view in many ecological studies that what is observed once is a general truth (but see the value of repeated experimentation in Collins, 1985).

Determination of the power of an experiment before it is planned is crucial if the biological realities of your research programme are going to be preserved. It requires considerable biological insight to be able to propose quantitative hypotheses, rather than vaguer qualitative ones. It does, however, cause much re-evaluation of the point and purpose of the experimental procedures before they are done. This is undoubtedly always a very good thing.

6.8 Alternative procedures

The assumption that the sample mean of a distribution should be normally distributed seems to frighten people into seeking alternative statistical procedures. Sometimes, the fright leads to a lemming-like rush to procedures that have several equally fierce assumptions, but these are lost sight of in the fear that data may not conform to normality!

Here, several alternatives to the use of *t*-tests are considered. First, there is a binomial test of paired data, then two tests equivalent to the unpaired situation. There are several different situations for which the latter tests may be used. In this discussion, the only hypotheses being considered are those to do with the equality of the locations of two populations (see Sections 3.6 and 6.3), as in the preceding examples.

The origins and details of these tests are discussed in full in numerous texts, of which those by Siegel (1953), Hollander & Wolfe (1973) and Conover (1980) seem to cover the ground fairly well. The last two, particularly Hollander & Wolfe (1973), have adequate and adequately detailed discussions of the assumptions underlying the use of non-parametric tests.

6.8.1 Binomial (sign) test for paired data

Another way to view paired data gathered to test 'before–after' hypotheses is to consider the direction of difference from before to after, rather than the magnitudes of the differences. The directions are shown in

Table 6.1, where data from such an experiment were analysed. If the null hypothesis were true and there is no consistent trend between the two sets of data, the differences among replicates are chance phenomena, due to individual variability among replicates or due to random errors of measurement. Therefore, the probability of the first reading being larger than the second should be the same as the probability of the first being smaller than the second. In contrast, if the null hypothesis is false, the difference from the first to the second set of readings is due to some systematic effect of a treatment or process. If the values increase, on average, there is a greater probability of the replicates in the second set being larger than those in the first set. The differences from the first to second reading in each pair will be more likely to be negative than positive. This leads to the following null hypothesis for the two-tailed case:

$$
\begin{aligned}
H_0:\ & \text{Prob (Before} - \text{After) is positive} \\
& = \text{Prob (Before} - \text{After) is negative} \\
& = 0.5
\end{aligned}
\tag{6.11}
$$

In the case of the desert rodents discussed in Section 6.1, the hypothesis predicted a decrease in weight. The one-tailed hypothesis and null hypothesis are:

$$
\begin{aligned}
\text{Hypothesis:}\ & \text{Probability (After} - \text{Before) is positive} \\
& > \text{Probability (After} - \text{Before) is negative} \\
H_0:\ & \text{Probability (After} - \text{Before) is positive} \\
& \leq \text{Probability (After} - \text{Before) is negative}
\end{aligned}
\tag{6.12}
$$

The null hypothesis can be analysed using the binomial expansion formula:

$$\text{Prob}(r) = \frac{n!}{r!(n-r)!} p^r (1-p)^{(n-r)} \tag{6.13}$$

where n is the number of experimental animals, r is the number of animals that show an increase from before to after the experiment (and can take on any possible value from 0 to n; note that 0! is 1) and p is the probability of one animal having a larger value before versus after (or after versus before) and $p = 0.5$ for this experiment if the null hypothesis is true (because animals should be equally likely to increase or decrease by chance during the experiment).

In our case, there are $n = 14$ animals, independently sampled in the population. The probability of the difference from before to after being positive is 0.5, if the null hypothesis is true. This is equal to the probability that the difference from before to after is negative. The observed number of positive differences is 11. The probability of getting such a large number or any larger number is obtained from Equation 6.13, to construct the one relevant tail of the distribution of the test statistic, r, the number of positive differences. The equation is evaluated for $r = 11$, 12, 13 and 14 positive differences in $n = 14$ trials with $p = 0.5$ chance of a positive occurrence in each trial.

In the example here, the probabilities are 0.022, 0.055, 0.0009 and 0.00006 for $r = 11$, 12, 13 and 14, respectively. The summed probability of getting 11 or more positive differences is smaller than 0.05. Thus, we reject the null hypothesis in favour of the alternative of a general decrease in weights during the summer. This is, of course, the same conclusion we reached using the t-test.

In a two-tailed case, because $p = 0.5$, the probabilities of any value of r are symmetrical and you double the probability calculated for positive differences to get the probability of getting a particular number of positive or negative values. If the present example were two-tailed, we would need to know the combined probability of getting 11, 12, 13 or 14 positive differences and 0, 1, 2 or 3 positive differences (which is the same as 14, 13, 12 or 11 negative differences) from before to after. The probabilities of 11, 12, 13 and 14 negative and 11, 12, 13 and 14 positive differences are identical. Thus, the relevant two-tailed probability is 0.058, which would not be significant.

The major advantage of this test over the t-test is that it requires no assumptions about the shape of the distribution of data being sampled, except that the distribution of proportions of positive or negative differences from before to after should be a binomial distribution. Where there are large variances among individuals, this test is also more likely to detect small overall differences from before to after an experimental treatment. These will tend to be missed by the t-test.

In general, however, if it is reasonable to assume normality, as required by the t-test and, where variances are relatively small, the binomial test is less powerful for a given size of sample (n) than is the t-test.

6.8.2 Other alternative procedures

There are useful alternative procedures for testing the null hypothesis that the means of two populations are the same. For symmetrical (non-skewed) distributions, a χ^2 contingency test can be done using the number of observations that are above or below the mean in each of the two samples. Samples must be larger than $n = 10$. The details are given by Hollander & Wolfe (1973) and Conover (1980).

This test has the advantage over the unpaired *t*-test that there are no assumptions about the normality of the distributions. There is, however, the necessary assumption that the two distributions being sampled are the same shape. Thus, except for their means (which are different if the null hypothesis is not true), the two distributions have the same frequency distribution (e.g. see Hollander & Wolfe, 1973; Conover, 1980). This is often expressed obscurely in books on so-called distribution-free statistics.

The Mann–Whitney ranks test can also be used for two independent samples. This test has other purposes (Siegel, 1953; Conover, 1980), but can be used to test hypotheses about the means of two populations. Again the details can be obtained in texts by Siegel (1953), Hollander & Wolfe (1973) and Conover (1980).

This test, again, makes no assumptions about normality of the distributions being sampled. It is, however, important to notice what the assumptions of the tests actually are. The crucial assumptions are that data are sampled independently within and between populations, which is identical with the assumption of the *t*-test. Conover (1980, p. 222) described the other assumptions and pointed out that the probabilities associated with the test are based on the underlying theory that only if the samples are identically distributed will every arrangement of data from sample A and data from sample B in a combined ranking be equiprobable. Thus, the distributions being sampled must be identical except for their mean. In other words, the variances, skewnesses, etc., of the two populations being sampled must be the same. This is at least as restrictive as the assumption that the two populations are normally distributed. Even if it were only considered to be important for the variances of the two populations, this assumption still requires that the two variances are equal.

Whenever the assumptions of normality are thought to be important, alternative test procedures should be considered. Alternatives, such as rank-order or other non-parametric tests are not valid alternatives if their assumptions are equally likely to have been violated. There are numerous examples in the ecological literature where someone has done analyses using the Mann–Whitney procedure, because the data did not have equal variances, so a test based on the *t*-test could not be used.

This is obviously silly. If one test is not valid because its assumptions are violated, so is another!

Apart from that, there are circumstances where rank-ordered data are the only type of data relevant. Then, the Mann–Whitney test is the logical choice. Otherwise, choose statistical procedures very wisely and be equally sure about their limitations rather than abandoning one for another without being aware of the problems of your second choice.

6.9 Are experimental comparisons of only two populations useful?

The general ecological answer to this question is 'No'. This is probably true for most branches of biology. There are rarely situations where it is possible to make valid comparisons between a single experimental and a single control group of replicates.

First, there are myriads of cases where the nature of replication requires more than two populations. These are discussed in Chapter 9, under nested designs of experiments, as the solution to the problem identified by Hurlbert (1984) as 'pseudoreplication'.

Much more importantly in the present context is the immense logical difficulty of having only one set of controls for some experimental treatment or manipulation such that the controls still represent the population specified in the observations, models and hypotheses. This was alluded to in the previous section, concerning experiments where the population being studied had to be changed to suit the hypothesis being investigated.

6.9.1 The wrong population is being sampled

Consider the experiments using two populations discussed so far. Most of them involve some manipulation or treatment that is also done to the controls. The experimental group then have this manipulation plus the treatment that creates the conditions specified in the hypothesis. Predators are excluded by use of cages or scarecrows. Suppose the scarecrows (or cages) also affect the prey.

Consider the lizards and ask what we should conclude if the lizards are frightened away by the presence of scarecrows and their controls? In experimental and control areas, there are now few lizards, regardless of the presence or absence of predatory birds. The conclusion from the statistical analysis of the final data from the study will probably be that predators do not make any difference to the numbers of lizards. This will probably occur because the numbers of prey are now so small that the influence of predators is extremely unlikely to be major. Unfortunately, concluding that predators have no effect is a poor outcome of the process. The predators may indeed have an effect as predicted in the hypothesis, but the conditions prevailing under the experimental regime are not those originally observed (i.e. densities of lizards are now smaller). Thus, the model is no longer relevant. It is no longer necessary to explain the sorts of observation made at the start of the study – they are not occurring in the experimental areas. More importantly, the experiment is irrelevant to the model, because of the artefact brought about by introduction of the experimental and control treatments.

In the case of some experiment involving the injection of such animals as mice, the hypothesis and model usually relate to animals that have not been picked up and injected. In both cases, the experiment has failed to provide the conditions specified in the hypothesis. It has failed to provide the conditions originally observed and which are explained by the model being examined experimentally.

What is required is therefore at least one other population of animals (or areas or plots or whatever) that is a true control and therefore not subject to any of the experimental manipulations, nor their procedural controls. These would serve to verify that the original prevailing conditions actually pertained during the course of the experiment. They will also serve to indicate whether there is an artefact due to the application of the control treatments. Consider the example of the scarecrow and its control (a pole without the flapping plastic). Construction of these may have frightened the lizards away. The presence of the poles may cause continuous vibration when the wind blows, driving lizards away. All sorts of things may be happening.

It is not in keeping with the whole logic of an experimental form of inquiry to dismiss these problems by saying that you know so much about the lizards and the habitat that these are trivial complaints. Unless you have data to support such a statement you should not be making it. You need information to support your hypothesis that there are no artefacts – an hypothesis based on some model about the way the lizards behave and about the effects of the poles on their behaviour. Without such information, there is no point in trying to use the hypothesis to help you to interpret the experiment. If you just 'know' all of these things, why are you going to all the trouble to do an experiment in the first place? You obviously already know all the answers!

Nevertheless, if there are potential artefacts, there are always going to be potential misinterpretations of the outcome of an experiment, however well and independently replicated it is.

The minimal experiment really involves the use of three experimental populations. As a result, there is no prospect that a statistical analysis involving only two populations is useful. The experimental design is inappropriate. This raises the possibility that the *t*-test might somehow be used for comparing more than two populations. This is considered below.

Some authors respond to the problem of potential artefacts in manipulative experiments by accepting the results of a comparison of areas that happen to differ in the variable required to be different by the hypothesis (e.g. Diamond, 1986). Thus, numbers of lizards would be compared in areas that happen, naturally, to have large and small numbers of predators. This so-called 'natural' experiment is not an experiment at all. Any difference in numbers of lizards, even if it is a difference consistent with the hypothesis, could be explained by numerous processes – not just predation. All attempts to interpret such comparisons are confounded (see also Section 6.2). These comparisons are better described as natural 'experiments' to draw attention to which word is suspect (Underwood, 1990). They are certainly natural. They are not experimental tests of specific hypotheses.

We need a specific example, so we will stay with the lizards even though it may seem far fetched to those who know about lizards. The contention is that the experimental manipulation and/or its control may cause a change in the numbers of lizards in the experimental and control plots. Thus, we need three treatments: (i) experimental replicates which have the scarecrows; (ii) controls or sham-experimentals with poles and so forth but no plastic scarecrows; and (iii) controls (i.e. true untouched controls) to which nothing is done until the end of the experiment, when the lizards are counted.

In Figure 6.4, I have illustrated the most important possible outcomes, in terms of differences among treatments, that may be found as a result of such an experiment. This should illustrate the point better than a specific numerical example. In Figure 6.4a, there are no differences among any of the treatments in terms of mean numbers of lizards. There are no effects of predators and no artefacts due to the poles used in the procedural controls. Thus, no differences would be detected by whatever analysis is used; no differences would have been detected by the *t*-test on the experimental and procedural control groups. A correct interpretation would be

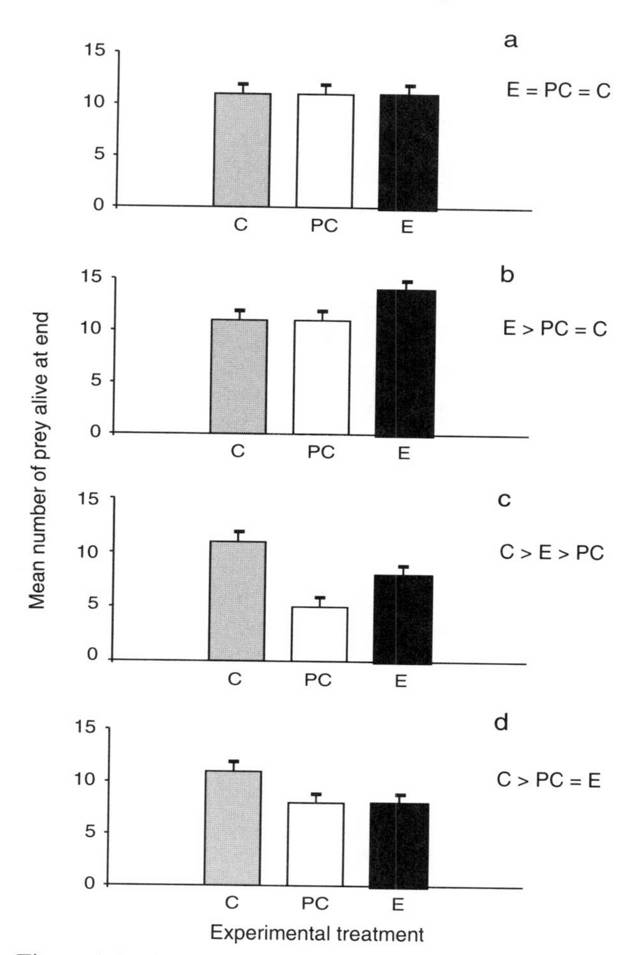

Figure 6.4. Different outcomes of an experimental removal of predators. There are three treatments: E have predators removed and a scarecrow. PC are procedural controls with controls for the poles used to make the scarecrows. C are untouched controls (see the text for other details). In (a), there are no effects of predation and no artefacts due to the poles. In (b), there is an effect of predation, but no artefacts. In (c), there is an effect of predation, but an artefact due to the poles. In (d), there is only an artefact and no effect of predation.

made, but this can really be the case only if potential problems due to artefacts of the manipulation are known to be absent (see below).

In the second case (Figure 6.4b), there is a major difference between the experimental and procedural control groups, which would, correctly, have been detected by the experimental design using only two treatments. In this case, the correct inference would be reached, because there is no artefact (there is no difference in the mean numbers of lizards between the procedural control and the true control groups). The experimenter, however, cannot know there is no artefact, so again the conclusion is not a sure one (see below).

In the third case (Figure 6.4c), there is again a difference between the experimental and procedural controls and it would have been detected by a *t*-test if these two treatments had been the only ones used. The inference that the predators affect the lizards as specified in the model is, however, difficult to accept. All we know as a fact is that the experimental treatment (removal of predators) caused an increase in the numbers of lizards, when these numbers were already reduced below the natural densities (as represented in the graph as the density in true control areas, C; these are not available in the experiment). Unknown to us, an artefact due to the poles in the two treatments caused a decline in numbers of lizards. Then, the predators reduced the numbers in the procedural control plots (PC in Figure 6.4c).

It is not a logical necessity that such an effect would have occurred if lizards were at their normal (i.e. larger) densities. It may be, for example, that when lizards are sparse, they come out into the open more often, because they are not threatened by larger competitors (if such things exist in this species of lizard). Therefore, the predators can catch them much more commonly than when the populations are intact and the larger competitors are also more abundant.

Thus, it is necessary that such results will be obtained in an experiment comparing only the experimental and procedural control treatments. It is, however, insufficient as evidence that the hypothesis is correct. In fact, the experiment has completely failed to test the model and hypothesis (the model is that predators reduce numbers of lizards at the observed densities; the hypothesis is that if predators are removed from areas with natural numbers of lizards, the numbers will end up being greater than in controls).

The final case illustrated in Figure 6.4d is that where there is no difference revealed by a comparison of the experimental and procedural controls, because, potentially, at reduced densities, the predators do not make any difference. It may be, for example, that birds do not forage where there are few lizards. It may be that, because their densities are small, the lizards have access to sufficient resources of food and shelter without being out in the open. Therefore they are rarely vulnerable to predators. Again, the experiment has failed to provide the conditions necessary to test the hypothesis – which was stated with respect to the normal, larger densities of lizards originally observed. Discovery of no influence of predators at an artificially small density of lizards is irrelevant to the model and hypothesis proposed.

In the last two cases (Figure 6.4c and d), the experimental design with only experimental and control treatments cannot reveal the artefact that causes the illogical conclusion. For such artefacts to be detected, untouched control areas must also be examined to demonstrate whether the conditions specified in the hypothesis (in this case, the originally observed densities) are maintained throughout the experiment. Therefore, the three treatments are necessary. Of course, some may argue that there need only be two treatments, the true control and the experimental treatment. Then, of course, it would be impossible to determine whether the result was due to an artefact of the existence of the poles, rather than the removal of birds. This is demonstrated by comparing the relevant treatments (E and C) in Figure 6.4d, which would certainly be incorrectly interpreted as an effect of predators if there were no procedural controls.

The logical structure of an experiment is always compromised by the lack of a proper series of complete controls. More groups (populations) need to be included in the experimental design. The use of *t*-tests to compare more than two populations is not possible without modification (see below) and is not efficient when modified. Therefore we shall need a new procedure to deal with virtually all experiments.

6.9.2 Modifications to the t-test to compare more than two populations

There are two problems associated with trying to use a statistical procedure defined in terms of two populations to compare more than two. First, there is a logical dilemma: consider the situation where there are three populations (E for experimental, PC for procedural control and C for control) being examined, as in the previous example. The simplest form of the null hypothesis is:

$$\mathrm{H}_0: \quad \mu_{\mathrm{C}} = \mu_{\mathrm{PC}} = \mu_{\mathrm{E}} \tag{6.14}$$

although, in the previous example, the contrast between the experimental and procedural control was one-tailed (it was expected that the presence of birds would decrease the numbers of lizards relative to controls). So, the specific case discussed has the following null hypothesis:

$$\mathrm{H}_0: \quad \mu_{\mathrm{C}} = \mu_{\mathrm{PC}} \leq \mu_{\mathrm{E}} \tag{6.15}$$

There are numerous tails to either form of the null. In fact, there are 12 alternatives in the first case and 11 in the second (work it out with a piece of paper and lots of equality and inequality signs!). It is difficult in logic to determine how a test statistic could be constructed that could have this number of tails. Dealing with the comparisons in pairs requires some statement about what sort of patterns should be expected to allow the comparison to work. None of this is specified in the null hypothesis and the whole procedure loses much of its logical integrity. Any more useful procedure must solve this problem (see Chapter 7 for the solution).

Even if we discount the logical problem, there is a statistical problem clamouring for attention. It is obviously possible to compare the means in pairs. If we simply ignore the logical dilemma we could test the following pairs:

$$\left.\begin{aligned} H_{01}&: \quad \mu_C = \mu_{PC} \\ H_{02}&: \quad \mu_{PC} = \mu_E \\ H_{03}&: \quad \mu_C = \mu_E \end{aligned}\right\} \tag{6.16}$$

So far, so good. There is, however, a problem to do with the probability of Type I error in these tests. In each test, let us assume that the conventional value of $\alpha = 0.05$ is used as the probability of Type I error. There are three tests and therefore the probability of Type I error over the set of three tests can be calculated using the binomial formula discussed earlier (Equation 6.13). In practice, the tests are not independent (each of the three sample means is used twice throughout the set of three tests), so the situation is even more confused than this discussion suggests. Nevertheless, the probability of making a Type I error somewhere in the set of three tests can be approximated using the number of events (i.e. errors) $r = 0$, the number of trials $n = 3$ and the probability of an event (error) occurring $p = 0.05$ in the binomial formula:

$$\text{Prob}\,(r = 0) = \frac{n!}{r!(n-r)!}\,p^r(1-p)^{n-r} = \frac{3!}{0!\,3!}(0.05)^0(0.95)^3 = 0.95^3$$

Thus, the probability of making no Type I errors is 0.95^3, which equals 0.86. Therefore, the probability of at least one Type I error across the three tests is all other possibilities (there must be 0, 1, 2 or 3 such errors in three tests). This is (1 minus the probability of no errors) and is 0.14 – nearly three times larger than previously determined before the experiment was done.

The tests require modification so that the Type I errors do not accumulate. One procedure is to 'Bonferronize' the probabilities; a new probability of Type I error (α_1) is calculated so that it produces 0.05 after the binomial formula has been applied. For k tests, the probability of Type I error for each test that will produce 0.05 over all tests is $\alpha_1 = \alpha/k$. For $k = 3$ tests, $\alpha_1 = 0.017$. This is the probability of Type I error necessary in each of the three tests to ensure that the set of three tests has the pre-determined probability (i.e. 0.05).

Thus, two things must be done. First, we would have to adjust the probabilities used in the test, as shown here. Second, in each of the individual t-tests, there is now considerably increased probability of Type II error, so that the individual comparisons are now less powerful. Both components suggest that something more efficient would be handy. This is taken up later in Chapter 7.

6.9.3 Conclusion

All of this discussion suggests very strongly that experiments with only two treatments are not usually much good. Experiments with more than two treatments require procedures of analysis different from the t-test used to compare the means of two populations.

So, here we are already in Chapter 6 and we have not yet met a situation where we can proceed with the design of an experiment – because the experiments have all turned out to be more complicated than the simplest designs and analyses. There is almost certainly a moral in this tale. The design of experiments that retain their logical structure is a complex business. Few, if any, experiments are logically interpretable *and simple*. This might be news to those who have read the ecological literature and seen many simple experiments. Next time, also look for the integrity of the logical structure!

7

Analysis of variance

7.1 Introduction

One of the most powerful and versatile tools in experimental design is the set of procedures known as analysis of variance. These start from a simple conceptual basis, but can be used in many different aspects of biological research. They were expressly designed to solve the problems of trying to compare more than two populations (see Section 6.9) and are used for tests about means of several populations. They are very suitable for planned, manipulative experiments and are also of widespread application in mensurative or comparative studies where predictive hypotheses are tested by collecting data from the appropriate parts of habitats, sizes of organisms or whatever.

The hypothesis underlying all analyses of variance is that some difference is predicted to occur among the means of a set of populations. The null hypothesis, reduced of any complexity (see several later models) is always of the form:

$$H_0: \quad \mu_1 = \mu_2 \ldots = \mu_i \ldots = \mu_a (= \mu) \tag{7.1}$$

where μ_i represents the mean (location) of population i in a set of populations. It is proposed as the null case that all populations have the same mean μ. Experiments with such a linear array of means are known as *single-factor* or *one-factor* experiments. Each population represents a single treatment or *level* of the factor.

Any departure from this null hypothesis is a valid alternative – ranging from one population being different from the other $(a - 1)$ populations to all a populations differing. Departure from the null hypothesis does not imply that all the populations differ.

The procedure is often thought to be complex and there is no doubt that some researchers have responded to this apparent complexity by doing vastly oversimplified experiments. Thereby, they have lost the available capacity to understand ecological variability in time or among sites in the outcome of variable processes. Analysis of variance is well suited to identifying interactive processes. These include variability in the outcome of processes such as predation, competition or recruitment and emigration, which often vary markedly from place to place and time to time.

The simplest way to understand analysis of variance is to deal with three different aspects of its rationale. First, consider the sorts of data that are collected in an experiment to test the null hypothesis indicated in Equation 7.1. These data can be manipulated, by simple operations, without any assumptions about the way they are distributed. The most useful manipulation is to analyse the data – to partition the available information into useful and identifiable components. This is explained below.

The second step is to consider an underlying theoretical model that might explain the data. This is useful to identify what is being measured by the samples. Again, it makes no assumptions about the frequency distributions being investigated.

Third, the components of the theoretical model are used to identify what has been measured in the partitioning of the data into their component parts. As a result, a very useful (and simple) test statistic can be identified to test the null hypothesis.

Because of its importance in the development of modern experimental design, it is worth examining these three steps in detail. It will pay off because all more complex designs use exactly the same operational steps. Understanding and interpreting the necessarily complex designs of experiments on variable biological processes is only likely if the simplest case is completely interpretable.

7.2 Data collected to test a single-factor null hypothesis

To test the null hypothesis in Equation 7.1, a representative sample is taken of each of the populations. In the simplest case, these samples are *balanced* – they are all of the same size, n. This causes the collection of n replicate data in each of a samples, as in Table 7.1. Each value X_{ij}, belongs to a particular population i and is the jth member of a sample of that population. The subscripts are a necessary pain, but don't panic. It is not complicated algebra.

The total variability among the numbers in the entire set of data can be measured by calculating how far the individual values (the X_{ij} values) are from $\bar{X}$, the overall mean of all data combined. Because of the usual

Table 7.1. Data for 1-factor analysis of variance

(a) Example of data; $a = 4$ experimental treatments, $n = 5$ replicates

	Treatment			
Replicate	1	2	3	4
1	414	353	50	413
2	398	357	76	349
3	419	359	28	383
4	415	396	29	364
5	387	395	27	365
Sample mean	406.6	372	42	374.8
Sample variance	184.3	465	452.5	601.2
Standard error	6.07	9.64	9.51	10.97

Overall mean = 298.85

(b) General layout

	Treatment					
Replicate	1	2	...	i	...	a
1	X_{11}	X_{21}		X_{i1}		X_{a1}
2	X_{12}	X_{22}		X_{i2}		X_{a2}
⋮						
j	X_{1j}	X_{2j}		X_{ij}		X_{aj}
⋮						
n	X_{1n}	X_{2n}		X_{in}		X_{an}
Sample mean	$\bar{X}_1$	$\bar{X}_2$		$\bar{X}_i$		$\bar{X}_a$
Sample variance	s_1^2	s_2^2		s_i^2		s_a^2

$$\text{Overall mean} = \bar{X} = \frac{\sum_{i=1}^{a} \bar{X}_i}{a} = \frac{\sum_{i=1}^{a} \sum_{j=1}^{n} X_{ij}}{an}$$

problem of deviations from a mean summing to zero (Section 3.8), the deviations are all squared. The total variation amongst the entire set of data (i.e. the sum of squared deviations from the mean over the entire set of data) is known as the total sum of squares (for fairly obvious reasons!).

$$\text{Total sum of squares} = \sum_{i=1}^{a} \sum_{j=1}^{n} (X_{ij} - \bar{X})^2 \tag{7.2}$$

noting how to calculate $\bar{X}$ in Table 7.1b. Where data are more (or less) variable, this will be a larger (or smaller) number.

Each sample also provides a sample estimate of the mean of the population from which it came.

7.3 Partitioning of the data: the analysis of variation

Without too much fuss or difficulty for comprehension, Equation 7.2 could/can be rewritten as:

$$\text{Total sum of squares} = \sum_{i=1}^{a} \sum_{j=1}^{n} [(X_{ij} - \bar{X}_i) + (\bar{X}_i - \bar{X})]^2 \tag{7.3}$$

because the $\bar{X}_i$ terms cancel each other out. So, without any alteration, this can be rewritten as:

$$\begin{aligned} \text{Total sum of squares} &= \sum_{i=1}^{a} \sum_{j=1}^{n} [(X_{ij} - \bar{X}_i) + (\bar{X}_i - \bar{X})]^2 \\ &= \sum_{i=1}^{a} \sum_{j=1}^{n} (X_{ij} - \bar{X}_i)^2 + \sum_{i=1}^{a} \sum_{j=1}^{n} (\bar{X}_i - \bar{X})^2 \qquad (7.4) \\ &\quad + 2\sum_{i=1}^{a} \sum_{j=1}^{n} (X_{ij} - \bar{X}_i)(\bar{X}_i - \bar{X}) \end{aligned}$$

using the well-known expansion of $(r + s)^2 = r^2 + 2rs + s^2$. This simple operation made R. A. Fisher the most generally contributory scientist of this century (i.e. the one who has had the most influence on so many fields of science). It is simple, but we aren't doing it for the first time!

It turns out that the right-hand, third term in Equation 7.4 equals zero (see Table 7.2) and so we have:

$$\text{Total sum of squares} = \sum_{i=1}^{a} \sum_{j=1}^{n} (X_{ij} - \bar{X}_i)^2 + \sum_{i=1}^{a} \sum_{j=1}^{n} (\bar{X}_i - \bar{X})^2 \tag{7.5}$$

The first of these two terms measures some function of variability *within* the samples – it is calculated as deviations of the data (X_{ij} values) from the mean of the sample to which they belong ($\bar{X}_i$ values). The second term obviously involves variability *among* the samples – it is calculated as the deviations of each sample mean from the overall mean. In principle (although sampling error makes this more complicated, as explained in Section 7.7 below), there should be no variability among samples if the null hypothesis were true. When it is true, the sample means would all be the same except for sampling error.

Table 7.2. Algebraic disappearance of the cross-product term in Equation 7.4

From Equation 7.4, the cross-product term is:

$$\sum_{i=1}^{a}\sum_{j=1}^{n}(X_{ij}-\bar{X}_i)(\bar{X}_i-\bar{X})$$

$$\bar{X}_i=\sum_{j=1}^{n}X_{ij}/n \quad \text{and} \quad \bar{X}=\sum_{i=1}^{a}\sum_{j=1}^{n}X_{ij}/an$$

$$=\sum_{i=1}^{a}\sum_{j=1}^{n}((X_{ij}\times\bar{X}_i)-\bar{X}_i^2-(X_{ij}\times\bar{X})+(\bar{X}_i\times\bar{X}))$$

$$=\sum_{i=1}^{a}\sum_{j=1}^{n}\left[\frac{X_{ij}\left(\sum_{j=1}^{n}X_{ij}\right)}{n}-\frac{\left(\sum_{j=1}^{n}X_{ij}\right)^2}{n^2}-\frac{X_{ij}\left(\sum_{i=1}^{a}\sum_{j=1}^{n}X_{ij}\right)}{an}+\frac{\left(\sum_{j=1}^{n}X_{ij}\right)\left(\sum_{i=1}^{a}\sum_{j=1}^{n}X_{ij}\right)}{an^2}\right]$$

Summing over n

$$=\sum_{i=1}^{a}\left[\frac{\sum_{j=1}^{n}X_{ij}\left(\sum_{j=1}^{n}X_{ij}\right)}{n}-\frac{\sum_{j=1}^{n}\left(\sum_{j=1}^{n}X_{ij}\right)^2}{n^2}-\frac{\sum_{j=1}^{n}X_{ij}\left(\sum_{i=1}^{a}\sum_{j=1}^{n}X_{ij}\right)}{an}+\frac{\sum_{j=1}^{n}\left(\sum_{j=1}^{n}X_{ij}\right)\left(\sum_{i=1}^{a}\sum_{j=1}^{n}X_{ij}\right)}{an^2}\right]$$

$$=\sum_{i=1}^{a}\left[\frac{\left(\sum_{j=1}^{n}X_{ij}\right)^2}{n}-\frac{n\left(\sum_{j=1}^{n}X_{ij}\right)^2}{n^2}-\frac{\left(\sum_{j=1}^{n}X_{ij}\right)\left(\sum_{i=1}^{a}\sum_{j=1}^{n}X_{ij}\right)}{an}+\frac{n\left(\sum_{j=1}^{n}X_{ij}\right)\left(\sum_{i=1}^{a}\sum_{j=1}^{n}X_{ij}\right)}{an^2}\right]$$

$$=0$$

This completes the algebraic component of the introduction to the analysis. The data have been *partitioned* or *analysed* into two components. The two sources of variability (within and among samples) add up to the total variability amongst all the data (Table 7.3). To do so is solely an

Table 7.3. Partitioning (i.e. analysis) of variation into identifiable components in a single-factor analysis of variance

Source of variation	Sum of squares
Among samples	$\sum_{i=1}^{a}\sum_{j=1}^{n}(\bar{X}_i - \bar{X})^2$
Within samples	$\sum_{i=1}^{a}\sum_{j=1}^{n}(X_{ij} - \bar{X}_i)^2$
Total	$\sum_{i=1}^{a}\sum_{j=1}^{n}(X_{ij} - \bar{X})^2$

algebraic operation. No assumptions have been made about the type of data, nor the sorts of frequency distributions from which they came. Each of these sources of variation has at least some intuitively interpretable meaning.

7.4 A linear model

The model developed to describe data from such an experiment is:

$$X_{ij} = \mu_i + e_{ij} \tag{7.6}$$

where μ_i is the mean of the population from which a sample is taken and X_{ij} is the jth replicate in the ith sample. e_{ij} is therefore a function of the variance of the population – it measures how far each X_{ij} is from the mean of its population. This is illustrated diagramatically in Figure 7.1. The X_{ij} values are distributed with variance σ_i^2 around their mean μ_i. The same data can obviously be replotted as e_{ij} values (as in Figure 7.1), centred around zero with variance σ_i^2.

In terms of the null hypothesis, all the populations have the same mean μ. If the null hypothesis is not true, the populations do not all have the same mean and it is assumed that they differ (at least, those that do differ) by amounts identified as A_i terms in Figure 7.2. There is a linear, additive difference between each population's mean and the overall mean of all populations.

Therefore:

$$X_{ij} = \mu + A_i + e_{ij} \tag{7.7}$$

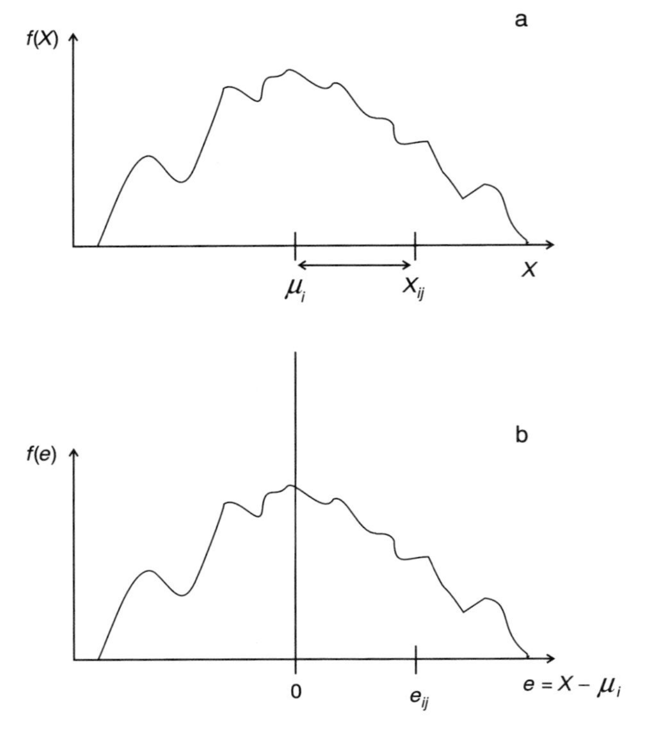

Figure 7.1. Diagram illustrating the error term in an analysis of variance. (a) Frequency distribution of a variable X, with mean μ_i and variance σ_i^2. X_{ij} is a particular value, at a distance e_{ij} from μ_i. (b) Frequency distribution of the same variable, transformed to a mean of zero (i.e. $X - \mu_i$). This is the distribution of e ($= X - \mu_i$), with mean zero and variance σ_i^2.

where μ is the mean of all populations, A_i is the linear difference between the mean of population i and the mean of all populations, if the null hypothesis is not true and e_{ij} is as before. Within any population, a particular individual reading (X_{ij}) differs from the mean of its population by an individual amount (e_{ij}). There is obviously no loss of generality in using this second equation (Equation 7.7) whether or not the null hypothesis is true. When the null hypothesis is true, all values of A_i are zero (compare Figure 7.2a and b). The original null hypothesis of equality of the means of the a populations (Equation 7.1) can be restated as:

$$H_0: \quad A_1 = A_2 = \ldots A_i = \ldots = A_a = 0 \qquad (7.8)$$

which is the situation illustrated in Figure 7.2a, where the null hypothesis was true.

From the linear model, without any new assumptions, the following can also be calculated for sampled data:

$$\bar{X}_i = \sum_{j=1}^{n_i} X_{ij}/n_i = \mu + A_i + \bar{e}_i \quad (7.9)$$

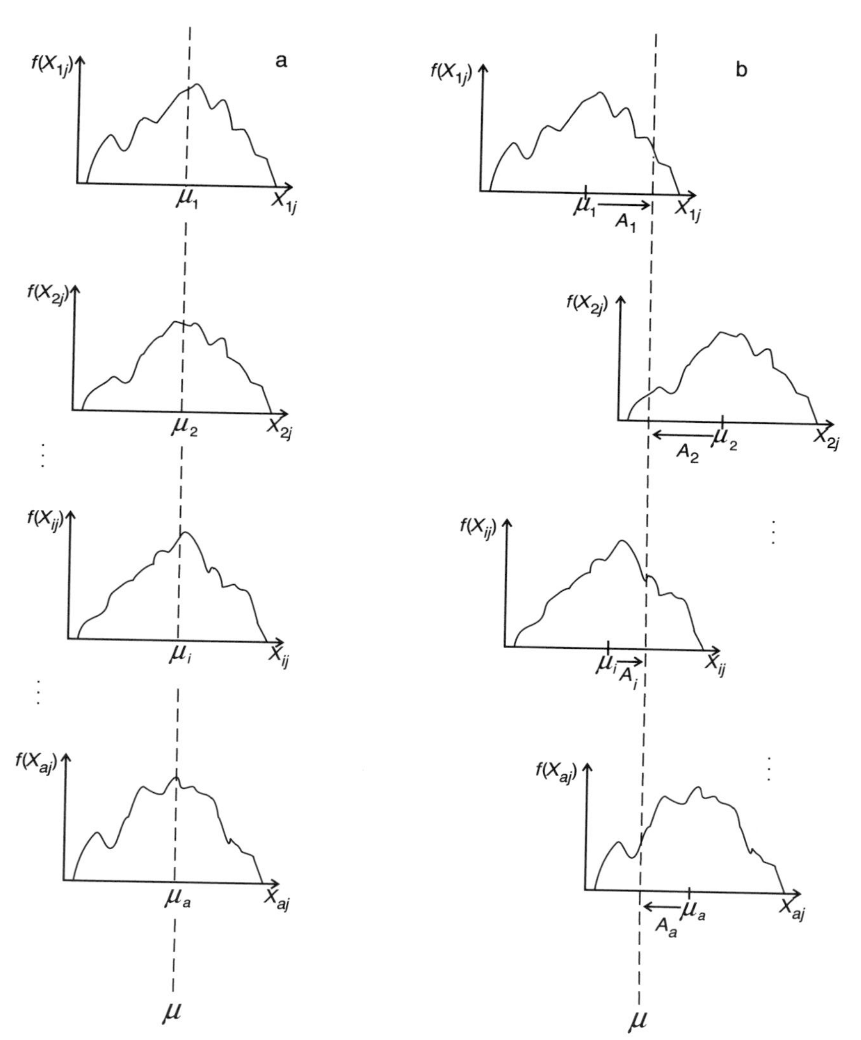

Figure 7.2. Illustration of A_i terms in the linear model for an analysis of variance. There are $1 \ldots i \ldots a$ populations of the variable X_{ij}, each with mean, μ_i. In (a), all populations have the same mean, μ; the null hypothesis is true. In (b), the null hypothesis is false and some populations have different means, which differ from μ, the overall mean, by amounts A_i.

Table 7.4. Analysis of data generated from a normal distribution with mean = $300 + A_i$ and variance = 1000 to show sampling error (e_{ij}). $\bar{A}$, the average of all A_i values is zero. $\hat{A}_i$ values estimate A_i values as the difference between $\bar{X}_i$ and $\bar{X}$

Example data: $\mu = 300$, $\sigma^2 = 1000$, $n = 5$, $a = 4$

		$A_1 = +30$		$A_2 = -20$		$A_3 = +40$		$A_4 = -50$	
$\mu_i = \mu + A_i$		330		280		340		250	
	e_{11}	+7	e_{21}	+67	e_{31}	+44	e_{41}	+65	
	e_{12}	−22	e_{22}	+7	e_{32}	−45	e_{42}	+42	
	e_{13}	−18	e_{23}	+22	e_{33}	−48	e_{43}	+29	
	e_{14}	+29	e_{24}	+17	e_{34}	−11	e_{44}	−25	
	e_{15}	−13	e_{25}	+26	e_{35}	−11	e_{45}	−33	
$\bar{e}_i$		−3.4		27.8		−14.2		15.4	$\bar{e} = 6.4$
	X_{11}	337	X_{21}	347	X_{31}	384	X_{41}	314	
	X_{12}	308	X_{22}	287	X_{32}	295	X_{42}	292	
	X_{13}	312	X_{23}	302	X_{33}	292	X_{43}	279	
	X_{14}	359	X_{24}	297	X_{34}	329	X_{44}	225	
	X_{15}	317	X_{25}	306	X_{35}	329	X_{45}	217	
$\bar{X}_i$		326.6		307.8		325.8		265.4	$\bar{X} = 306.4$
$\hat{A}_i$		+20.2		+1.4		+19.4		−41.0	$\sum = 0$
s_i^2		452.3		530.7		1374.7		1807.3	
$\hat{A}_i$ is wrong by		−9.8		+21.4		−20.6		+9.0	$\sum = 0$
Which is $\bar{e}_i$		−3.4		+27.8		−14.2		+15.4	
$\bar{e}$		+6.4		+6.4		+6.4		+6.4	
So $\bar{e}_i - \bar{e}$		−9.8		21.4		−20.6		+9.0	

where $\bar{X}_i$ is the mean of a sample of n_i replicates (allowing the size of sample to vary from one population to another); A_i is, as before, the difference of that population from the overall mean if the null hypothesis is false and $\bar{e}_i$ is the average of all deviations of e_{ij} from their population's mean. If the sample represents the population perfectly, $\bar{e}_i$ would be zero (the mean of e_{ij} values in Figure 7.1 is zero). In fact, $\bar{e}_i$ will not be zero because of sampling error and its magnitude will be a function of the variance of the population σ_i^2 and the size of the sample n_i, as described in Section 5.2, where the standard error was discussed.

The origin and meaning of $\bar{e}_i$ is illustrated in Table 7.4, which is a fabricated analysis of four populations, each with the *same* variance (which is important later – see below). The means of the populations are different and known to us (I made them up!). They differ by the amounts identified as A values in the Table. Samples of $n = 5$ from each population allow calculation of the e terms, because we know the means, unlike what happens in the real world. The means of these in each sample ($\bar{e}_i$ values) are not exactly zero because of sampling error. Therefore, the estimates of means of the populations ($\bar{X}_i$ values) are not exact measures; nor are the estimates of differences among populations (A_i values) and the overall mean (μ). The latter is wrong because of the average sampling error ($\bar{e}$) over all four populations. These data were randomly generated from normal distributions with the specified means and variances. Obviously, where variances are large and/or samples are small, sampling error can be quite large. The data in Table 7.4 provide some visible, numerical reality to the algebra being manipulated here. The details are discussed in Section 7.9.

Finally, the mean of all data, over all samples, can also be calculated as:

$$\bar{X} = \sum_{i=1}^{a} \sum_{j=1}^{n_i} X_{ij} \Big/ \sum_{i=1}^{a} n_i = \mu + \bar{A} + \bar{e} \tag{7.10}$$

where $\bar{X}$ is the mean of all sampled data, $\bar{A}$ is the average of the deviations of means of the populations from the overall mean (see Sections 7.20 and 8.2 for clarification of $\bar{A}$). $\bar{e}$ is the average of the averages of individual deviations from the means of the populations due to sampling error in each sample.

7.5 What do the sums of squares measure?

These derivations for each sample and for the entire set of data can be used to determine what we measure in the analysis, or partitioning of data done earlier (Section 7.3). The two sources of variation can be interpreted by substitution of the terms by their equivalent expressions from the linear model. So, variation within samples (if samples are balanced, i.e. of the same size) is:

$$\sum_{i=1}^{a} \sum_{j=1}^{n} (X_{ij} - \bar{X}_i)^2 = \sum_{i=1}^{a} \sum_{j=1}^{n} [(\mu + A_i + e_{ij}) - (\mu + A_i + \bar{e}_i)]^2$$

$$= \sum_{i=1}^{a} \sum_{j=1}^{n} (e_{ij} - \bar{e}_i)^2 \tag{7.11}$$

This can be simplified by noting that, for each population sampled, the sample variance s_i^2 estimates the population's variance σ_i^2. So:

$$s_i^2 = \frac{\sum_{j=1}^{n} (X_{ij} - \bar{X}_i)^2}{(n-1)} = \frac{\sum_{j=1}^{n} (e_{ij} - \bar{e}_i)^2}{(n-1)} \tag{7.12}$$

and:

$$\sum_{j=1}^{n} (e_{ij} - \bar{e}_i)^2 = (n-1)\, s_i^2 \text{ and estimates } (n-1)\, \sigma_i^2 \tag{7.13}$$

From this, the variation *within* samples is (from Equation 7.13):

$$\sum_{i=1}^{a} (n-1)\, s_i^2 \text{ and estimates } \sum_{i=1}^{a} (n-1)\, \sigma_i^2 \tag{7.14}$$

The variation *among* samples is more complex, but is, from Equations 7.9 and 7.10:

$$\sum_{i=1}^{a} \sum_{j=1}^{n} (\bar{X}_i - \bar{X})^2 = \sum_{i=1}^{a} \sum_{j=1}^{n} [(\mu + A_i + \bar{e}_i) - (\mu + \bar{A} + \bar{e})]^2$$

$$= \sum_{i=1}^{a} \sum_{j=1}^{n} [(A_i - \bar{A}) + (\bar{e}_i - \bar{e})]^2 \tag{7.15}$$

This is expandable (take my word for it if you don't like algebra) as:

$$\sum_{i=1}^{a} \sum_{j=1}^{n} (A_i - \bar{A})^2 + \sum_{i=1}^{a} \sum_{j=1}^{n} (\bar{e}_i - \bar{e})^2$$

$$+ 2 \sum_{i=1}^{a} \sum_{j=1}^{n} (A_i - \bar{A})(\bar{e}_i - \bar{e}) \tag{7.16}$$

Now we shall make an *assumption* that each population is sampled independently and, in each population, every member of a sample is taken *independently*. This requires that there is *no* influence of the individual replicates in one sample on the sample taken for other populations. There is considerable discussion of this crucial point in Section 7.13, with examples demonstrating what it means in practice. For now, make the assumption!

It has several consequences, of which the most useful one here is that, *if the populations are sampled independently*, the last right-hand term in Equation 7.16 is zero. Another way of looking at this is to view the product of the two terms $(A_i - \bar{A})$ and $(\bar{e}_i - \bar{e})$, each of which is the deviation of a term from its mean, as a measure of correlation. If two variables measured as deviations are uncorrelated, their summed product (i.e. $(A_i - \bar{A})(\bar{e}_i - \bar{e})$) is zero. If they are independent, they must be uncorrelated. So, the variation among samples is:

$$\sum_{i=1}^{a} \sum_{j=1}^{n} (\bar{X}_i - \bar{X})^2 = \sum_{i=1}^{a} \sum_{j=1}^{n} (A_i - \bar{A})^2 + \sum_{i=1}^{a} \sum_{j=1}^{n} (\bar{e}_i - \bar{e})^2 \tag{7.17}$$

Now, the second term on the right-hand side of Equation 7.17 can be interpreted very easily. For any sample of a variable X with mean μ and variance σ^2, the sample mean is often normally distributed with mean μ and variance σ^2/n (see Section 5.1). Thus, a series of a samples, each of size n, give sample means $\bar{X}_1, \bar{X}_2 \ldots \bar{X}_i \ldots \bar{X}_a$ and these have an overall mean $\bar{X}$. If these sample means are derived from independent observations, they have a variance σ^2/n. This was described and illustrated in Section 5.1 and see Figure 5.2. So, the variance of a set of means of samples is:

$$\sum_{i=1}^{a} (\bar{X}_i - \bar{X})^2/(a-1) \text{ and estimates } \sigma^2/n \tag{7.18}$$

From this, it follows directly that:

$$n\sum_{i=1}^{a} (\bar{X}_i - \bar{X})^2/(a-1) \text{ estimates } \sigma^2 \tag{7.19}$$

which is the same as:

$$\sum_{i=1}^{a}\sum_{j=1}^{n} (\bar{X}_i - \bar{X})^2/(a-1) \text{ estimates } \sigma^2 \tag{7.20}$$

and thus still estimates the variance of the population being sampled.

By analogy, the same algebra holds true for samples from different populations, provided that the distributions all have the same variance. The same argument will hold for samples of any variable, including the e_{ij} terms in the linear model in Equation 7.16.

Provided that the sample means are normally distributed and conform to the central limit theory (as above and in Section 5.1) and provided that the a populations being sampled all have equal variances, this algebra can be applied to the data in the linear model. Therefore, *under the assumption of equality of variances of the populations*:

$$\sum_{i=1}^{a}\sum_{j=1}^{n} (\bar{e}_i - \bar{e})^2/(a-1) \text{ estimates } \sigma_e^2 \tag{7.21}$$

where σ_e^2 is the variance of each (all) of the populations.

And:

$$\sum_{i=1}^{a}\sum_{j=1}^{n} (\bar{e}_i - \bar{e})^2 \text{ estimates } (a-1)\,\sigma_e^2 \tag{7.22}$$

Table 7.5. Analysis of variance with one factor. Sums of squares estimate terms from the linear model for samples of equal sizes

Source of variation	Sum of squares (SS)	SS estimates
Among samples	$\sum_{i=1}^{a}\sum_{j=1}^{n}(\bar{X}_i - \bar{X})^2$	$(a-1)\,\sigma_e^2 + \sum_{i=1}^{a}\sum_{j=1}^{n}(A_i - \bar{A})^2$
Within samples	$\sum_{i=1}^{a}\sum_{j=1}^{n}(X_{ij} - \bar{X}_i)^2$	$a(n-1)\,\sigma_e^2$
Total	$\sum_{i=1}^{a}\sum_{j=1}^{n}(X_{ij} - \bar{X})^2$	

This allows the interpretation of the calculation in Table 7.5 where Expression 7.22 has been substituted into Equation 7.17. We are now nearly at the end of this detailed interpretation of an analysis of variance.

The assumption that all the populations have the same variances (σ_e^2) modifies the formulae used to interpret the variation within samples (as in Equation 7.13). If the variances of all the populations (σ_i^2 values) are equal, they are all equal to a variance, which we can call σ_e^2. Therefore, if all the variances are equal (all σ_i^2 values $= \sigma_e^2$), Equation 7.11 can be rewritten, so that:

$$\text{Variation within samples} = \sum_{i=1}^{a}\sum_{j=1}^{n}(X_{ij} - \bar{X}_i)^2 \text{ measures } \sum_{i=1}^{a}(n-1)\,s_i^2$$

$$\text{and estimates } \sum_{i=1}^{a}(n-1)\,\sigma_e^2 = a(n-1)\,\sigma_e^2 \qquad (7.23)$$

The results so far are summarized in an analysis of variance table in Table 7.5, which brings together the sums of squares and what they estimate.

7.6 Degrees of freedom

The next thing we shall need is the degrees of freedom associated with each of the calculations. Remembering the operational rule for calculating these (as in Section 3.10), the results are in Table 7.6. For the total sum of squares, there is a total of n data in each of the a samples and therefore an data in the set, of which we need to know the overall mean $\bar{X}$. The degrees of freedom are therefore $(an - 1)$. For the variation among

Table 7.6. Analysis of variance with one factor. Mean squares estimate terms from the linear model (see Table 7.5)

Source of variation	Sum of squares (SS)	Degrees of freedom	Mean square	Mean square estimates
Among samples	$\sum_{i=1}^{a}\sum_{j=1}^{n}(\bar{X}_i - \bar{X})^2$	$(a-1)$	$\dfrac{\sum_{i=1}^{a}\sum_{j=1}^{n}(\bar{X}_i - \bar{X})^2}{(a-1)}$	$\sigma_e^2 + \dfrac{\sum_{i=1}^{a}\sum_{j=1}^{n}(A_i - \bar{A})^2}{(a-1)}$
Within samples	$\sum_{i=1}^{a}\sum_{j=1}^{n}(X_{ij} - \bar{X}_i)^2$	$a(n-1)$	$\dfrac{\sum_{i=1}^{a}\sum_{j=1}^{n}(X_{ij} - \bar{X}_i)^2}{a(n-1)}$	σ_e^2
Total	$\sum_{i=1}^{a}\sum_{j=1}^{n}(X_{ij} - \bar{X})^2$	$an-1$		

samples, there are a sample means ($\bar{X}_i$ values) and we must know the overall mean $\bar{X}$); hence, there are $(a-1)$ degrees of freedom. The degrees of freedom within samples is self-evident from these considerations (in each sample, there are n data, of which we know the mean, so there are $(n-1)$ degrees of freedom in each of the a samples). Altogether there are $a(n-1)$ degrees of freedom within samples.

7.7 Mean squares and test statistic

The final manipulation of the algebra is to divide the sums of squares for the two sources of variation by their degrees of freedom (because this is ultimately convenient for constructing a useful test statistic). These numbers are known as mean squares, as in Table 7.6. Why they are called mean squares baffles me too!

The final term of the estimate provided by the mean square among samples is very easily related to the null hypothesis. If, as the null requires (Equation 7.8), all A terms are equal to zero, $\bar{A}$ (the average of all A_i values, see Equation 7.10) is zero and so is $(A_i - \bar{A})$ for all populations. So, when the null hypothesis is true:

$$H_0: \quad \sum_{i=1}^{a}\sum_{j=1}^{n}(A_i - \bar{A})^2 = 0 \tag{7.24}$$

is also true. Therefore, when this null hypothesis is true, the mean square among samples estimates only the variance of the populations σ_e^2 and so does the mean square within samples (see Table 7.6). When the null hypothesis is true and there are no differences among the means of a set of populations, we have:

$$\text{H}_0\text{: Estimate of } \sigma_e^2 \text{ from mean square among treatments}$$
$$= \text{Estimate of } \sigma_e^2 \text{ from mean square within treatments} \quad (7.25)$$

The mean squares are composed only of squared numbers (i.e. $(A_i - \bar{A})^2$ cannot be negative), so the only possible alternative to the null hypothesis is:

$$\text{H}_\text{A}\text{: Estimate of } \sigma_e^2 \text{ from mean square among treatments}$$
$$> \text{Estimate of } \sigma_e^2 \text{ from mean square within treatments} \quad (7.26)$$

Entirely equivalent formulations are:

$$\text{H}_0\text{: } \frac{\text{MS}_{\text{among}}}{\text{MS}_{\text{within}}} = 1 \quad \text{and} \quad \text{H}_\text{A}\text{: } \frac{\text{MS}_{\text{among}}}{\text{MS}_{\text{within}}} > 1 \quad (7.27)$$

This, although you may not recognize it, is a familiar calculation. When the null hypothesis is true, each of these mean squares is a sampled variance (each is an estimate of σ_e^2). The ratio of two sampled variances in the two-tailed case, where either could be smaller or larger, was used to determine whether two samples had the same variance (in Section 6.4.2). Now it must be assumed that the populations being sampled are normally distributed. If so, the distribution of the ratio of two estimates of a variance when the null hypothesis that they are equal is true is known as the F-ratio. It is therefore a convenient test statistic for the present null hypothesis, although its use in analysis of variance is one-tailed because of the above null and alternative hypotheses (Equations 7.27). Note that the F-distribution does not have a mean (or expected value) of exactly 1, as implied by Equation 7.27. So, technically, the null hypothesis is strictly that the ratio mean square among treatments divided by mean square within treatments can be expected to be distributed as a central F-ratio distribution. There is, however, no loss of understanding if we use the short-hand version in Equation 7.27.

7.8 Solution to some problems raised earlier

Notice that this rigmarole has solved the logical and statistical problems that arise when more than two populations are being compared (raised in Section 6.9). If the region of rejection of the test statistic is chosen by having the probability of Type I error $\alpha = 0.05$, there is only one test to evaluate all potential differences among means. Therefore, there is no difficulty of excessive Type I error as there would have been in a series of pair-wise t-tests.

Also, the test has converted the null hypothesis to a one-tailed situation. There is now no potential logical problem for interpretation due to an excessive number of tails (as there would be in any series of pair-wise comparisons).

Finally, although there are many issues still to be resolved (e.g. about the underlying assumptions), this procedure extends without much new theory to cover an enormous variety of scenarios, models and experimental designs. It is the basis of most modern experimental design.

7.9 So what happens with real data?

A set of data has been prepared in Table 7.4 to demonstrate what happens. The data came from distributions that conform to the linear model (i.e. all four populations have means that are the sum of μ and different A_i terms for each population). I made the four treatment effects – the four values of A_i – add up to zero to simplify things (this is discussed in Section 8.3). The four samples of $n = 5$ data were randomly generated from normal distributions with the appropriate means and with $\sigma_e^2 = 1000$ for every population.

In an experiment, you would get only the X_{ij} values (i.e. you do not know μ nor the A_i values). You use the data to estimate all the parameters in the model (i.e. μ and the A_i values) as is also shown in Table 7.4. Note that algebra alone has preserved the fact that the estimated values of A_i values still add up to zero (this matters later – see Section 7.20). Otherwise, the estimates are not particularly close (that for treatment 2 is +1.4 instead of the −20 originally used to generate the data). The estimate of μ, i.e. $\bar{X}$ (the average of all the data) was 306.4, which is not bad for the original 300 put in.

The differences between estimates and the original values are due to sampling errors, which have two different effects. In each sample, the average of the data is not exactly the mean of the population, because the average of the e_{ij} terms in each sample is not exactly zero. It would only be zero if each sample perfectly represented the population from which it came. $\bar{e}_i$ terms – the average deviation between the sample mean and the true mean for each population (see Equation 7.12) – are shown in Table 7.4.

Second, the overall mean $\bar{X}$ is not exactly the original mean μ because of the error in each sample. The difference $(\bar{X} - \mu)$ is $\bar{e}$, the average of all error terms (e_{ij} values) which is, of course, also the average of all deviations of sample means ($\bar{X}_i$ values) from population means ($\mu + A_i$ values). When samples are the same size, $\bar{e}$ is therefore the mean of $\bar{e}_i$ terms (see Equation 7.10).

Each estimated treatment effect ($\hat{A}_i$) is calculated as the difference between a sampled mean ($\bar{X}_i$) and the overall mean of the data ($\bar{X}$). $\bar{X}_i$ is subject to error ($\bar{e}_i$); $\bar{X}$ is subject to error ($\bar{e}$). Thus, the estimate of A_i is subject to these errors combined (as in Table 7.4).

This demonstrates the self-evident fact that when small samples are taken from populations with large variances, statistical estimates are not very precise. Nevertheless, it also demonstrates in a visible manner what the estimates are and how they are formed. These data might also cause caution in the interpretation of small experiments!

7.10 Unbalanced data

For the single factor analysis, there is no need for the sizes of samples to be equal or balanced. Before taking unbalanced samples, there ought to be some thought about why it is appropriate. The null hypothesis is about a set of means. It is not obvious from anything in the way it is stated that justifies or explains why some of them should be sampled more precisely (see Section 5.7.2 for discussion of the relationship between precision and size of sample). It is not easy to see why some treatments should have more effort spent on them to achieve greater precision. I doubt whether any such reason ever exists and sizes of samples just 'happen' without prior thought. Therefore, 15 replicates of some control are taken, but only three replicates of the experimental treatments because these require a great deal of work to set up, or to maintain.

Often, however, samples become unbalanced due to loss of data, death of animals, arrival of washed-up whales rotting the plot, plane crashes, human interference and, perhaps, 'excesses of custodial zeal' (Frank, 1975). There are several ways to proceed in order to complete the analysis.

If there are few replicates in nearly every treatment and only the odd one is missing here or there, do the analysis anyway. Use all the data you have because you have very few data! If there are numerous replicates (say, $n \geq 20$) and only an occasional one is lost, do the unbalanced analysis because this amount of difference is small and the effect on precision is minuscule when n is large.

What should be avoided is the situation where some samples have large and others have small n. This leads to great imbalance in the precision of sampled means. More importantly, it leads to a difference in power to detect differences among samples when multiple comparisons are done. If five treatments have $n = 5, 5, 10, 10, 20$, respectively, there will be greater precision in a comparison of treatments 3 and 4 with treatment 5 than a comparison of treatments 1 and 2 with treatment 3 or 4.

It is usually simpler, in addition to being better, to use balanced samples.

7.11 Machine formulae

Although we have derived the formulae for calculating sums of squares as measures of variation among data, they should not be used! These formulae involve considerable rounding errors in their numeric results, because means are usually rounded somewhere in the calculations. Instead, algebraically identical formulae that introduce little rounding should be used (as in numerous textbooks and commercial software).

7.12 Interpretation of the result

Having done the analysis of variance, the value of F obtained allows you to determine whether to reject or retain the null hypothesis of no difference among means. If it is rejected, you know that at least one mean differs from the others. You do *not* know that all the means differ. Consequently, the next step is usually a procedure to determine the alternative to the null hypothesis. This is a thorny problem for some experimental procedures and some designs.

In general, however, the sample means are compared in some form of multiple comparison. All multiple comparisons have the same purpose – to group the means so that the specific alternative to the null hypothesis can be determined (if it is discernible with the available data). The actual procedures are discussed in the next chapter (Section 8.6).

The analysis of variance is a procedure to allow a simple test of a logically complex null hypothesis. Once it is determined that the null should be rejected, subsequent tests are necessary. These come in a variety of forms depending on the nature of the alternative hypothesis that has been proposed.

The analysis of variance for a single factor is an important analysis. It forms the basis for all more complex procedures and is the key structure for the design of more complex experiments. The linear model (Section 7.4) can be extended to include terms for many more complex components of experiments and has become a major underpinning of the design of many types of biological experiments.

7.13 Assumptions of analysis of variance

In the development of the analysis, several assumptions become necessary. These were not to do with the algebra of partitioning the data (Section 7.3). They were made necessary by the interpretation of the partitioned data in terms of the linear model and the construction of a test statistic. Very useful and informative discussions of these assumptions are available in papers by Bartlett (1947), Cochran (1947), Eisenhart (1947), Box (1953, 1954a,b) and Scheffé (1959).

The assumptions identified *en route* were:

1. Independence of data within and among samples, which appeared in Section 7.5 and Equations 7.15 and 7.16 to remove an annoying term in the algebra. This term was a measure of the correlation of the deviations of data in each sample from the mean of the sample and the difference of a sample mean from the overall mean. If this correlation is zero, the nuisance term disappears, simplifying everything (Equation 7.17). We assume the term to be zero in order to make things simple.
2. Homogeneity of variances for all populations, which appeared in Section 7.5. This assumption allowed a simple definition of the calculated mean square within treatments (Table 7.6) and enabled Fisher (1928) to determine the distribution of the test statistic (F-ratio). This distribution, under the null hypothesis of no difference among the means of the populations, is known *if* and *only if* the populations have similar variances.
3. Normality of data in each distribution. This assumption was, again, necessary to allow construction of the test statistic's distribution (Section 7.7).

In addition, a further assumption (the summation assumption) becomes necessary in the use of analyses of variance to estimate the magnitudes of differences among treatments. This is described later (Section 7.20), but is an assumption necessary for interpretation of patterns of difference among populations once they have been found. It is not an assumption that affects, in any way, the validity of the statistical test itself.

These assumptions and their effects on the analysis and its interpretation are considered in detail in the following sections. They are dealt with in order of their importance with respect to the havoc they cause if they are not true. Thus, non-independence is more deleterious to the orderly, logical use of an analysis of variance than is heterogeneity of variances. Both are much more problematic than non-normality of data. The effects of violations of the assumptions and how to deal with the assumptions are important components of designing and interpreting experiments. These effects must be understood so that the logical framework of making inferences from experimental data can be monitored throughout the experiments.

7.14 Independence of data

This problem – ensuring that data are independently sampled – is so great that many biological and ecological studies are invalidated by a failure to solve it. This has, unfortunately, not prevented the publication of results of such experimental studies without any consideration of the consequences of the problems for interpretation of the analyses caused by non-independence of the data (Hurlbert, 1984; Underwood, 1981). As a result, subsequent sampling and experiments are done using published methods and designs even though these are flawed. It is therefore crucial to break the self-perpetuating cycle of the use of invalid procedures.

The best way to consider non-independence of data and its effects on interpretation is by examples of the four different types (see Table 7.10). Each of these can be avoided or managed, provided attention is paid to the necessity of removing them from experimental designs.

The onus of creating independence always falls on the experimenter. Within limits set by the needs of the stated hypotheses, independent data can almost always be obtained by application of more thought, more effort and more money. The externally imposed limitations of the last two of these (funding and logistical constraints) may explain why so many non-independent data are collected and analysed.

The most intriguing thing about dealing with potential problems of non-independence is that, first, you do not get much help from books on experimental design. For example, Hairston (1989) did not discuss the problem. There has, however, been much better realization of the problems in the literature on behavioural experiments and there is an excellent consideration of the practical problems in Martin & Bateson (1993).

Second, and much more importantly, it involves considerable thought about and knowledge of the biology of the system being investigated (see below). This makes it a topic well suited to biologists to solve. It makes it a problem that biologists should want to solve. It is not the responsibility of professional statisticians to understand why biological data are likely to be non-independent in any particular experiment. It is necessary for the biologist or ecologist to explain the potential problems, in order to receive better advice from a statistical consultant.

There are, essentially, four types of non-independence.

7.14.1 Positive correlation within samples

Consider a behaviourally induced form of non-independence affecting estimation of growth of small lobsters. In the first case considered, the lobsters are completely independent of one another – they are not influenced by the presence of other lobsters. If they are allowed to feed whenever they choose, the number of foraging excursions each animal makes and the period spent foraging each time will not be influenced by other animals. This was simulated (Figure 7.3, Table 7.7) by sampling 15 lobsters emerging to feed during a period of 12 hours. Each animal emerged at a random time and spent a random period of 15–25 minutes feeding. It then went back into shelter for a minimal random interval of a further 15–25 minutes before it could emerge again. Each animal was modelled to produce an average number of excursions of 10 per animal, each lasting an average of 20 minutes. Therefore each animal spent an average of 200 minutes out feeding and a minimum of 200 minutes in shelter after feeding in each 12 hours (720 minutes). It therefore has a 10/320 (=1/32) chance of emerging to feed per minute when not feeding nor being made to shelter.

The simulation results in the frequency distribution of numbers of excursions in Figure 7.3a and of periods of feeding in Figure 7.3d, as summarized in Table 7.7a. Because the lobsters were modelled independently, it makes no difference whether they were kept together in one aquarium or were kept separately in 15 different tanks.

Now consider the result of sampling a group of lobsters that *are* influenced by the behaviour of other animals. Suppose the lobsters behave as described above when they are isolated. When they are kept in the sight of each other, however, they are inquisitive and tend to come out to feed whenever they see another lobster out feeding. In addition to initiating their own bouts of foraging, they are more likely to emerge to feed when another lobster is feeding.

This was modelled for a group of 15 lobsters by increasing the probability of emerging when in shelter. To illustrate the point, each lobster

Table 7.7. Results of sampling feeding by lobsters ($n = 15$), as described in the text and illustrated in Figure 7.3.

(a) Statistical estimates from samples

	No. of feeding excursions per lobster			Length of feeding excursion (min)			Total time spent feeding per lobster		
	Mean	Variance	n	Mean	Variance	n	Mean	Variance	n
Independent data (Figure 7.3a and d)	10.40	1.97	15	20.56	8.62	156	214	1112	15
Positively correlated (Figure 7.3b and e)	17.40	0.38	15	20.44	7.91	260	356	111	15
Negatively correlated (Figure 7.3c and f)	4.80	17.30	15	16.44	60.00	72	79	9575	15

(b) Results of 200 simulated experiments to test the null hypothesis of no difference among three diets ($n = 9$ lobsters in each diet), as described in the text and illustrated in Figures 7.4 and 7.5. The mean and variance of time spent foraging are from samples of nine lobsters; three samples were taken from each of 200 simulations to represent the three diets for each experiment.

	Time spent foraging				Proportion of F-ratios significant at		
	Mean	Variance ($n = 600$)	Mean F-ratio ($n = 200$)	Mean square among treatments ($n = 200$)	$P = 0.25$	$P = 0.10$	$P = 0.05$
Independent data	199.2	879.0	1.09	904.2	0.25	0.11	0.04
Positively correlated	340.9	162.0	6.21	899.0	0.66	0.50	0.42
Negatively correlated	100.2	6767.9	0.06	162.0	0	0	0

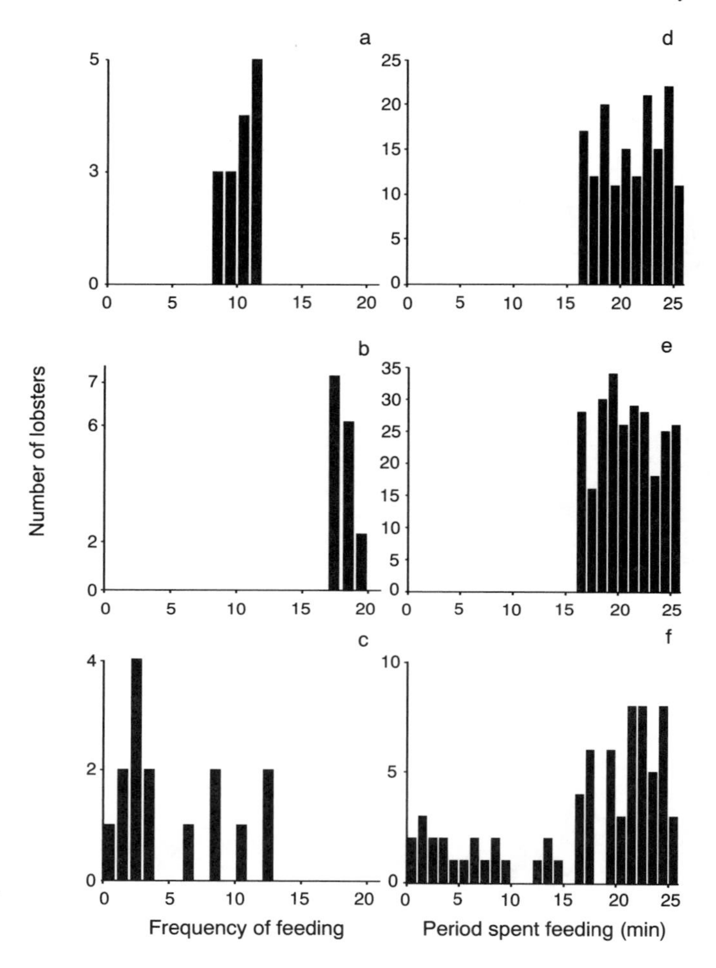

Figure 7.3. Feeding behaviour by samples of 15 lobsters in aquaria. (a), (b), (c) Numbers of foraging excursions (per 12 hours); (d), (e), (f) Frequency distributions of length of feeding bouts (over all 15 lobsters). (a), (d) are independently sampled; (b), (e) are lobsters with positively correlated behaviour; (c), (f) are lobsters with negatively correlated behaviour. For explanation, see the text.

was made 20 times more likely to emerge whenever another lobster was out feeding (i.e. 20/32 chance per minute). It would then stay out for a period between 15 and 25 minutes before returning to shelter for 15–25 minutes.

As would be expected as a result of the influence of the other animals, each lobster came out to feed more often than did the independent animals described earlier (Figure 7.3b). The length of feeding excursion was not influenced by the presence of other lobsters (Figure 7.3e), but, as a result of more frequent feeding, the total time spent feeding was greatly increased (Table 7.7a).

There are two consequences of the *positive* correlation among the lobsters. For number of foraging bouts and total time spent feeding, there was the predictable increase for lobsters that were not independent of one another. Much more importantly, there was a substantial *decrease* in the variance of these variables (Table 7.7a). This will occur whenever there is a positive correlation among the replicates in a sample. It is a general consequence of positive non-independence among units in a sample.

The consequences for statistical tests on a sample are very drastic. In the case just considered, the increased mean time spent feeding (and therefore, probably, the rate of growth of the animals) was increased. Any comparison of such factors as diets, temperatures, rates of flow, etc., on rate of growth would be under the conditions caused by non-independent behaviour. Only if any differences were evident at enhanced rates of growth would they be detected by experiments. If growth is already maximized by the increased tendency to forage caused by the influence of other animals, different diets will not alter it. But growth at the rate shown by independently sampled lobsters would, perhaps, have revealed differences due to these factors.

More generally, however, the influence of positive correlation among replicates is to cause excessive Type I error in statistical tests on differences among samples. This is best illustrated by simulating an experiment on the growth of lobsters in tanks, to test the null hypothesis that there is no difference in mean rate of growth on three diets. Diets represent a fixed factor with three levels. For the sake of simplicity, assume that there are no differences from one aquarium tank to another due to such factors as position in the laboratory (i.e. amount of light, rate of flow of water) or any other uncontrolled source of variation. This is unlikely in the real world and requires replicated aquaria to ensure that any differences among tanks can logically be attributed to differences in diets (if they exist), rather than to confounded influences (see Chapter 9). Nine juvenile lobsters are kept in each tank and fed with an excessive amount of food each day.

Now, suppose that there is no difference in growth due to different diets and that growth is directly related to time spent feeding. The experiment was simulated for independent lobsters exactly as described above. The mean time spent foraging per individual in 12 hours over all 200 experiments is 199 minutes (i.e. estimates 200 minutes as specified in the model with 10 excursions per lobster, each of an average of 20 minutes). The mean square among treatments in each experiment also measured the variability among lobsters because there were no differences among diets (see Section 7.7). In Table 7.7b, the average variance among lobsters (i.e. within treatments) was very similar to that among treatments. Consequently, the average F-ratio for testing the null hypothesis of no difference among diets was 1.09, satisfactorily close to the value expected when the null hypothesis is true (see Table 7.7b). Note the comment at the end of Section 7.7, which explains why the average value of F-ratios is not exactly 1. The probability of Type I error was as predicted by the null distribution of F with 2 and 24 degrees of freedom (shown for $P = 0.25$, 0.10 and 0.05 in Table 7.7b). The distribution of F-ratios from the 200 simulated experiments is shown in Figure 7.4. Thus, when independently sampled, the experimental results conform properly to the theory of the analysis of variance used.

In complete contrast, when the experiments were simulated in exactly the same way but creating positive correlation among the lobsters as described before (Table 7.7), the effect on the analyses was disastrous. There were no differences among diets – the null hypothesis was still true. The positive correlation among replicates decreased the variation within samples (shown as the variance in time spent foraging). The mean time spent foraging increased, but, because this did not differ among diets, it had no influence on the outcome. The mean square *among* treatments continued to measure the intrinsic variability among lobsters as though they were behaving independently. This happens because the estimate of variability among treatments comes from the means of each sample and the relationships between the variance of sample means and the variance of the distributions being sampled (as explained in Section 7.4 and Table 7.4).

The result is dramatic. Because the estimated variation within samples is much smaller than that found when lobsters are independently sampled, the variability among samples is relatively large (even though it is correctly estimated). The resulting F-ratios are much larger than they should be if data were independently sampled. The average F-ratio in 200 experiments and the proportions larger than the critical values (at $P = 0.25$, 0.10 and 0.05) are all excessive (see Table 7.7b). The distribution of F (Figure 7.4) is markedly shifted to larger values and there is, consequently, massively excessive probability of Type I error. Even though there were no differences among diets, 84 of the 200 experiments caused rejection of the null hypothesis (at $P = 0.05$; this is a probability of Type I error of

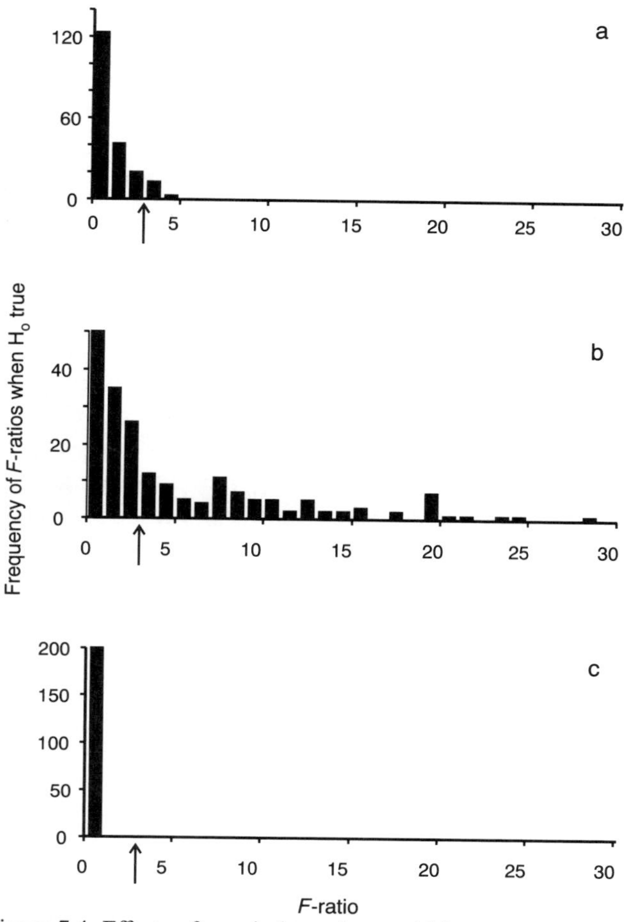

Figure 7.4. Effects of non-independence within treatments. Data are F-ratios sampled from 200 experiments on growth of lobsters on $a = 3$ diets, with $n = 9$ lobsters per diet. In each case, the null hypothesis is true; there are no differences among diets. In (a), lobsters feed independently. In (b), lobsters feed in a positively correlated manner and in (c) in a negatively correlated manner (as described in the text). The arrows indicate the critical value of $F = 3.40$ (at $P = 0.05$ with 2 and 24 degrees of freedom).

0.42, as in Table 7.7b). Positive correlation among replicates has caused increased Type I error, making it increasingly likely to detect apparent differences among treatments than are predicted by chance when the null hypothesis is true.

Thus, the biology of lobsters – their response to each other when in tanks – has serious outcomes for statistical tests on data derived from animals in groups. Whether this problem occurs in a particular experiment is not easy to determine (see Section 7.14). It is always better to assume that non-independence will occur and that animals, plants, nests, home-ranges, etc., are *not* independent unless you sample them to make them so. In the case of the lobsters described here, the problem should have no practical consequences because there should be replicate tanks for each diet (see Chapter 9) to ensure that controlled, uncontrolled and unknown differences from tank to tank that are not related to differences due to the diets would not be interpreted as influences of different diets. Thus, the data being compared for a test of the null hypothesis that there are no differences among diets are mean rate of growth for each aquarium, not for each lobster. The aquaria can, by appropriate design of the experiment, be sampled independently (but see Chapter 9).

If it is important to keep animals, or plots or whatever experimental units, separate because the null hypothesis requires it, keeping them together will almost certainly create some form of non-independence. Data will be independently sampled only by having independent replicates. If, however, the null hypothesis is about groups of animals, or experimental plots or clumps of plants, it is essential to have measures of the average results in replicated groups or set of plots or clumps, not from individual members of groups or clumps. This always results in more work because data from individuals cannot be used unless it can be demonstrated that they are independent. As explained in Chapter 9, the degrees of freedom and power of tests are often much smaller when the individuals are not the replicate units. There is, however, a compromise when nested analyses are used in conjunction with *post hoc* pooling (see Section 9.7).

Experimental biologists must *always* be on their guard against non-independence. The requirement for independent data is not a unique property of analysis of variance (see Section 6.4.1), nor of parametric versus distribution-free tests. It is a necessity due to underlying probability. It is, however, a biological problem. Biological (behavioural and ecological) processes cause non-independence among replicates.

7.14.2 Negative correlation within samples

Non-independently sampled data may also be negatively correlated. Again, a simple example will illustrate this. Let us reconsider the lobsters discussed in Section 7.14.1, only this time have a more aggressive species. Some of the animals come out to feed whenever they choose, regardless of whether others are out feeding or not. But they come out doing the equivalent of a crustacean snarl and swagger. As a result, more timid lobsters already feeding retreat back into shelter and do not come out again until the more aggressive (or larger, or whatever the cue is) lobsters are all back in shelter.

This was modelled, as before, by dividing 15 lobsters into three groups. The first five were considered aggressively dominant and fed as independent individuals. The second group were intimidated by the first group (thus, ceasing feeding and not coming out to feed if a member of the first group was feeding). The final group were very timid and were intimidated by both of the first groups. A single sample modelled in this way showed marked negative correlation among the individuals. As a result, the number of feeding excursions per individual and the mean length of excursion both decreased (as required by the model) compared with independently sampled lobsters (Table 7.7; Figure 7.3). The total time spent feeding was very dramatically reduced. Of much more consequence to statistical tests, the sampled variances of each of these variables were very much *larger* than for independent lobsters. Obviously, if some lobsters feed as they will, but others are prevented from doing so, there will be more scatter in the numbers of excursions and their durations than if all lobsters did as they wished.

In experiments with such lobsters, keeping them separate will cause different behaviour from that which occurs when they are kept together. As a result, the estimated means and variances of some measured variables will differ. This requires the experimenter to determine which form of measurement is appropriate for the null hypothesis being examined and then ensuring the chosen experiment (individuals or groups?) is designed. There is, however, still the statistical problem of the greater estimates of variance in non-independently, negatively correlated sets of replicates. This is illustrated by simulating, as before, 200 experiments to test the null hypothesis that growth of lobsters is unaffected by the choice of one of three diets. The timid lobsters are used, with $n = 9$ replicates in one tank with each diet, ignoring differences among tanks (i.e. ignoring potential confounding) as in Section 7.14.1. The experiments were run with no differences due to the different diets.

The influence of negative correlation among replicates within treatments is very clear. F-ratios testing for differences among total times spent feeding are, on average, much smaller than for independently sampled lobsters, because the variance among replicates was so inflated (Table 7.7b; Figure 7.4). The variability among treatments was also smaller than for independent lobsters because the mean time spent foraging was reduced. Differences among diets were therefore, on average, smaller than for independent lobsters. No observed F-ratios exceeded the critical values for F at $P = 0.25$, 0.10 or 0.05. The rate of Type I error was therefore zero. The probabilities of Type I error and the observed distribution of F-ratios (Figure 7.4) are nothing like those specified in tables, even though the null hypothesis is true.

What this means in practice, however, is that the probability of Type II error (retaining incorrect null hypotheses) is inflated. In other words, if there were differences due to the three chosen diets, they would have to be much larger to be detected in experiments with negatively correlated lobsters than with independently sampled ones. Or, as an alternative view, if a certain magnitude of difference actually occurred among the diets, it would be much more likely to be detected as significant in experiments with independent lobsters than with samples of negatively correlated individuals. The *power* of the experiment to detect differences is dramatically reduced when replicates are negatively non-independent.

As before (Section 7.14.1), the experimenter has to take control of this situation, by being wise about the biology of the system being investigated and by being aware of the problem.

7.14.3 Negative correlation among samples

The previous two situations were non-independence due to correlation among replicates within samples (or treatments). There can also be correlation *among* the treatments. This occurs in numerous ways due to spatial or temporal relationships, or the use of the same animals or plants repeatedly for different experimental treatments. To illustrate the problems, some very simple cases are discussed here.

First, there can be negative correlations among experimental treatments. Suppose that two species of prostrate plants have been observed to have very patchy distributions at a small spatial scale. Patches of either species are about 5 to 50 cm in diameter. The overall impression is that one of the species is more abundant (i.e. occurs with greater percentage cover of the ground than the other). The model to explain this observation is that there really is a greater cover of species A than of species B (as opposed to models that there is only an illusion or a mistake – see Section 5.10). An appropriate hypothesis is that careful, quantitative sampling of the cover of the two species will reveal a greater abundance of A than of B and the null hypothesis is H_0: $\mu_A \leq \mu_B$. This is to be tested by sampling with quadrats divided into a grid of 100 points. The number of points lying over each species will provide an estimate of their covers.

There are two ways to proceed. In one, a sample of quadrats is thrown representatively in the study-area and both species are counted in each quadrat. In the other, a sample of quadrats is thrown and the cover of species A recorded. Then a second set of independently placed quadrats is sampled to estimate the cover of species B.

In the second case, the estimates of mean cover of the two species are independent. Consider the situation where there is no difference in mean cover between the two species and there is no bare space (i.e. $\mu_A = \mu_B = 50\%$). Independently thrown quadrats (two samples each of $n = 10$ quadrats) provide unbiased estimates of these means, which can be compared by an analysis of variance (or equivalently by a *t*-test).

Such data were simulated in 1000 'experiments' for a situation where the null hypothesis is true and there was no difference between the mean covers of the two species. The results are illustrated in Figure 7.5, where the distribution of observed *F*-ratios conforms closely with that expected by chance. In contrast, if the two species are sampled in the same quadrats, there is obviously going to be extreme negative correlation between them. Where one covers a great deal of space in a quadrat, the other can only have a small cover and vice versa. This tends to maximize the difference in cover between the two species per quadrat. If, by chance, several quadrats happen to have above-average covers of species A, the mean cover will be overestimated, but the mean of species B must be underestimated because the two estimates are tied together by the method of sampling.

As a result (Figure 7.5), there will be more samples showing significantly large apparent differences in mean cover than should be expected by chance. This was simulated 1000 times as summarized in Table 7.8 and Figure 7.5. As predicted, there was a much larger apparent difference between the covers of the two species and observed *F*-ratios were larger than expected (Figure 7.5), causing a much increased probability of Type I error (Table 7.8). The null hypothesis would have been erroneously rejected in 18% of experiments – between three and four times the rate of error from independently sampled data.

This is not a special feature of the situation where the two species are absolutely negatively correlated because their covers sum to 100% in each quadrat. To demonstrate this, the two situations (sampling with independent sets of quadrats or sampling both species in the same quadrat) were simulated another 1000 times with mean cover of the two species 40%. Thus, there was, on average, 20% bare space in the quadrats.

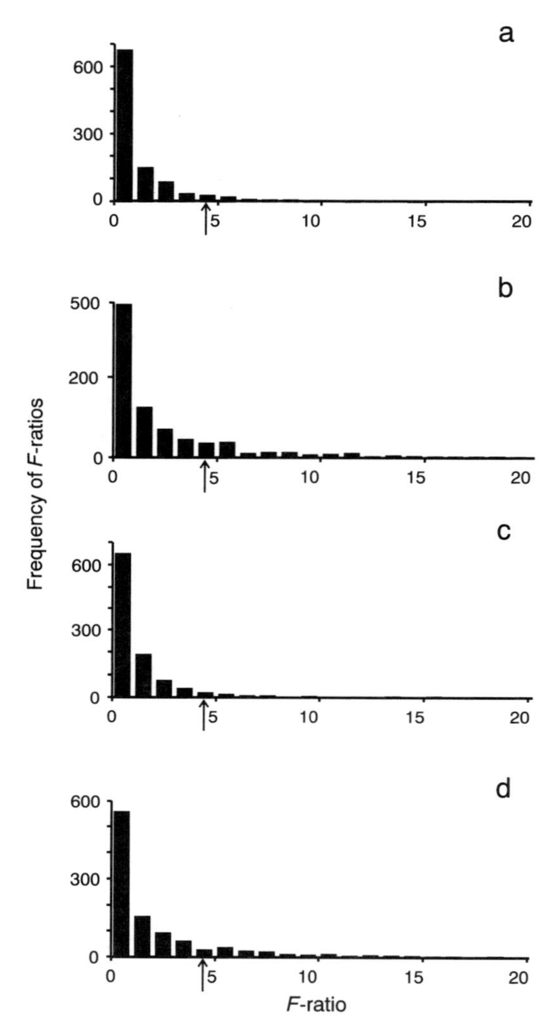

Figure 7.5. Effect of negative correlation in comparisons of mean percentage covers of two species. In each case, data are frequency of F-ratios in 1000 simulated experiments comparing mean cover of the two species with $n = 10$ quadrats. In (a) and (c), independent quadrats were sampled for each species (i.e. 20 quadrats were examined). In (b) and (d), the percentage covers of the two species were recorded in the same quadrat. In (a) and (b), mean cover of each species is 50%. In (c) and (d), mean cover of each species is 40%. Arrows indicate the critical value of $F = 4.41$ ($P = 0.05$ with 1 and 18 degrees of freedom).

Table 7.8. Non-independence among treatments: negative correlation in the cover of two species of plants. In (a) and (b), mean cover of each species is 50%; in (c) and (d), mean covers are 40%. In (a) and (c), data were independent – covers were estimated in different quadrats for each species. In (b) and (d), the covers of the two species were estimated in the same quadrats. Each situation was simulated 1000 times with samples of $n = 10$ quadrats. Means, variances and F-ratios were averaged from 1000 samples (see Figure 7.5)

	Cover of Species A		Cover of Species B			Proportion of F-ratios significant at	
	Mean	Variance	Mean	Variance	Mean F-ratio	$P = 0.10$	$P = 0.05$
$\mu_A = \mu_B = 0.5$							
(a) Independent	50.04	122.41	50.01	123.40	1.19	0.10	0.06
(b) Negatively correlated	49.89	123.81	50.02	123.80	2.64	0.25	0.18
$\mu_A = \mu_B = 0.4$							
(c) Independent	40.11	120.41	40.06	119.70	1.15	0.09	0.05
(d) Negatively correlated	40.08	120.93	39.94	122.14	1.95	0.20	0.13

This might be thought to solve the problem because each species can, theoretically, be independently distributed. Not all the space is occupied, so the amount of ground available to be occupied by one species is not just what is left free by the other species.

Despite this reasoning, many quadrats will still show negative correlation between the two species. Whenever one is more abundant than 50% cover, there can only be less than 50% cover of the other. Wherever one species is present in large amounts, the other can only have a small cover and vice versa.

The negative correlation again increased the probability of Type I error (Table 7.8), causing an increased chance of rejecting the null hypothesis in any single experiment. The effect of non-independence among samples was not as great as in the previous case (note the average F-ratios and rates of Type I error in Table 7.8 and the generally smaller values of F in Figure 7.5). Nevertheless, the problem remained large.

In both cases, there was a major difference from the situation described earlier when there was negative correlation *within* samples. Here, the variance in each sample (i.e. among replicates) is not affected by the presence of correlations. It is the difference among treatments that is erroneously measured. Increased average F-ratios were caused by increased magnitudes of the mean square among treatments. The variances for each species were very similar when estimated from separate, or from the same, quadrats (Table 7.8).

The situation described here for non-independent sampling could very simply be remedied by treating the data as paired (percentage cover of A, percentage cover of B) for each of $n = 10$ quadrats, resulting in a paired t-test with nine degrees of freedom (see Section 6.1). This solution turns the negatively correlated data into a single variable. The solution would, however, not be available if there were several species or a more complex sampling design.

The only satisfactory solution to the problem of non-independence among experimental treatments is to do the extra work necessary to ensure the availability of independent sampling.

7.14.4 Positive correlation among samples

The final form of non-independence that is of major concern to biologists is also among treatments, where there is a pattern of positive correlation. This will be illustrated by an experimental manipulation of prey of some predatory species that operates in a very patchy habitat.

The natural history of the predators includes the fact that the intensity or frequency or duration of predation varies from place to place. Observations on prey of two species have suggested greater survival of one species where another, perhaps preferred or more easily consumed, species is also present. The model to explain this correlation is that where species 2 is abundant, predators choose them or otherwise spend most of their time eating them. Consequently, survival of species 1 is enhanced. Alternative models would explain the observations as being, for example, the result of differential rates of predation on species 1 that happen to coincide with patterns of distribution of species 2. In other words, alternative models are not based on the notion that the presence of species 2 *causes* the observed reduced mortality of species 1.

A relevant hypothesis from the proposed model is therefore that reduced survival of species 1 will be predicted to occur where species 2 are experimentally removed than in similar areas where species 2 remain. This hypothesis is quite different from those derivable from alternative models, which do not predict a change in survival of species 1 when species 2 are removed.

The null hypothesis is that the removal of species 2 will result in no change or an increase in survival of species 1. To test this, several replicate plots are established from which individuals of species 2 are removed. Control plots are not manipulated. In the real world, there would need to be procedural controls for disturbances caused by removal of species 2 (blank shots fired, extra poisoned carcasses left lying around, crushed remains of animals littering the landscape, dug-up patches representing the removal of plants of species 2, etc., as appropriate) without removing the prey items. Here, however, we will keep it simple.

A typical, independently sampled experiment would look something like Figure 7.6a. Control and experimental plots are scattered over the study-site. Unknown to the experimenter, the study-site is actually a set of patches with different, natural rates of predation. The variation among replicates of each of the treatments will be a measure of the differences from patch to patch, as sampled for each treatment, by a set of replicates scattered across the patches.

The outcome of such experiments is illustrated in Table 7.9 and Figure 7.7 for two situations. In the first case, there is no difference in mortality due to predators when the second species of prey was removed. The hypothesis and model were wrong; the null hypothesis should have been

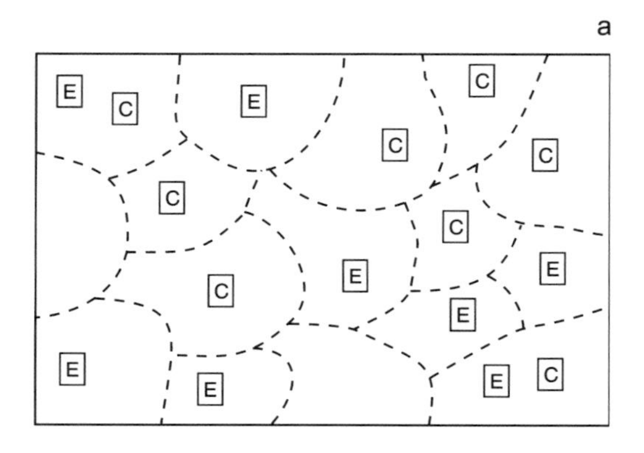

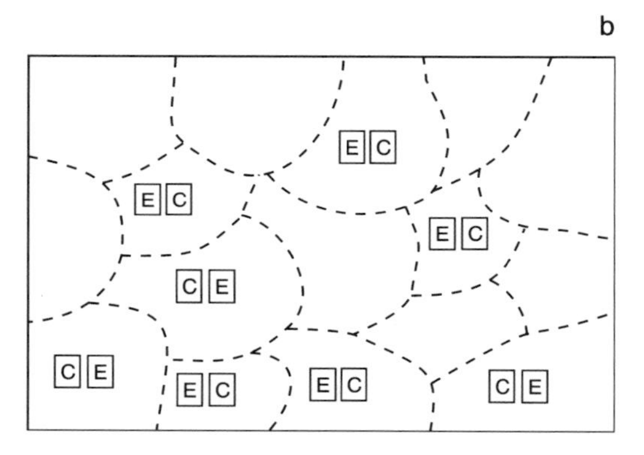

Figure 7.6. Comparison of control (C; untouched) and experimental (E; prey species 2 removed) plots to test an hypothesis about effects of predation on prey species 1. There are eight replicates of each treatment. In (a), replicates are scattered randomly across the study-site, independently within and between treatments. In (b), a replicate of each treatment is placed close together (in the mistaken belief that this 'reduces' variability). In both diagrams, the dotted lines indicate patches of habitat with different intrinsic rates of predation – the arrangement, boundaries and other details of which are not known to the experimenter.

retained. In 1000 simulations, the null hypothesis was rejected, as expected, in 49 experiments (i.e. the probability of Type I error was about 0.05; see (a) in Table 7.9).

In contrast, when the model and hypothesis were correct and there was a difference in survival when species 2 were present, the independently sampled experiment sometimes detected it. The power of the experiment to detect such a difference (Section 8.3) was, however, small (0.31 in (c) Table 7.9), but this is illustrative and not supposed to be demonstrative or career-building!

Table 7.9. Non-independence among treatments: positive correlation in the number of prey of species 1 surviving in experimental areas (where prey of species 2 are removed) and control areas (where they remain). In (a) and (b), there is no effect of removal of species 2. In (c) and (d), removal of species 2 causes increased mortality and therefore fewer survivors of species 1. In (a) and (c), the replicate plots of control and experimental treatments are scattered independently. In (b) and (d), there are pairs of one control and one experimental plot. Each situation was simulated 1000 times with samples of $n = 8$ plots. Means, variances and F-ratios were averaged from the 1000 simulations. Variance among patches (see Figure 7.6) was 600; variance among replicate plots within patches was 100

	Control: No. of species 1 per plot		Experimental: No. of species 1 per plot			Proportion of F-ratios significant at:	
	Mean	Variance	Mean	Variance	F-ratio	$P = 0.10$	$P = 0.05$
No effect							
(a) Independent	180.2	685.4	180.3	702.9	1.19	0.10	0.05
(b) Positively correlated	179.9	719.4	180.0	714.1	0.19	0.002	0.0
Increased mortality							
(c) Independent	180.1	683.2	160.0	719.2	3.96	0.64	0.31
(d) Positively correlated	180.3	716.4	160.2	720.5	3.10	0.37	0.18

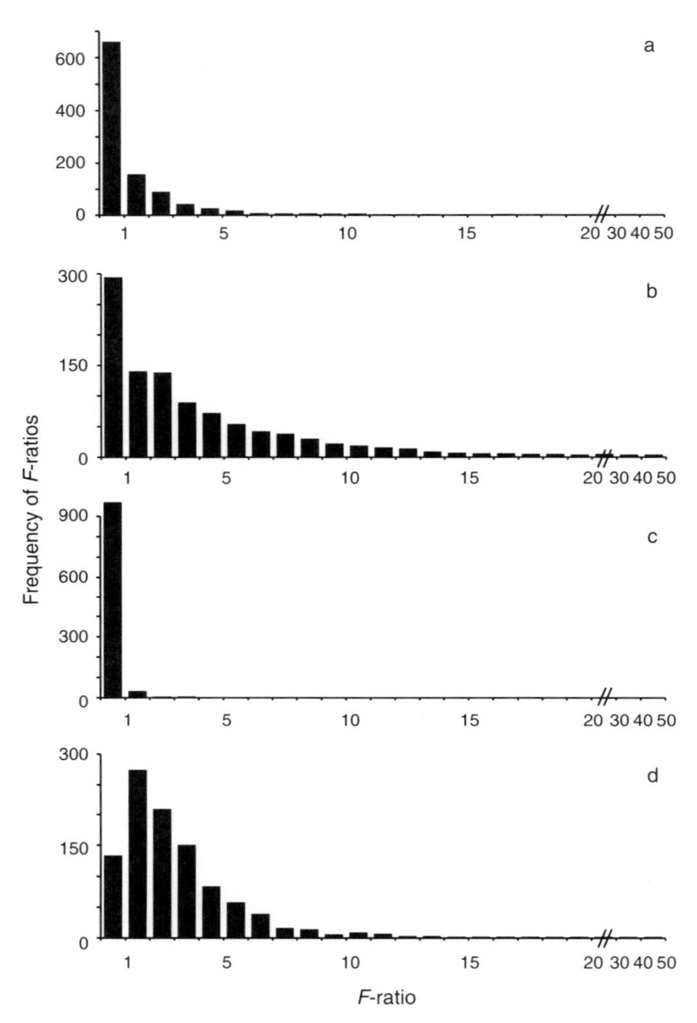

Figure 7.7. Effect of positive correlation among treatments. Data are frequency distributions of F-ratios from 1000 simulated experiments to compare numbers of a prey species surviving in areas where another prey is removed or left intact. $n = 8$ plots of each treatment. In (a) and (c), there is no effect of the experimental treatment on mean survival. In (b) and (d), survival is reduced where the alternative prey is removed. In (a) and (b), experimental plots are sampled independently. In (c) and (d), the plots are near each other in pairs (see Figure 7.6 and Table 7.9).

The experiment was then simulated using the alternative design (see e.g. Menge, 1976) of having the control and experimental units close to each other (Figure 7.6b). This resulted in them being in pairs, each pair being (unknown to the experimenter) in one of the patches. The variability among replicates within samples is identical, on average, with that in the previous design (compare (a) and (b) in Table 7.9). The replicate plots in each treatment are, after all, still scattered across a random sample of the patches. What is different in this arrangement is that there is no difference from one treatment to another that represents differences among patches.

The point is best revealed by considering the situation when the null hypothesis is true and there is no effect of the experimental treatment. There are, however, differences among all the plots. The difference between one plot and another of the same treatment in the survival of prey species 1 is due to two components. One is the difference caused by one plot being in a patch with a particular intensity of predation and the other plot being in a different patch. In addition, there is also a difference that would occur between one plot and another due to minor variability in the same patch. Thus, no individual plot provides an error-free estimate of mortality for the entire patch. This is smaller than the differences among plots – otherwise the original observation that predation is patchy would not have been made. So, variation from replicate plot to replicate plot within a treatment includes variation from patch to patch and any variation from one spot to another within a patch.

Where there are no differences between the treatments, exactly the same components of variation occur in the observed differences between plots from different treatments. So, if there are no differences among treatments, variation among replicate plots in the same treatment is the same as the average variation among plots measured between treatments. This was explained algebraically in Section 7.7.

In the second experimental design, the variability among replicates of the same treatment is exactly as before. It consists of variations from patch to patch and the individual differences among plots in a patch. In contrast, for the control and experimental plots in the same patch, the only difference is the smaller, within patch variation. The variation among treatments no longer includes any component measuring variability among patches! The effect of this is that variability among treatments, if the hypothesis is false and there is no average difference due to the experimental treatments, will be smaller than that among replicates within treatments. A more diagrammatic representation of this point is presented later (Section 12.1 and Figure 12.1).

The consequence to the analysis of variance is that the F-ratio of the mean square among treatments divided by the mean square within treatments is no longer expected to be approximately 1 if the null hypothesis is true. It is expected to be an unknown amount *smaller* than 1. If there are differences among treatments (the hypothesis is correct), the F-ratio will always be distributed with a smaller mean than is expected when data are sampled independently.

This is illustrated by 1000 simulations of experiments as before, but with one replicate plot of each treatment placed close together, as in Figure 7.7c. When there was no difference in survival of species 1 due to removal of species 2 ((b) versus (a) in Table 7.9, Figure 7.7c versus a), the F-ratios were smaller and the probability of Type I error was much less than that stated by a table of F-ratios.

The simulations were then done using this non-independent design, with the same difference due to removal of species 2 used before (i.e. an increase of 20 individuals per plot surviving where species 2 was removed). The mean value of the F-ratio over 1000 simulations was reduced and the probability of detecting the difference (the power of the experiment) declined to 0.18 ((d) in Table 7.9). The distribution of F-ratios (Figure 7.7d) demonstrates the general decreases due to positive correlation among treatments.

As with the previous cases, the responsibility for doing the experiment with due regard to independence of treatments is entirely the experimenter's. It is not appropriate or necessary to argue that, in the case considered here, we do not know there are patches or, because we do not know about patchiness, we do not need to worry. What we are responsible for is the demonstration that the spatial and temporal patterns of arrangement of the experimental treatments are the same as the arrangement of replicates within the treatments. Then, any non-independence outside the control of the experimenter will be equal in both estimates of variation, i.e. among and within treatments.

Obviously, if there is a potential for positive correlation among treatments, discovering significant differences among treatments in a statistical analysis means that the problem of non-independence did not cause an error (Table 7.10). Positive correlation among treatments causes increased probability of Type II error (retaining a null hypothesis when it should be rejected). Therefore, if the null hypothesis *is* rejected, a Type II error cannot have occurred, despite the potential for non-independence.

Table 7.10. Consequences for interpretation of experiments of non-independence among replicates (within treatments) or among treatments (variance σ_e^2 is explained in Section 7.5)

Non-independence within treatments	Non-independence among treatments
Positive correlation	
σ_e^2 within samples underestimated	σ_e^2 among samples underestimated
F-ratios excessive	F-ratios too small
Increased Type I error	Increased Type II error
Spurious differences detected	Real differences not detected
Negative correlation	
σ_e^2 within samples overestimated	σ_e^2 among samples overestimated
F-ratios too small	F-ratios excessive
Increased Type II error	Increased Type I error
Real differences not detected	Spurious differences detected

There will, however, be problems for estimation of the relative magnitude or the 'importance' of experimental treatments (discussed in Section 7.18). The assumptions underlying that analysis require that the variation among replicates within samples (σ_e^2 in Section 7.7) is equally measured among and within treatments. Non-independence within or among treatments guarantees that this assumption is not correct (Table 7.10).

7.15 Dealing with non-independence

It is not always possible to do formal tests on experimental results to determine whether there are patterns of non-independence. The sort of issues raised by non-independence among treatments can best be dealt with by two considerations. First, a considerable amount more thought should be given to the biology and processes operating in the system being studied. Being aware of the need to avoid non-independence is the most important step. Do not use the same plots, animals, tanks, nests, etc., in more than one treatment if they are the replicated units.

At some scales, there is no prospect of eliminating correlations. As a simple example, patients attending a local hospital may get assigned independently of each other to several pharmaceutical treatments in an experiment on different drugs. It may be, however, that all of them come from an area affected by the toxic output from a faulty factory chimney. As a result, unknown to the experimenters, the data are all, in some sense, correlated. This means that results really apply only to the population being sampled. The careful definition of this population and proper caution in the extrapolation of results to other situations is the only appropriate procedure. This was discussed in more detail in Section 3.4. Then, of course, the specificity of the results caused by unknown factors, such as the factory chimney, will not cause errors of interpretation. If you want to generalize to other areas and other populations, you *must* test the hypothesis that results elsewhere will differ and demonstrate that this notion is wrong (by failing to reject the null hypothesis). Then, any unknown causes of non-independence affecting the entire set of data will not affect interpretations.

Within treatments, among replicates, it is sometimes possible to test for correlations. Where large numbers of repeated readings are taken on the behaviour, physiology or growth of individuals, these can be examined by temporal autocorrelations. The analysis can be used to determine at what temporal scales data are sufficiently independent to be used as independent replicates (see, for example, the details in Swihart & Slade, 1985, 1986). The same procedures can be used where there are numerous spatial replicates, such as transects of continuously monitored plankton or oceanographic variables.

These are, however, not the usual style of ecological experiments where there are few replicates. Probably the only way to ascertain whether there is correlation *within* samples is to determine from pilot studies or the available literature (if any exists) the variance among replicates for independently sampled data. If, for example, there have been previous studies of the sort of organism, in the sort of experiment you are doing in which variability among replicates has been measured independently, you could compare your potentially non-independent estimates with published values. If they are similar, you will not have problems with analysis of your data. Often, however, there are no published data that are relevant or they are also potentially non-independently sampled.

Alternatively, you should do a pilot experiment with a few replicate quadrats, plots, individuals, whatever, treated independently and a few others treated as you planned. Thus, a few animals are grown on their own and another few together. If the variance in growth among the individuals is different for the two sets, there is an effect of keeping them together due to non-independence in their growth. For situations like those described for the estimates of cover of species in quadrats, the existence of correlation in each quadrat can be shown by calculating the sample correlation coefficient for the species, in pairs.

Otherwise, it is very difficult to determine whether there actually is independence in data that have been collected in a non-independent manner. The best solution is to avoid the problem by planning and doing experiments so that data are independently collected.

7.16 Heterogeneity of variances

The second assumption (Section 7.13) was that variances within samples were equal. This has consequences for the validity of the analysis when there are marked departures from homogeneity of the variances. The problem arises because samples from populations with different variances but with the same mean can differ in more ways than samples from populations that have the same variances. This is illustrated in Figure 7.8. On the

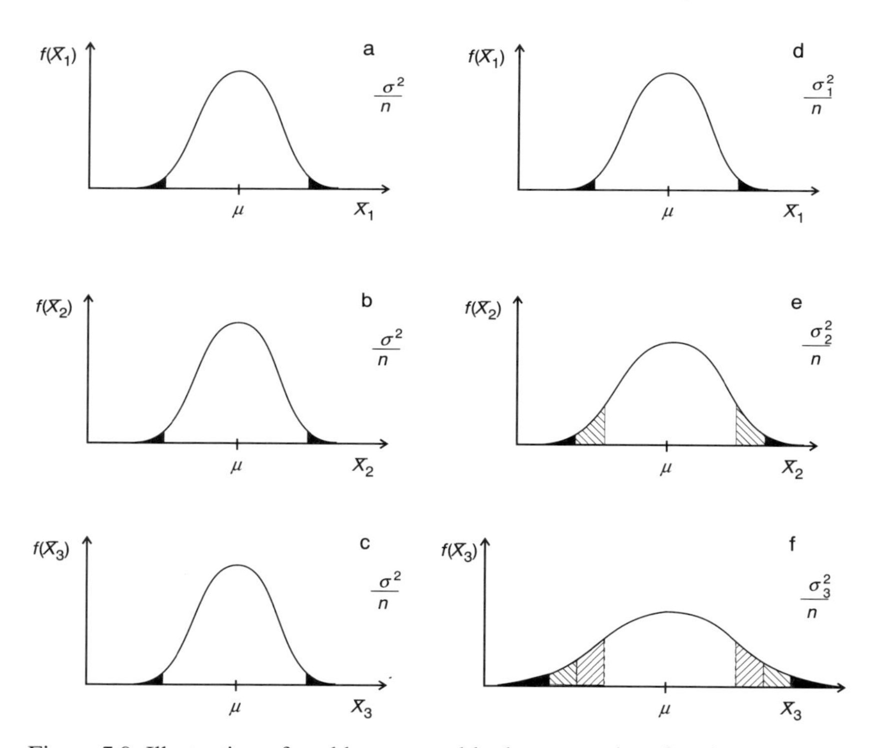

Figure 7.8. Illustration of problems caused by heterogeneity of variances. Frequency distributions of sampled means of populations are shown for (a), (b), (c): three populations with the same mean (μ) and variance (σ^2), repeatedly sampled with size of sample n. Confidence limits (95%) are shown. In (d), (e) and (f), the frequency distributions of sample means from three populations with the same mean (μ), but different variances (σ_1^2, σ_2^2, σ_3^2) are shown, with 95% confidence limits. The shaded areas indicate regions in which sampled means would cause rejection of the null hypothesis of no difference among means (for details, see the text).

left ((a)–(c)) are three populations with the same means and variances. If samples of size n are taken from each population to test the null hypothesis of no difference among the three means, an analysis of variance will reject the null hypothesis 1 in 20 times by chance, if the chosen probability of Type I error is 0.05. This will occur whenever the sample mean from one population happens, by chance sampling, to be outside the 95% confidence limits for its population of sample means (i.e. at least one sample of the three available in an experiment is in the black areas of Figure 7.8a, b or c). Of course, if all three samples happen, by chance, to be in the same tail (i.e. all are in the right or all are in the left tails), the sampled means will be similar and the F-ratio will be small. In numerous repeated experiments, the 5% of F-ratios that are significant represent the occasions when sampled means fall in the tails of these distributions.

In Figure 7.8d–f, the same situation is illustrated, but the three populations being sampled do not have the same variance. Consequently, the frequency distributions of possible sample means do not have the same variances (see Section 5.1). Now there are many more possible samples of population (e) that will have significantly different means from those in samples from population (d). Any sample mean in the shaded area of frequency distribution (e) will differ statistically from any sample mean within the confidence limits of frequency distribution (d). Obviously, the probability of getting a sample mean from frequency distribution (e) that differs from a sample mean from distribution (d) is much greater than that specified by the 5% tails calculated as confidence intervals (i.e. the black areas). Samples from population (f) will differ much more often from (e) and more often from (d) than would be expected by chance if the population had the same variances.

The distribution of the F-statistic was calculated under the assumption that all populations have the same variance. When they do not, the probability of Type I error (rejecting the null hypothesis even though the populations have the same mean) is much greater than that specified for any chosen critical value. This is true only when sampling is balanced (i.e. all the samples have the same number of replicates). When samples are not the same size, heterogeneity of variances can cause decreased probability of Type I error.

It turns out that, for balanced samples, this problem is serious for use of analysis of variance when one of the sampled populations has a larger variance than the others. It is much less of a problem, or no problem, when the variances differ from one population to another, but there is no single exceptionally large variance (an example was illustrated in

Underwood, 1981). It is not a problem when one variance is smaller than the others (e.g. Box, 1953). Also, Box (1953) showed that the effects of heterogeneity of variances are much worse if the sizes of samples differ from one population to another. This suggests very strongly that samples should be balanced (see also Section 7.10).

Because of the problem of increased probability of Type I error caused by heterogeneity of variances, it is appropriate to examine the sample estimates of variances before proceeding to the analysis of variance.

7.16.1 Tests for heterogeneity of variances

Several tests have been proposed and are widely used in the ecological literature. One is Bartlett's (1937) test, which has proved popular, largely because it can be used in cases where sample sizes differ. This is undesirable (see above). Its popularity is not matched by its usefulness. There are innumerable circumstances where Bartlett's test indicates a significant departure from homogeneity of variances, but the degree of heterogeneity is not sufficient to cause excessive Type I error in an analysis of variance. Box's (1953) description is still the best – such a preliminary test on variances to check the validity of an analysis of variance is akin to 'putting to sea in a rowing-boat to find out whether conditions are sufficiently calm for an ocean liner to leave port'.

The robust analysis of variance will not be swamped by many conditions of heterogeneity of variance that cause the fragile Bartlett's test to be sunk. Bartlett's test is particularly sensitive to non-normality of data, which has little effect on analysis of variance (see Section 7.19).

Levene's (1960) test examines the absolute values of the deviations of every replicate from the mean of its sample. Where the variances are different, the absolute deviations differ, on average, among samples. The deviations can be tested, for differences, by an analysis of variance (Levene, 1960), but this already assumes that the variance of the deviations is the same in every treatment!

Scheffé (1959) invented a test that is insensitive to non-normality, but did not recommend it for general use. Hartley (1950) proposed a procedure that uses as a test statistic the ratio of the largest to the smallest sampled variance. This suffers from the problem that the test statistic is large when the smallest variance is small, even if all other variances are the same. Yet, under such circumstances, there is little or no problem for the subsequent analysis of variance. Hartley's (1950) test is also sensitive to non-normality of the data.

As a result, Cochran's (1951) test is probably the most useful. It uses as a test statistic the ratio of the largest variance to the sum of the sampled variances. The latter is obviously related to the average variance, so this test is specific for one excessively large variance

$$\text{Cochran's } C = \frac{\text{largest } s_i^2}{\sum_{i=1}^{a} s_i^2} \tag{7.28}$$

where s_i^2 values are sample estimates of variances of the a populations sampled. The frequency distribution of C has been tabulated when the null hypothesis is true and the variances *are* equal and populations are normally distributed. The table of C involves a, the number of treatments or populations and $(n - 1)$, the degrees of freedom in each sample. Note that n must be the same for all samples.

If Cochran's test is significant, there is evidence of a serious potential problem for any ensuing analysis of variance. There are several things to do once heterogeneity of variances has been identified. These are discussed below, including the fact that heterogeneous variances lead to excessive Type I error in analyses of balanced samples, so non-significant results of an analysis are perfectly acceptable.

7.17 Quality control

One of the first things to consider when experimental data have heterogeneous variances is 'why?' After all, when the experiment started, the samples allocated to the various treatments were supposed to represent the same original population (see Sections 6.4.2 and 6.5). Thus, there is no obvious reason to expect heterogeneous variances. Their discovery may, therefore, be worth noting as an outcome of the experiment – the treatments differ because they cause different patterns of variation among replicates rather than differences among means. The causes of different variances are at least as interesting for interpreting nature as are the causes of different means.

In general, however, as with many other things in nature, many biologists and ecologists view heterogeneity of variances as a problem, rather than as information. The first step in dealing with the discovery of heterogeneous variances is to determine whether this is unusual. You may have experience with the type of data you are analysing and your previous experience suggests that there is normally no heterogeneity of variances. For example, you may have sampled the concentration of chlorophyll

Table 7.11. Quality control in assessment of heterogeneity of variances. The data are variances among numbers of the intertidal snail *Littorina unifasciata*, in samples of 10 quadrats (4.5 cm × 4.5 cm) at ten times during five years. The variance at time 6 is excessive. See text for further information

Time	Variance
1	26.9
2	29.7
3	46.7
4	35.7
5	63.5
6	466.2
7	42.3
8	36.6
9	37.0
10	45.2

per gram of leaf in numerous previous studies and never found heterogeneity of variances among different samples. In this case, you should give careful thought to the individual replicates in the sample that generated the excessive variance. If one replicate in that sample is very different from the others and that one is large, you should consider whether or not it is valid. Sometimes (as in the example of snails, below), the replicate value is unlikely, or impossibly large. Perhaps someone put a thumb over the cuvette before it was measured in the spectrophotometer? If you have evidence from numerous previous studies of the usual sizes of variances and range of data to be expected from sampling chlorophyll in leaves, you have independent evidence that your data are *wrong*. The offending variance is caused by an erroneous reading.

Sometimes, heterogeneity of variances occurs because of mistakes with data. As a real example from a long study of populations of intertidal snails, *Littorina unifasciata* (A. J. Underwood & M. G. Chapman, unpublished data), Table 7.11 shows the variances of samples ($n = 10$ quadrats) of a population at ten times of sampling over a period of several years. One variance (at time 6) is very much larger than the others. Altogether, there are 100 quadrats sampled in these data. The range of numbers of snails per quadrat was 0 to 35, except for one quadrat at time 6, which was entered in the data-set as 225 snails. So, not only is there an exceptional (and very much larger) variance in only one of the ten samples, but also there is one extreme replicate out of 100 sampled. This one turned out, by checking the original field data-sheets, to be an error of transcription – the number of snails was really 22. Thus, heterogeneity of variances alerted us to an error in the data, because normally we do not have such heterogeneity in this type of sampling and because we have ample previous or other data to be sure that something was wrong. In passing, it is worth noting the effect of the error on the mean number of snails at time 6. The error caused the mean of the ten quadrats to be 38.2 snails rather than the correct value of 17.9. This would have had serious consequences for any analysis and its interpretation!

So, quality control can be a very useful approach to understanding the data. If there is no obvious correction that can be made, but evidence of an error, the erroneous datum should be omitted from the analysis. For example, in the case of the analysis of chlorophyll, there may be no error in recording the data, but an error in the measurement. If there is sufficient independent evidence that there is an error, omit the offending replicate. Replace it with the average of the other replicates. This dummy number has the property that it keeps the set of data balanced (see Section 7.10), but does not alter the mean of the valid data, nor alter their variance (the deviation of the mean from the mean is, of course, zero). Analyse the data and adjust the degrees of freedom for the within samples sum of squares by subtracting one (for the missing or dummy replicate).

Sometimes, heterogeneity of data is caused by a complete set of replicates being wrong. For example, in an experiment involving samples of insects in sweep-nets, one whole set of replicates may have been taken by an inexperienced assistant. This, of course, would be poor practice – observers should, as far as is practicable be assigned randomly or evenly across experimental treatments to avoid confounding the differences among samples with differences among observers. Nevertheless, if this has occurred and the single sample is the one with excessive variance, provided you have adequate evidence it is potentially erroneous, it can be eliminated.

Suppose you have sufficient previous data to mistrust the estimated variance *and* you know that the single sample was collected in an untrustworthy way, you could eliminate the variance from the analysis. Calculate the sum of squares for that sample (i.e. the variance of the sample multiplied by its degrees of freedom) and subtract it from the sum of squares within samples in the analysis of variance. Then subtract the degrees of freedom for that sample from the degrees of freedom within samples.

Now you can calculate a new mean square within samples that does not use the large variance from the odd sample. Note that this is probably an acceptable option only where you have many samples. For example, if the entire experiment involves 20 samples, this procedure can alter the mean square by only, at most, 5% of the data. If, however, you have only four samples, this procedure will have a very large effect (it alters 25% of the estimated variances).

Also, this procedure assumes that the *mean* of the outlying sample is still an appropriate one to use. This may, of course, be illogical. If there is doubt about the variance because of the way the data were sampled, there should equally be doubt about the validity of the mean. Often, it would be better to eliminate an untrustworthy sample altogether.

Finally, the above discussion assumes that common sense prevails. You cannot remove offending replicates or samples from your data just because they cause heterogeneity of variances. There must be justifiable cause based on the previous, extensive sets of data available from similar work to demonstrate that an error has been made.

7.18 Transformations of data

An important procedure for dealing with heterogeneity of variances is to transform the data to some scale in which there is no heterogeneity. This is only effective if there is a relatively constant relationship between variances of samples and their means, so that where there are differences among means, there will be differences among the variances. Furthermore, to retain the possibility of being able to interpret the transformed data, transformations must be *monotonic*. This means that the transformation leaves the means of different samples in the same rank order (the largest is still the largest; any mean larger than another in untransformed scale will still be larger than the second one when transformed).

The general procedures for transforming data to remove heterogeneity of variance are described in full by Snedecor & Cochran (1989) and Winer *et al.* (1991). There are, however, three general classes of ecological or biological data that potentially cause heterogeneity of variances. For these types of data, there are standard transformations that are widely used to solve the problem. In all cases, unless there are other reasons to do so, transforming the data should only be done to solve the problem of heterogeneous variances. Routine transformation of data is pointless. The procedure should therefore be to test for heterogeneity of variances (say, by Cochran's test). If significant heterogeneity is found, transform the data (see below) and then test the variances of the transformed data. If there is no significant heterogeneity, proceed to analyse the transformed data. If there is, stop and think (see below, Section 7.18.4).

Sometimes, it is appropriate to transform data anyway – regardless of heterogeneity of variances, but because data are in an inappropriate scale. For example, the distances moved by animals during a random walk are often exponentially distributed (Underwood, 1977). This suggests that the underlying processes influencing rates of movement and lengths of periods of movement are probably in some logarithmic scale. Routinely, therefore, it makes sense to use a logarithmic scale for the distances moved, so they are transformed to logarithms for analysis.

Similarly, it can be argued that growth of animals or populations over relatively short periods is probably exponential or approximately so. Again, logarithmic transformation of the data is appropriate.

Here, the three commonest transformations of biological data are considered.

7.18.1 Square-root transformation of counts (or Poisson data)

Many types of biological data, particularly those involving frequencies or counts per unit are distributed approximately as Poisson distributions. It is a property of these distributions that their variances equal their means. So, in any experiment where means differ, so will variances, thus compromising an analysis of variance.

This occurs commonly in data that are counts per leaf or nest or quadrat or net, particularly where the mean numbers are small. If there is heterogeneity of variances, the appropriate transformation is the square-root of the data. A preliminary step is to plot the variances against the means, which should be an approximately linear relationship (as in Figure 7.9a). In the case illustrated, the data are numbers of shrubs in quadrats. There is a very good chance that these will be scattered as a Poisson distribution, unless there is something in their natural history that causes aggregation or unless they compete for resources and are spaced further apart than would happen by chance. The latter would cause similar variances in different sites. Here, the variances in ten sites are significantly heterogeneous, but the plot of variances against means is roughly linear. Transformation to $\sqrt{(X+1)}$ removed the heterogeneity (Figure 7.9b). The data can now be analysed in this transformed scale.

Various authors have provided different versions of this transformation to improve the removal of heterogeneity when data are close to zero (e.g.

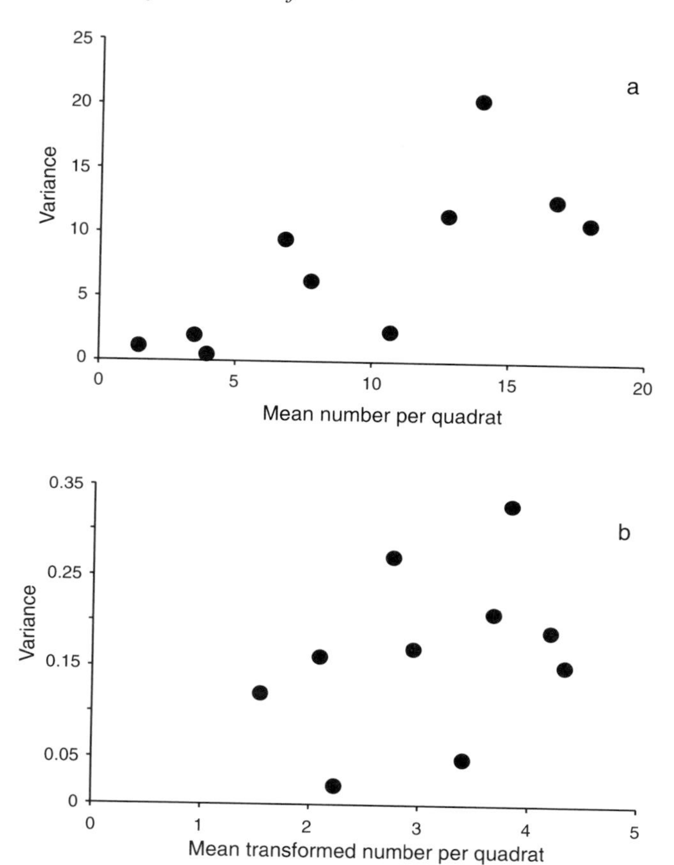

Figure 7.9. Square-root transformation of data. (a) Sample variances and mean numbers of animals per quadrat from samples in ten sites; (b) sample variances and means after transformation of data to $\sqrt{(X+1)}$.

Tukey, 1957; Mosteller & Youtz, 1961). Also, this transformation will be effective wherever sample variances are proportional to sample means (the plot of variances on means is roughly linear), even if the data are not from Poisson distributions.

7.18.2 Log-transformation for rates, ratios, concentrations and other data

Many types of data, particularly those that are ratios of two variables, are highly skewed to the right. In such distributions, where the mean is large, the variance is very large. Such data are often approximately log-normally distributed (as described in Chapter 3). Consider, for example, the measurements of the ratio of number of prey eaten per predator in plots of some habitats. Both components of the ratio are variable. The ratio cannot, however, be less than zero (which occurs when no prey are eaten). It can be very large – when a few predators are massively voracious in some plot. There is no theoretical limit to the number of prey eaten. As a result, in areas where predators are individually very voracious or active, they will cause a large mean value of the ratio and occasional plots will have very large values. Where the mean is smaller (because there are fewer prey or because the predators are less active or efficient) it is quite unlikely that individual plots will have large values. Thus, the variances will increase very radically as the mean increases, leading to marked heterogeneity of variances.

This situation can be discerned in a graph of the standard deviations (i.e. square-root of the variance) against the means (Figure 7.10a). Where such a plot is approximately linear, transforming all the data to logarithms will usually remove the heterogeneity of variances. In the case illustrated, the data are concentrations of chlorophyll per weight of leaves. This is, in essence, a ratio of two variables even though converted to concentration per gram. The concentration and the weight vary from leaf to leaf. Small leaves can have very large concentrations of chlorophyll, so there is the potential for very large concentrations even where the mean is quite small. These are skewed distributions. The standard deviations are reasonably linearly related to the means. Transformation to natural logarithms (Figure 7.10b) removed any trace of relationship between the variances and the means, thereby removing the heterogeneity of variances.

There are two things to note about log transformations. First, there is no difference in the effect of transformations according to the base of logarithm used. All logarithms to one base are a constant multiple of those at any other base. So, in transformed scale, the relationship between standard deviation and mean is constant, whatever the base. It is, however, important to report what logarithmic transformation is used so that others may compare their results with yours.

The second issue is what to do when some of the data have the value zero. The logarithm of zero is minus infinity. It is customary to add a small amount (1 or 0.1) to the numbers before transformation. The problem with this is that the transformation is no longer necessarily strictly monotonic. There can be changes in relative (rank) order of means of samples depending on the magnitude of the data and the size of the constant added to them. In practice, it seems that samples (or

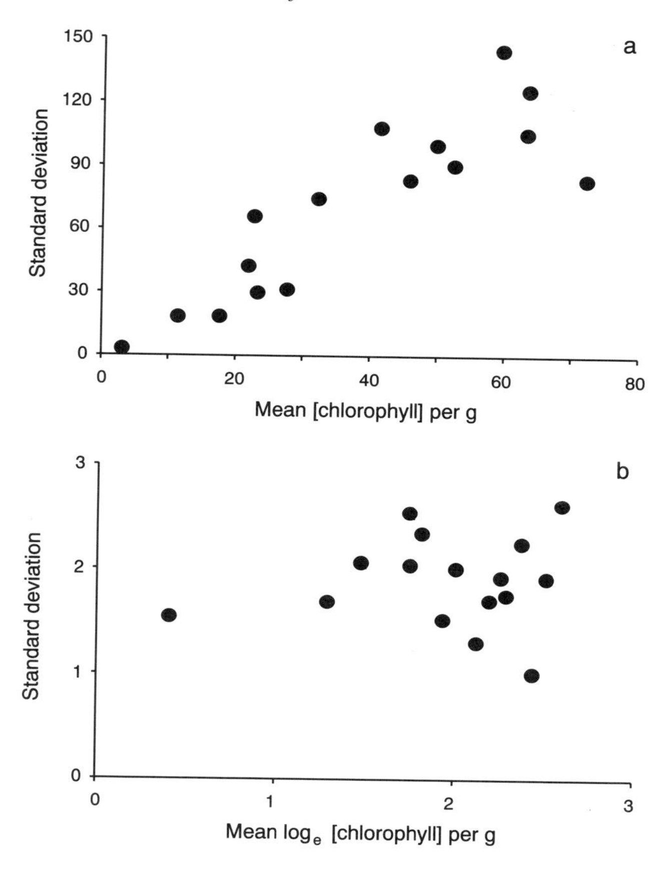

Figure 7.10. Logarithmic transformation of data. (a) Standard deviations and mean concentrations of chlorophyll per gram of leaf in 16 samples; (b) standard deviations and means after transformation of data to $\log_e X$.

treatments) that change in order relative to each other are not statistically significantly different from one another, so no interpretational difficulties occur.

There does, however, remain the problem that addition of a constant will cause different relative relationships of means from those where logarithms are used directly. Thus, addition of a large constant (say, 1) to all data may be appropriate when the data are mostly in the range 10 upwards. It will make little difference to the transformation. If some of the samples are small numbers (<0.1), but others are large (say, >1), with occasional zero values, their relative magnitudes will now be made very different by addition of 1. The smaller numbers are much more affected. Transformations involving logarithms cause several difficulties in other contexts (for a discussion of scale, see Doran & Guise, 1984; and the comments in Box & Cox, 1964). McArdle *et al.* (1990) and McArdle & Gaston (1992) have more recently discussed problems of additions of constants before log transformation of counts of animals in various types of ecological studies.

7.18.3 Arc-sin transformation of percentages and proportions

Where data are percentages or proportions, they are often binomially distributed. As a result, variances are larger where means are near 0.5 (or 50%) than where means are small or large (near 0.1 (10%) or 0.9 (90%)). This can lead to heterogeneity of variances where means are different. An example of percentage cover of plants is shown in Figure 7.11a. In areas with an approximately 50% average cover, the variance among quadrats is greater than elsewhere. In this case, the appropriate transformation is one that 'spreads out' the data towards the end of the possible range of values (i.e. near 0 or 1 for proportions or 0 and 100 for percentages). In contrast, the transformation should do little to the values near 0.5 or 50%. In this way, the variances of transformed data near the ends of the range of possible values will be increased to match the variance in the middle of the range. The appropriate transform is the arc-sin of the square-root of the proportion (i.e. the angle which has its size equal to the square-root of the data). Thus, the transformed data, X', are:

$$X' = \sin^{-1} \sqrt{X} \tag{7.29}$$

where X values are the original data, as proportions (i.e. divided by 100 if percentages). Again, this has the effect of removing heterogeneity of variances (Figure 7.11b).

7.18.4 No transformation is possible

Often, biological data being what they are, there are no simple ways of dealing with heterogeneous variances. The problem is not solved by the approach of checking variances against published values, or prior experience, or the use of monotonic transformation.

Under these circumstances, you should note that, in large experiments, analyses of variance are robust to departures from the assumptions. In other words, the validity of the test and the probabilities associated with

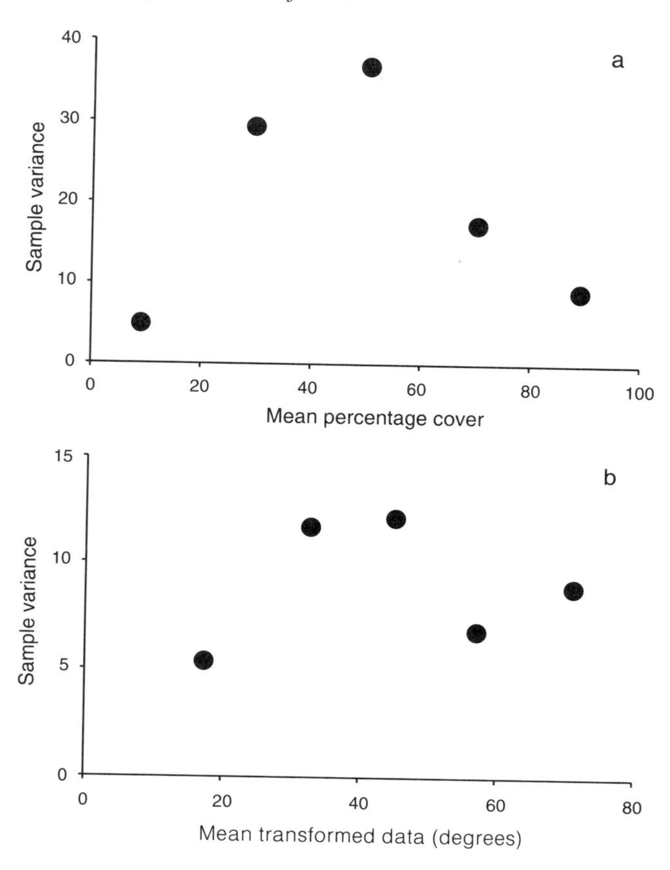

Figure 7.11. Arc-sin transformation of data. (a) Variances and mean percentage covers of plants in samples from five sites; (b) variances and means after transformation of data to $\sin^{-1} \sqrt{(X/100)}$.

the F-ratio distribution are not affected much by violations of the assumption. This is particularly true where data are balanced (i.e. sizes of samples are all made the same) and where samples are relatively large. 'Relatively large' is one of those delightfully vague terms, but having more than about five treatments, with n more than about 6 seems satisfactory from simulations. The problems of heterogeneity, in terms of statistical tests were discussed in Sections 6.4.2 and 7.16. Obviously, where samples are large, there is little chance of getting all of a sample from one tail of its distribution. So, there is no increased probability of differences occurring among samples simply because one has a larger variance.

Where there are many samples or treatments, there can be only a small effect of any one sample variance on the estimated variances provided by the mean square within treatments. So, large, balanced experiments, particularly with large samples in each treatment, will not cause problems for interpretation of an analysis with heterogeneous variances.

For small heterogeneous experiments, the best advice is to do the analysis anyway. If it turns out that there are no significant differences among treatments, this is a valid conclusion for balanced samples. It is valid because heterogeneity of variances leads to increased probability of Type I error. Therefore, if no significant differences occur among the means, you cannot make a Type I error, which is an erroneous *rejection* of a null hypothesis.

The only problem occurs where there are small samples with heterogeneous variances and their means differ according to the analysis of variance. This may be an erroneous result and should be dealt with cautiously. The best practice would be to treat it as an exploratory or pilot experiment and to use it to plan a larger, more reliable test as the first step of furthering the model being worked on (as described in Chapter 2). There are some reliable models for analysing sets of data with heterogeneous variances, but these are available only for the simplest experimental designs (see Weerahandi, 1995).

Choosing, instead, a non-parametric statistical procedure (such as the Kruskal–Wallis test discussed in Section 8.5) will not solve the problems caused by heterogeneity of variances (see also Section 6.8.2).

7.19 Normality of data

The assumption that data are normally distributed is not very important even where there are quite small samples (Box, 1953). The analysis of variance is quite robust to non-normality – in other words, its outcome and interpretation are not affected by the data being non-normal. Again, this is particularly the case where experiments are large (there are many treatments) and/or samples of each treatment are large. It is also the case where samples are balanced.

If it is anticipated that data may be from highly skewed distributions, log transformation (Section 7.18.2) will remove the skewness and make the data much more normally distributed. So, transformation may be appropriate even if variances are not significantly heterogeneous.

The major type of non-normal data that seems to cause problems is sampled from multimodal frequency distributions (as in Section 5.1).

These would best be handled by constructing better models and hypotheses to deal with the various modes as different distributions and therefore to sample them independently and not as part of a single distribution.

7.20 The summation assumption

A final assumption must sometimes be made about analyses of variance, where it is intended to estimate the magnitudes of effects of various experimental treatments or a variable. The assumption is not at all relevant where the purpose, as is usual for ecologists, is to examine the differences among populations and to test hypotheses about these differences.

If the actual effect of a given treatment is to be estimated properly in an analysis of variance, it is important that the effects of each of the treatments sum to zero. In algebraic terms, using the linear equation for an analysis, as in Equation 7.7, we must assume that all treatment effects sum to zero, or:

$$\sum_{i=1}^{a} A_i = 0 \tag{7.30}$$

where A_i is the effect of treatment i as in Equation 7.7. This must be the case where the null hypothesis is true (Equation 7.8) and must be assumed to be the case when the null hypothesis is false.

As done in an earlier review (Underwood, 1981), it is easier to explain the relevance of this assumption by showing what happens when it is *not* met. Consider a set of experimental treatments to put fertilizers on fields to test the hypothesis that there are differences in yield of some crop under the different regimes. The average yield of this fictitious crop is 10.8 bushels per acre (or other obscure agricultural measure). So, for the entire population of plots to be sampled, $\mu = 10.8$, although this is not known to the observer. There are four fertilizer treatments and their effects have been chosen for this illustration to disobey the summation assumption (see Table 7.12). Again, this is not known to the experimenter.

Now, assume that the samples taken in the experiment are perfectly representative of the populations being sampled. The sample means will then be exactly the means of the populations. Of course this will not happen in nature, but it makes no difference to what follows and allows the point to be demonstrated much more simply. So, the experimenter measures the sample mean from a set of replicates for each of the four

Table 7.12. Violation of the assumption that effects of treatments sum to zero

The true mean $\mu = 10.8$

Treatment (i)	1	2	3	4	
Effect (A_i)	+5.3	+3.6	+1.1	−2.0	$\sum A_i = 8.0$
Sample mean ($\bar{X}_i$)	16.1	14.4	11.9	8.8	$\bar{X} = 12.8$
Estimated effect $(\hat{A}_i) = \bar{X}_i - \bar{X}$	+3.3	+1.6	−0.9	−4.0	$\sum \hat{A}_i = 0$

Estimated difference				Real difference
$\hat{A}_1 - \hat{A}_2$	=	1.7	=	$A_1 - A_2$
$\hat{A}_1 - \hat{A}_3$	=	4.2	=	$A_1 - A_3$
$\hat{A}_1 - \hat{A}_4$	=	7.3	=	$A_1 - A_4$
$\hat{A}_2 - \hat{A}_3$	=	2.5	=	$A_2 - A_3$
$\hat{A}_2 - \hat{A}_4$	=	5.6	=	$A_2 - A_4$
$\hat{A}_3 - \hat{A}_4$	=	3.1	=	$A_3 - A_4$

treatments ($\bar{X}_i$ values as in Table 7.12). Unknown to the experimenter, these are exactly what should be measured without sampling error (i.e. $\mu + A_i$) for each of the four treatments.

Apart from testing whether these are different (by analysis of variance), they can be used to estimate the value of using any of the fertilizers. The effects of the treatments can be estimated as $\hat{A}_i$, the difference between the mean of a treatment and the overall mean $(\bar{X}_i - \bar{X})$. These estimates are *all* wrong. For treatment 3, the estimate is for a negative effect when it is really positive. The estimates of magnitude of effects of treatments are all inaccurate because the assumption of summation was not true.

Note, however, that the differences between treatments are properly preserved (Table 7.12). If you want to know how much more yield you would get with one treatment compared with another, the experiment is perfectly valid even though the assumption is violated. Thus, to test the null hypothesis that there are no differences among the four means, the data can be used whether or not this assumption is true.

In practice, all that is needed to solve the problem for purposes of estimation is to include a proper control – plots with no fertilizer treatment. The mean over all treatments will be the same as the mean of the control if the assumption is correct. If it is not, the overall mean will differ from the mean of the control by an amount that can be used to correct the estimates of effects of each treatment. This is not illustrated here because it is simple algebra. The important point is that uncontrolled experiments are difficult to interpret.

For testing null hypotheses, this assumption has no relevance. It is, however, important to realize that it has been discussed entirely with reference to what is know as a *fixed* factor. This concept and the different types of experimental factors are discussed in the next chapter.

8

More analysis of variance

Having waded through a detailed introduction to analyses of variance, there are other crucial issues to sort out. These are the definitions of two types of experimental treatments, the power of an analysis of either kind and how to determine the alternative to the null hypothesis if the analysis of variance causes you to reject it.

8.1 Fixed or random factors

One very important consideration in an analysis of variance is to be sure whether experimental treatments represent *fixed* or *random* factors. This makes little practical difference for a one-factor analysis, but will turn out to matter a great deal for extensions beyond one factor. The distinction was made explicit by Eisenhart (1947) in his thoughtful consideration of different models in experimental designs.

There are many ways of considering this issue but, in the end, they all boil down to you, the experimenter, also being very thoughtful about your hypotheses. The relevant definition of an experimental factor is a function of the way hypotheses are stated and *not* a property of the analysis. As an immediate consequence of this, you should always know which sort of factor you have, because you cannot interpret the experiment without the hypothesis being clear. As will be demonstrated below, the statement of an hypothesis requires definition of whether or not the experimental treatments are fixed or random.

The difference revolves around whether you are concerned with specific, identifiable required treatments (which would make a fixed factor) or whether you are interested in a general problem, some components of which are included (representatively) in the experiment.

Whether or not an hypothesis relates to a fixed or a random factor is a crucial distinction, despite this being denied in some texts on the topic. Mead (1988), in particular, has sometimes been cited as having stated that it does not matter whether factors are fixed or random and therefore experimenters need take no notice of the difference. Of course, Mead (1988) never implied this and, although he stated that the distinction was a minor one, he made it clear that it was necessary because of the 'consequent change in the interpretation of the experimental results'. Because of my insistence that we spend more time focussing on the hypotheses and purposes of an experiment and its logical structure, I must insist that the consequences for interpretation need to be understood before the experiment is planned. It makes a large difference to interpretations, to the mechanics of analyses of all but the simplest designs and to methods that might help to optimize the allocation of resources in experimental designs.

It is, however, true that there are better ways of describing analytical frameworks (Nelder, 1977; see below). Because some distinction is necessary in logic, hypotheses and analyses and because the contrast between 'fixed' and 'random' is widely discussed in the literature, I shall keep the more usual terminology.

Here is a specific example to illustrate the nature of the difference. In studies of the rate of consumption of seeds of several species of shrubs, there is reason to believe that seeds are not all equally palatable or chosen by the consumers. There is variation from plant to plant in the rate at which seeds are consumed, but it has been proposed in a model of how the ecology of the plants works that there are also differences among species. This hypothesis can be tested by taking samples of seeds of several species and measuring percentage survival after a relevant period in the field. The null hypothesis is that there will be no difference in mean percentage survival among species. The species chosen are the experimental factor and the samples of seeds are the replicates. Decide that four species will be examined, out of the large number in the study-area.

There are two ways you might decide which species to choose. These 'ways' encapsulate the different sorts of hypothesis. In one case, there is no reasoning behind which species to choose. The hypothesis is very general ('species differ') and you are examining four species in order to get a range of species. You have no particular reason for choosing any particular species. You choose four essentially at random out of those available. In this case, the species are randomly chosen to represent the population of the species in the area. Any set of four species will do the job – the hypothesis simply considers that there are differences among species. Any randomly chosen set of species could be examined.

The hypothesis simply says 'there are variations among species' and therefore the difference between the null hypothesis ('no difference among species') and the hypothesis is whether or not there is significant variation. If there is, the null hypothesis will be rejected – but you would also expect to reject it for any set of four species. In other words, the experimental factor consists of four representative species and any other sample of four species should be similarly representative. If one set of four shows variation, so should another set.

Contrast this with the following scenario. There are numerous species, but you choose only certain ones to examine. You look at A and B because they are the ones most discussed in the literature. You choose species C because it is the most abundant in the area and therefore getting seeds will be easy. Finally, you choose species D because you are concerned about its long-term conservation and so you would like to include it in your study.

Now you no longer have a random factor. You have, in the way you are now phrasing the hypothesis, created a fixed factor. The levels of the factor, i.e. the particular species chosen, are fixed by special characteristics they have (or that you use to describe them). You are no longer representing all species – you are specifying a particular set of species and they *cannot* represent any others. You have defined species to consist of 'the two most discussed in the literature, the most abundant species and one about which you have conservatory concern'. No other species can be used in the experiment because you are now hypothesizing that the mean rate of disappearance of seeds is different, on average, for *these* four species. It is a much more precise experiment, but it will be less general than the random factor considered before. By including a particular suite of species, no extrapolation to other species is possible. The ones chosen are the complete set of species specified – they are not a sample of a more generally defined population of species.

The discerning reader will, at this point, object to this example. In the fixed scenario, the species chosen were not included in the experiment because of some condition imposed by the hypothesis. I agree – in this particular case, it is not a choice and the species included should be chosen simply to represent the variety of species in the area. If time and resources were available, there would be no reason not to examine *all* the species. Given constraints of time, budget and patience, only a few will be used in the test. These were chosen arbitrarily as a random factor because the original observations and model did not, in any way, specify which species mattered – the fixed-factor experiment is not a logical choice.

The example does, however, indicate that there are two different types of factor. In one, a fixed factor, all the relevant levels (i.e. relevant to the hypothesis being tested) are included in the experiment. The experiment provides a precise rationale for rejecting or retaining the null hypothesis as it applies to those treatments.

The other type of factor, a random factor, includes only a sample of relevant levels. The population of levels or treatments to which the hypothesis applies is much larger than the number of levels included. The experiment provides a less precise answer. Its results apply to a sample; they are, in a sense, a sampled answer. As with any sample, the answer represents the answer for the population and is not exactly the answer for all possible levels that might have been examined. But this loss of precision is compensated for by having a more general answer. The outcome of the experiment should, all other things being equal, be the outcome that would be obtained in experiments on other samples of levels that might have been included.

There are several other ways of considering this concept. One is the operational definition used by Simpson *et al.* (1960). They advised that a useful rule-of-thumb is to decide, if you were going to repeat the experiment as a new test of the hypothesis, would you have to use the same levels of the treatment? In the case of the species of plant, when considering them as a random factor, there would be no reason to choose the same species again. Any other sample of species would do. For the fixed-factor situation you *must* use the same four species, otherwise you would clearly be testing a different hypothesis.

This rule works fine until it comes to experimental factors that involve time. You cannot do an experiment to compare the intensity of competition amongst lizards in three different years and then, later, decide to do it again using the same years!

Some other examples will help to clarify the differences between fixed and random factors. Consider a test of the hypothesis that abundance of a population changes from time to time. For example, under the model that disturbances lead to fluctuations in numbers, this would be a relevant hypothesis provided that the interval between times is long enough for disturbances to occur. At each time, several replicate samples are taken to estimate mean abundance. These are compared by a one-factor analysis of variance with times as the factor. The times chosen can be entirely random. All that matters is to determine whether abundance varies. Someone else testing this hypothesis can use completely different times of sampling and should expect to get a similar answer.

Contrast this with the situation where change in abundance of a population is explained as a result of particular processes operating after a disturbance, so that there must be an increase in numbers between the second and third year after a disturbance. Now, the hypothesis predicts a particular pattern in abundance *and* when it should occur. The time elapsed since the disturbance is important and samples must be taken at the correct times. This fixes the times and the hypothesis also fixes how many times must be sampled. In an analysis, time is now a fixed factor. Anyone else doing such an experiment must use the same times of sampling after a disturbance. The only times that matter for the hypothesis must be sampled in the experiment.

You will find other examples given by Simpson *et al.* (1960), Underwood (1981) and discussions of the consequences by Eisenhart (1947) and Wilk & Kempthorne (1955). One of the clearest distinctions between the relevant hypotheses and their relationships to analyses was by Burdick & Graybill (1992). If the hypotheses concern differences among the means of the populations in the different levels of a factor, the design should ensure that the levels represent a fixed factor and the analysis will be a fixed effects model. In contrast then, the primary focus is an hypothesis about the variance among the levels of some factor in an experiment (as will be the case in the nested analyses in Chapter 9), the analysis must use a random effects model and the factor must be considered random.

Winer *et al.* (1991) discussed an operational definition of whether a factor is fixed or random in terms of the following algebra of the fraction of a population sampled. If a factor (A) in an experiment consists of X total levels and the number in the experiment is x levels, the proportion of the population that is not sampled is:

$$1 - \frac{x}{X} \tag{8.1}$$

This measures, in a sense, the remaining ignorance about the population of levels. If x is small relative to X, we know little about most of the population. From this Winer *et al.* (1991) defined:

$$\left.\begin{aligned} \text{A fixed factor is when } \left(1 - \frac{x}{X}\right) &= 0 \\ \text{A random factor is when } \left(1 - \frac{x}{X}\right) &\rightarrow 1 \end{aligned}\right\} \tag{8.2}$$

Thus, when there are only five species to be considered in some hypothesis (e.g. the null hypothesis concerns mean rate of growth of the five species in a particular genus), $X = 5$. If all five are examined in the experiment because the hypothesis requires it, $x = 5$. The factor *is* fixed (all relevant levels are in the experiment) and Equation 8.2 equals zero. If, however, the five species are arbitrarily chosen out of many, X is large and x is small relative to it. Equation 8.2 tends towards 1. You should now pick the five species at random out of those available.

There are several issues about this that need thought. First, the operational definitions used above tend to allow you to be wise after the event. It is, however, not sensible to wait until you have allocated the levels of the treatment into the experiment before deciding whether it is a fixed or random treatment. It is your job to sort this out when you propose the hypothesis. Once again, tedious though it is to keep saying it, you need to put enough thought into this because the hypothesis determines whether a factor should be fixed or random.

Second is the issue of ensuring that there is proper replication. For example, suppose you explain variations in numbers of wading birds seen feeding from one place to another by the model that their densities differ among habitats. You hypothesize that there will be different densities in a mud flat, a sandy beach and an area of salt marsh. These three habitats are a fixed factor – you need all three in the experiment or you cannot test the null hypothesis of no difference. Even though wading birds may live in other places, you are not hypothesizing about them. So $X = 3$ and $x = 3$ and Equation 8.2 is zero. You must, nevertheless, now ensure that sampling is done in several representative lumps of each habitat. If censuses are done in marked out, replicated grids on one area of each habitat, you should only conclude that three sampled areas differ – not that the three habitats differ (see Section 6.2).

Another point is not to confuse the definition of fixed versus random with the definition of the relevant population (see also the discussion in Section 3.4). Consider the example of species and seeds discussed above. You might not want to consider all the species in the area. You design the hypothesis in terms of those species that are producing seeds in spring, when you are going to do the experiment. That, of course, restricts the relevance of the experimental results to the population of all species that have seed in spring. The species included in the experiment can (and probably should) still be a random subset of such species – even though the definition of relevant species was determined by a restrictive definition.

One final point. Suppose there are only a few possible levels of a factor (i.e. X is quite small in Equation 8.2). For example, there are only nine species of plants setting seed in the area used for the experiment on mortality of seeds. Four are in the experiment (you couldn't test all nine because of lack of resources). Thus, Equation 8.2 scarcely identifies zero or unity.

This is not really a problem, except to show the limitations of Equation 8.2. The species chosen should be picked randomly. Now, they cannot be a fixed factor. Differences or similarities among the four should represent any of the possible samples of four from nine species. That still represents ($9 \times 8 \times 7 \times 6/4 \times 3 \times 2 \times 1$; i.e. many) possible experiments you might have done! Treat it as a random factor; don't worry about the sampling fraction. Alternatively, use $(1 - x/X) \neq 0$ as the definition of random. Either way of defining a random factor will ensure that you do not make errors in analysis or interpretation when the differences between fixed and random factors alter the outcome of analyses (see Chapters 10 and 11). You will have correctly identified the factor as random.

Nelder (1977) discussed the definition of 'complete' (i.e. $x = X$ in Equation 8.2) versus 'incomplete' ($x < X$ in Equation 8.2) sampling. He wished to replace the terms 'fixed' and 'random' to provide a more general and simpler framework for the algebraic formulation of the underlying analytical models. This also allowed a much simpler understanding of such situations where a subset of treatments is chosen non-randomly from those available. 'Complete' sampling referred to a 'fixed' effects model. 'Incomplete' sampling was related to random effects sampling.

Other examples of fixed and random factors will appear with factorial and general experimental designs (Chapters 10 to 12). For now, recognize that they are different.

8.2 Interpretation of fixed or random factors

Obviously, the definition and interpretation of experimental factors depend entirely on the null hypothesis and hypothesis being examined. There is, however, a major difference between the interpretations that are appropriate for random as opposed to fixed factors. Consider the fixed case first. The whole notion is that there is a series of levels of a factor chosen because they are needed by the null hypothesis. In the simplest type of experiment, there are, for example, treatments chosen because of the need for controls. Consider the type of experiment where there is one manipulation, one 'sham' manipulation and an undisturbed control (see Section 6.6 for a complete discussion).

The three treatments are necessary in order to interpret the result. This is the essence of a fixed factor. There can be choices whether or not the pattern of differences (if the null hypothesis is rejected) can be articulated before the experiment. This leads to different types of comparison if the analysis causes the null hypothesis to be rejected (see Section 8.6.3 below). Nevertheless, all of the treatments are equally necessary and the alternative to the null hypothesis must involve a consideration of all the levels of the factor.

In contrast, a random factor is *sampled* in the experiment. A number of levels (places, sites, species, whatever) is chosen to represent those available and defined as relevant by the null hypothesis. Differences among them, so that the null hypothesis is rejected, indicate that there are, in general, differences among samples of the levels of the factor. The individual treatments do not matter. Comparisons among the specified levels used in the experiment are neither necessary, nor, usually, useful. Discovering that place A differs from places B, C and D and that C differs from D cannot be particularly informative. After all, to test the null hypothesis of no differences among places, the particular places A, B, C and D might not even have been examined because they might not have been sampled as randomly chosen levels of a factor. Thus, the information that there *are* differences is all that is needed. The specific pattern of differences does not help in any interpretation – it was not part of an hypothesis.

This fundamental difference helps to explain why the algebra differs in analyses of fixed and random factors. In the discussion of the linear model for a one-factor analysis, two things were relevant to this. First, the mean square among treatments includes a term measuring differences among treatments and one measuring variance within treatments (see Section 7.7 for details). Second, the terms representing differences among treatments were assumed to sum to zero – or it was assumed that the algebra would set them to zero (see Section 7.20).

In the linear model for an analysis (as in Equation 7.7 in Section 7.4):

$$X_{ij} = \mu + A_i + e_{ij} \tag{8.3}$$

where A_i is the effect of treatment i if the null hypothesis is false. In the discussion of the meaning and derivation of the analysis, differences among treatments were estimated as:

$$n \sum_{i=1}^{a} (A_i - \bar{A})^2 \tag{8.4}$$

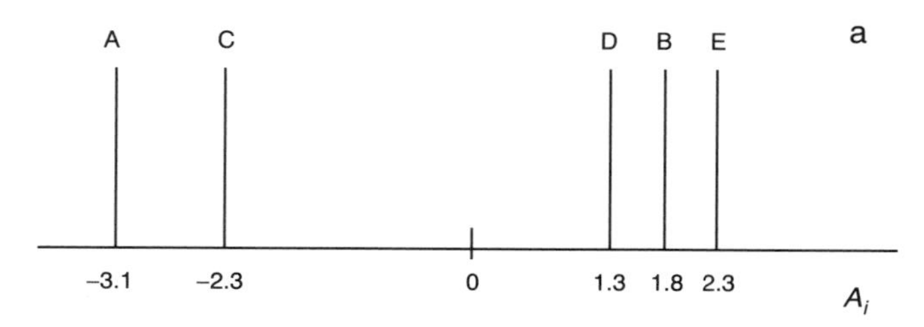

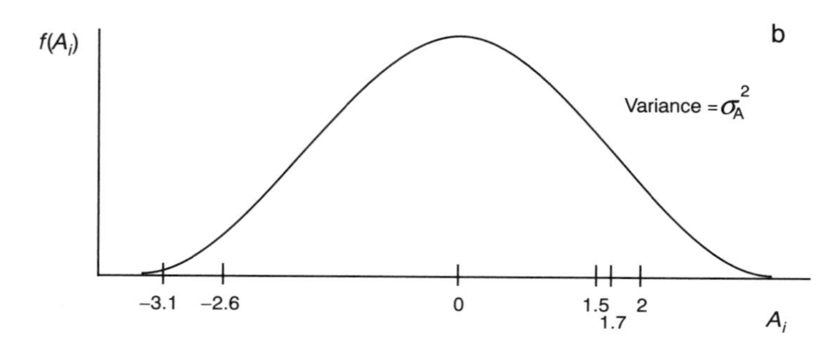

Figure 8.1. The difference between a fixed and a random factor. The levels of the fixed factor (a) are discrete, defined values. If the null hypothesis is false, a set of particular values (A_1 to A_5) applies. These are shown for a situation where $a = 5$. Note that $\sum_{i=1}^{a} A_i = 0$. For the random factor (b), there is a distribution of possible effects or levels. This is shown as a normal distribution with mean zero and variance, σ_A^2. A particular set is obtained for the experiment as a *sample* of this distribution (in this case the sample mean is $\bar{A} = -0.1$, $\sum_{i=1}^{a} = -0.5$).

The principle of a fixed factor is that all possible values of A_i are included in the experiment. The effects of different levels of a fixed factor are like the diagram in Figure 8.1a. There are five species of bird investigated in an experiment to test the null hypothesis of no difference in the rate of consumption of seeds among species. The five species were chosen because they all eat seeds (i.e. the relevant population of species must be seed-eaters) and these five were the only ones in the study-area. All five were chosen in order to be able to compare all the species in the area. If the null hypothesis is not true, the rates of consumption of the five species differ from the average over all species, as shown by the five values of A_i in Figure 8.1a. Three species (B, D, E) feed faster than average; species A and C are slower than average.

Contrast this with a similar experiment about weights of ovaries of adult lizards in fields. There are numerous fields. More small lizards are seen in some fields and lizards are known to differ in size from one place to another. As a general model, larger lizards have larger ovaries and are more fecund (perhaps?). From this, the differences in numbers of juveniles can be explained as a function of different sizes (and therefore weights of ovaries) of adults.

Therefore it is hypothesized that ovaries differ in weight among fields. The null hypothesis is that mean weight is the same in all the fields. Several fields are chosen randomly and a sample of ovaries of lizards in each field is weighed to test the null hypothesis. Now, the values of A_i that enter the experiment are a sample from the population of values belonging to all of the fields. The null hypothesis requires the fields all to have $A_i = 0$. What we have is a sample of the distribution of all A_i values. Such a distribution is illustrated as a normal distribution in Figure 8.1b. The distribution is of deviations of average weights of lizards in fields from average weight over all the fields for a large, normally distributed population of such deviations.

If the null hypothesis were true, all values of A_i are zero. Thus, the mean and variance (σ_A^2) of the population of A_i would both be zero. If the fields differ, A_i values are not all the same, but their mean must, on average be zero, otherwise they are not deviations from the overall mean weight. Now, however, $\sigma_A^2 > 0$. The relevant null hypothesis is that $\sigma_A^2 > 0$. The mean square among treatments furnishes an estimate of this variance because:

$$\frac{n\sum_{i=1}^{a}(A_i - \bar{A})^2}{(a-1)} \text{ estimates } n\sigma_A^2 \tag{8.5}$$

This can easily be verified by considering the familiar calculation of a sample variance (s^2) as an estimate of a population's variance (σ^2) as explained in Section 3.9. Here is a sample of a values of a population with variance σ_A^2. Hence, the estimated variance is as calculated.

This estimated variance has *two* potential sources of error. The first, as for the fixed factor, is due to sampling error within samples (σ_e^2). A_i values are not estimated exactly because samples are not exactly the same as the populations being sampled. The second source of error is the fact that the A_i values are not *all* examined in the experiment. There is a sample of them. So, the parameters of the population of A_i values are *estimated* from samples. Their variance (σ_A^2) is estimated from the sample of a levels of the factor. Their mean ($\mu = 0$, see Figure 8.1b) is also estimated.

Thus, the fixed and random cases differ in what they measure and how they measure it. In the fixed case, except for sampling error, differences

Table 8.1. Algebraic conventions to represent differences between fixed and random factors

$$X_{ij} = \mu + A_i + e_{ij}$$

where X_{ij} is jth replicate in ith treatment (ith level of factor A; $i = 1 \ldots a$), A_i is difference between ith level of factor A and overall mean of all levels (μ), e_{ij} is the deviation of replicate j in ith sample from the mean of that population.

Fixed factor:
By definition:

$$\sum_{i=1}^{a} A_i = 0$$

(see Section 7.6).

Analysis of variance	Mean square estimates
Among treatments	$\sigma_e^2 + \dfrac{n\sum_{i=1}^{a}(A_i - \bar{A})^2}{(a-1)}$ or $\sigma_e^2 + nk_A^2$
Within treatments	σ_e^2

where k_A^2 indicates fixed differences, all sampled in the experiment.

Random factor:

$$E\left(\sum_{i=1}^{a} A_i\right) = 0$$

Meaning you *expect* $\sum_{i=1}^{a} A_i = 0$ on average, over many experiments, but in a single experiment, A_i values as sampled may not sum to zero.

Analysis of variance	Mean square estimates
Among treatments	$\sigma_e^2 + n\sigma_A^2$
Within treatments	σ_e^2

where σ_A^2 is the variance of the population of A_i values sampled in your experiment.

among A_i values are measured exactly. In the random case, differences among A_i values provide a sampled estimate of the variance σ_A^2 of the entire population of possible levels. If a different experiment is done, with a different set of fields, the A_i values would be different, but they would still furnish a proper estimate of σ_A^2. Thus, the random factor provides a general statement about differences (if there are any) among any set of levels. The fixed factor provides a specific measure of the differences among the chosen levels.

The difference between the two types of factors starts to matter enormously when there is more than one factor in an experiment (see Chapter 10). It also greatly influences calculations of power (see Sections 8.3 and 8.4). Because of this, it is worth documenting the algebraic differences between fixed and random factors, as in Table 8.1.

Of course, the largest differences between the two types of factor occur in your interpretation of them. This is entirely a function of the hypothesis you propose.

8.3 Power of an analysis of a fixed factor

8.3.1 Non-central F*-ratio and power*

The theory and principles of determination of the power of an experiment were discussed in detail for a mensurative experiment in Chapter 5. Power of a fixed-factor analysis of variance is influenced by the probability of Type I error, the size of sample, n, the effect size and the variability of the populations being sampled.

The basis for determining power in the case of an analysis of variance of one fixed factor is demonstrated by the expected values of the mean squares used in the F-ratio (Section 7.7). In terms of what these mean squares measure (Table 7.6), F is calculated as:

$$F = 1 + \frac{\lambda}{(a-1)}, \quad \text{where } \lambda = \frac{n \sum_{i=1}^{a} (A_i - \bar{A})^2}{\sigma_e^2} \tag{8.6}$$

and n is the number of replicates of each of a populations sampled, with σ_e^2 as their common variance. A_i values are the differences among treatments if H_0 is false.

When the null hypothesis (H_0: $\mu_1 = \mu_2 = \ldots = \mu_i \ldots = \mu_A$) is true, A_i values are all equal to zero and therefore $\lambda = 0$. F is then distributed as the central F-distribution, the one tabulated with $(a-1)$ and $a(n-1)$ degrees of freedom (see Table 7.6) and a parameter ϕ set to zero. The general class of F-ratio distributions is:

$$\left.\begin{array}{l} F_{[(a-1),\, a(n-1),\, 0]} \text{ is the central } F \\ F_{[(a-1),\, a(n-1),\, \phi]} \text{ is a non-central } F \end{array}\right\} \tag{8.7}$$

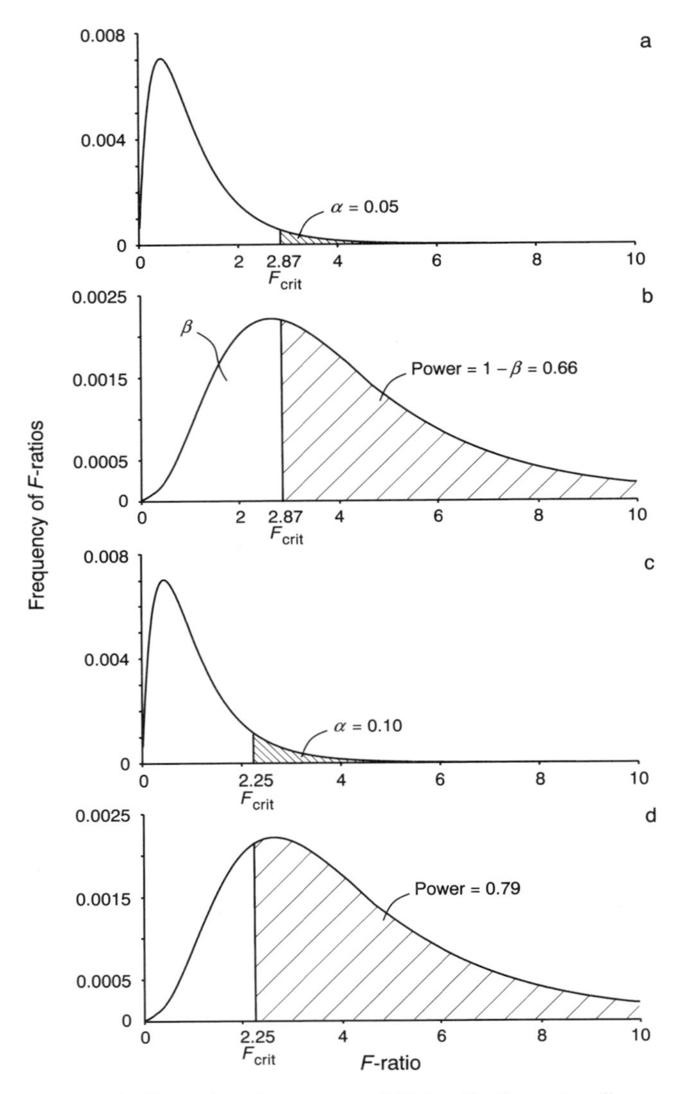

Figure 8.2. Central and non-central F-distributions. In all cases, there are $a = 5$ treatments, with fixed values of $A_i = 0$ when the null hypothesis is true ((a) and (c)) and +6, +4, −2, −3, −5 $(\sum_{i=1}^{a}(A_i - \bar{A})^2 = 90)$ when it is false ((b) and (d)). There are $n = 5$ replicates in each treatment and $\sigma_e^2 = 200$ in all treatments. In (a) and (b), F_{crit}, with 4 and 20 degrees of freedom and $\alpha = 0.05$, is 2.87. In (a), the probability of Type I error (α) is the shaded region to the right of F_{crit}. In (b), the probability of Type II error (β) is all values of F to the left of F_{crit}. Power is the shaded region $(1 - \beta)$. (c), (d) are the same as (a), (b) except probability of Type I error is 0.10, therefore $F_{\text{crit}} = 2.25$ and power is greater.

where $\phi = \sqrt{(\lambda/a)} = \sqrt{\left[n \sum_{i=1}^{a} (A_i - \bar{A})^2 / a\sigma_e^2\right]}$

Thus, if the null hypothesis is true, the calculated value of F for a given experiment comes from the central F-distribution, tabulated with the requisite number of degrees of freedom. If, instead, there is a particular alternative to the null hypothesis, i.e. particular magnitudes of A_i values, these set λ and therefore ϕ and F comes from the specified non-central F-distribution with the appropriate number of degrees of freedom.

This is illustrated in Figure 8.2a. For an hypothetical experiment, with $a = 5$ treatments and $n = 5$ replicates, the central F-distribution is shown. The critical value of F for α, the probability of Type I error, set to $P = 0.05$ is shown. Any observed F in the shaded region to the right of F_{crit} would cause rejection of the null hypothesis. If the null hypothesis is true, the probability of this occurring by chance has been chosen to be $P = 0.05$.

Under a particular, designated alternative hypothesis (H_A), the specified values of A_i allow calculation of λ and therefore ϕ (this is illustrated below). If this alternative hypothesis were correct, the calculated value of F would come from the non-central F-distribution in Figure 8.2b. The probability of getting a value of F smaller than F_{crit} is shown as β. This, of course, is the range of values of F from the non-central distribution that would cause you to retain the null hypothesis (they are all smaller than the value that would cause you to reject it). β is the probability of Type II error (failing to reject the null hypothesis when it is false). So, the power of the test is $(1 - \beta)$ and is the shaded portion of the non-central F to the right of the critical value in Figure 8.2b.

8.3.2 *Influences of* α, n, σ_e^2 *and* A_i *values*

Now it is possible to examine the various influences on power of an experiment (Figures 8.2 to 8.4). If a different value of α is chosen, F_{crit} moves and immediately influences the probability (β) of Type II error and power. An increase in α must cause an increase in power. This is common sense (if the probability of incorrectly rejecting a true null hypothesis is greater, null hypotheses are rejected more often; therefore they are less likely to be retained). It was also the case for the t-test (Chapters 5 and 6). It is illustrated in Figure 8.2b, d.

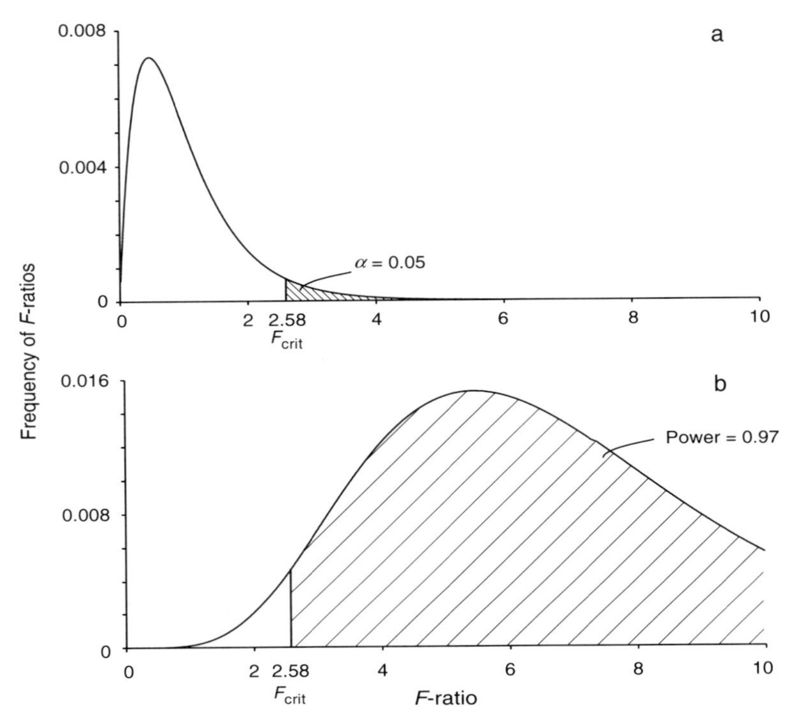

Figure 8.3. Effect of increasing size of sample (n) on power of analysis of variance. (a) and (b) are identical with Figure 8.2a and b, respectively, except $n = 10$, not 5. F_{crit}, with 4 and 45 degrees of freedom is 2.58. Note the greater power for the specified alternative hypothesis (compare (b) with Figure 8.2b).

An increase in sizes of samples also increases power. This is, again, common sense – a larger experiment with more replication should increase your capacity to detect differences. Examination of Equations 8.6 and 8.7 shows directly why increasing n increases λ and ϕ and therefore power, as illustrated in Figure 8.3.

The influence of the intrinsic variability of the experimental populations is also important. More variable data lead to less powerful experiments. This is intuitively clear and note that σ_e^2 is a divisor in Equation 8.6. The larger σ_e^2 is, the smaller is the non-centrality parameter and, thus, the less difference between the central and non-central F-distributions (Figure 8.4a).

Finally, the larger the effect sizes, the more likely it is that they will be detected and therefore the greater the power of an experiment. Again, this is clear in the determination of ϕ and λ (in Expression 8.7) and illustrated in Figure 8.4d.

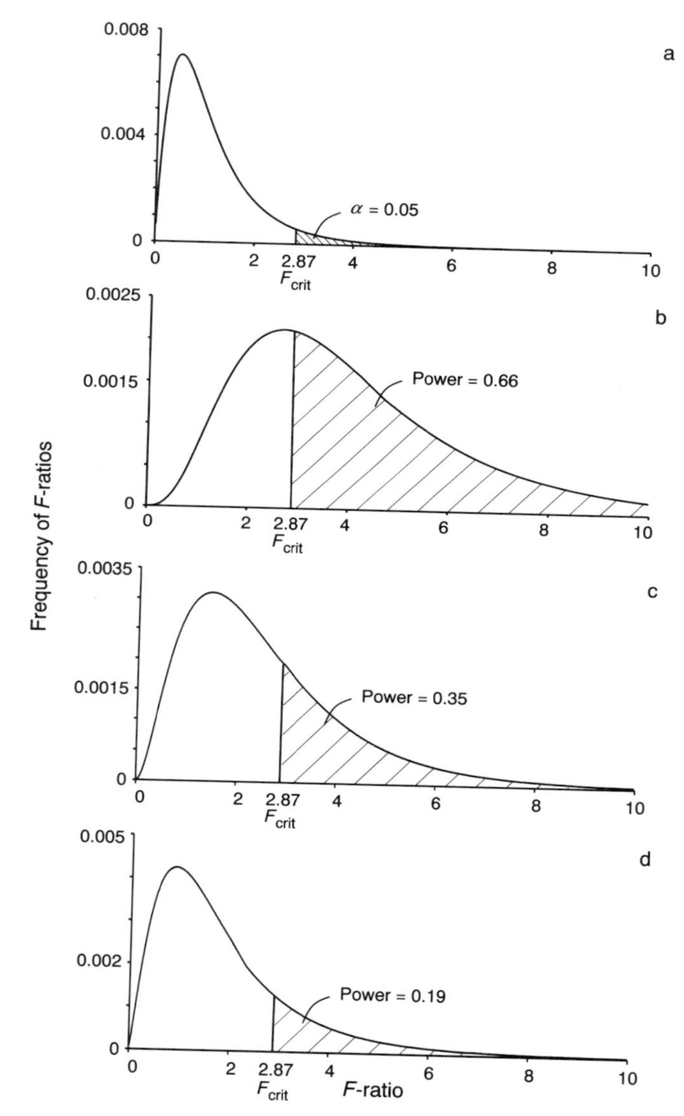

Figure 8.4. Influence of variance and effect size on power of analysis of variance. (a), (b) are identical with Figure 8.2a and b, respectively. (c) The same as (b), except that $\sigma_e^2 = 400$, not 200. Note the decreased power where there is greater variance within populations. (d) The same experiment, but the alternative hypothesis has smaller effect sizes. A_i values are +3, +2, −1, −2, −2 $(\sum_{i=1}^{a}(A_i - \bar{A})^2 = 22)$. Power to detect the alternative hypothesis is therefore smaller than in the case of larger effect sizes (compare with (b)).

Theoretically, the number of levels of the factor included in the experiment might influence its power (a is included in the determination of the non-centrality parameter ϕ, in Expression 8.7). In logical practice, however, a cannot vary or be considered to be different in a particular experiment. A fixed factor must have all possible levels that are relevant to the hypothesis included in the experiment. Thus, a is determined in defining the null hypothesis. It can't be altered or the hypothesis and null hypothesis are arbitrary and the factor would *not* be fixed!

So, power is intuitively comprehensible in relation to the design of the experiment. For a chosen level of α, pre-determined sizes of sample n and estimated variance within populations (estimated σ_e^2), the power of an experiment to detect a specified alternative to the null hypothesis can be calculated. Or, the experiment can be designed to have n large enough for some specified power to be achieved. The influences of various components of an experiment on its power are summarized in Table 8.4, below.

8.3.3 Construction of an alternative hypothesis

As with the previous discussions of power (in Chapters 5 and 6), the whole exercise hinges on determining, in advance, the minimal alternative to the null hypothesis that matters. This is the minimal differences among treatments that, according to the logic of the model and hypothesis you have proposed, should be detected to enable you to interpret sensibly a rejection of the null hypothesis.

Here is an example. Suppose you have observed a difference in mean abundance of prey from place to place of about 30 individuals per unit area. Your model to explain this is that predators are removing about 30 individuals from some places. You hypothesize that removal of predators will increase densities by 30. In the experimental areas, you propose to remove predators by shooting them. Controls will be untouched areas. There will also be procedural controls where disturbances are similar and shotguns have been fired, simulating removal of predators.

Your null hypothesis is that there will be no difference among treatments. The alternative that matters is that there should be no effect of the procedure (the difference between controls and procedural controls is zero), but there should be an average difference of +30 between the experimentals and controls. Thus:

A_1 = Effect on controls = 0

A_2 = Effect on procedural controls = 0

A_3 = Effect on experimentals = +30

In order to have these fixed effects (it is clearly a fixed factor – there are no other relevant treatments) sum to zero (see Section 8.2), these will be $A_1 = -10$, $A_2 = -10$, $A_3 = +20$, relative to the mean numbers of prey observed in the experiment. Control and procedural control plots will have fewer than the overall mean; experimental plots will have 30 more individuals. So, these are the effect sizes for which we want to know the power of the experiment. For this experiment, the alternative hypothesis is that $\sum_{i=1}^{a} (A_i - \bar{A})^2 = 600$.

If we had an estimate of the variability among replicates in such experimental treatments (which we should be able to calculate from the original observations; see Section 6.7), we could now calculate the power of the experiment for any chosen amount of replication (n). Better still, we could determine an appropriate amount of replication to achieve adequate power. The details of this sort of calculation are explained in Cohen's (1977) book. Assume for the example that follows that the variance among replicate plots (i.e. estimated σ_e^2) will be about 100.

The issue of how to determine adequate power revolves around the consequences of making a Type I or Type II error (as discussed in Section 5.11). Suppose we decide to set α and β equal to each other. We are willing to do the experiment with an equal chance of falsely rejecting (α) or falsely retaining (β) the null hypothesis. We also choose to keep these small (say, $P = 0.05$). Thus, we have chosen the power to be 0.95.

In this case (see Expressions 8.7):

$$\phi = \sqrt{\left[\frac{n\sum_{i=1}^{a} (A_i - \bar{A})^2}{a\sigma_e^2}\right]} = \sqrt{\left(\frac{600n}{300}\right)} \tag{8.8}$$

Consultation of tables of non-central F-distributions (Winer *et al.*, 1991) or the tables in Cohen (1977) leads to the determination of power, as in Table 8.2.

Provided that a sensible case can be made for specifying a relevant alternative to the null hypothesis, the power of an experiment can easily be determined in advance, for a fixed-factor experiment. To help with the calculations, there are, as mentioned above, tables of non-central

Table 8.2. Calculation of power in an experiment with a fixed factor. The experimental treatment (E) is removal of predators. C are untouched controls; PC are procedural controls (e.g. scarecrows, partial fences, etc.)

Anticipated effect sizes
$A_{\mathrm{C}} = -10,\ A_{\mathrm{PC}} = -10,\ A_{\mathrm{E}} = +20$

$$\sum_{i=1}^{a}(A_i - \bar{A})^2 = 100 + 100 + 400 = 600$$

Estimated variance $(\sigma_e^2) = 100$
Probability of Type I error $(\alpha) = 0.05$

$$\lambda = n\sum_{i=1}^{a}(A_i - \bar{A})^2/\sigma_e^2 = 600n/100 = 6n$$

$$\phi = \surd(\lambda/a) = \surd(6n/3) = \surd(2n)$$

Size of sample (n)	2	3	4	5
Non-centrality parameter (ϕ)	2.00	2.45	2.83	3.16
Probability of Type II error (β)[a]	0.58	0.21	0.05	0.01
Power $(1 - \beta)$	0.42	0.79	0.95	0.99

If $n = 4$, power of 0.95 could be achieved.

[a] Interpolated from tables of non-central F and power in Winer *et al.* (1991).

F-distributions and power curves for different sample sizes (see, for example, Winer *et al.*, 1991). The most complete tables are those in Cohen (1977), but Winer *et al.* (1991) also provided an algorithm approximating the probabilities of non-central F-distributions. Several computer packages will enable calculations of power.

8.4 Power of an analysis of a random factor

8.4.1 *Central* F*-ratios and power*

When a random factor is being considered in an experiment, there are two different types of variability included in the mean square among treatments (see Section 8.2). The first is the intrinsic variation among replicates within samples, which estimate variances in the populations being sampled. The second is the variability caused by the fact that the levels of the factor in the experiment are, themselves, a sample and their average will not exactly represent the average of the population of levels about which the hypothesis is proposed (i.e. zero if the null hypothesis is true).

Because of the difference from estimates obtained when investigating a fixed factor, random factors have different distributions of F-ratios. In the case of a fixed factor, F-ratios are distributed according to F with $(a-1)$, $a(n-1)$ degrees of freedom and non-centrality parameter ϕ. When the null hypothesis is true, $\phi = 0$ and the distribution is a central F-distribution. F is, thus, distributed as $F_{[(a-1), a(n-1), \phi]}$.

When the factor is random, the distribution of F is (see Table 8.1):

$$F = \frac{\text{Mean square among}}{\text{Mean square within}}, \quad \text{which estimates } \frac{\sigma_e^2 + n\sigma_{\mathrm{A}}^2}{\sigma_e^2} = 1 + n\theta \qquad (8.9)$$

where $\theta = \sigma_{\mathrm{A}}^2/\sigma_e^2$. Winer *et al.* (1991) provided a proof that, provided data are normally distributed, have homogeneous variances and are independently sampled, with balanced samples, $F' = F/(1 + n\theta)$ is distributed as a central F-distribution, where F' is the distribution of the ratio of mean square among samples divided by mean square within samples if the null hypothesis were false and the variance among levels of factor A is actually σ_{A}^2. F is the corresponding central F-distribution if the null hypothesis were true. When the null hypothesis *is* true and there are no differences among levels of factor A, F' reverts to the central F-distribution $F_{[(a-1),a(n-1),0]}$. The power of such a test is therefore determinable using the central F-ratio distribution. For example, for any test where F_{crit} is the critical value with $(a-1)$ and $a(n-1)$ degrees of freedom, at $\alpha = 0.05$, the power of the test can be determined using F_{crit} and the ratio of $\sigma_{\mathrm{A}}^2/\sigma_e^2$ chosen as the specified alternative to the null hypothesis. α, the probability of Type I error, when the null hypothesis is true, is the probability that F exceeds F_{crit}. Thus:

$$\alpha = \text{Prob}\,(F \geq F_{\text{crit}}) = \text{Prob}\,(F \geq F_{\text{crit}[(a-1), a(n-1), 0]}) \qquad (8.10)$$

The probability of Type II error is the probability of not exceeding this value, when a specified alternative is true:

$$\beta = \text{Prob}\,(F < F_{\text{crit}}) = \text{Prob}\left(F < \frac{F_{\text{crit}[(a-1), a(n-1), 0]}}{1 + n\theta}\right) \qquad (8.11)$$

Therefore, (see Figure 8.5) power is $(1 - \beta)$:

$$\text{Power} = \text{Prob}\left(F \geq \frac{F_{\text{crit}[(a-1), a(n-1), 0]}}{1 + n\theta}\right) \qquad (8.12)$$

The procedure is explained by Cohen (1977), Winer *et al.* (1991) and, in the context of assessments of environmental impacts, by Underwood (1993). An example is given in Table 8.3.

Table 8.3. Calculation of power in an experiment with a random factor. The experiment compares densities of insects per plant in randomly chosen fields (a random factor). $a = 5$ fields will be sampled. Previous studies suggest that, where abundances differ, differences among fields have been estimated (as σ_A^2) to be about 200. Power is the probability of exceeding F' in the central F-distribution with the specified numbers of degrees of freedom for each size of sample.

Anticipated effect size (σ_A^2) = 200
Estimated variance among plants (σ_e^2) = 100
Probability of Type I error (α) = 0.05

$\theta = \sigma_A^2/\sigma_e^2 = 2.0$

Size of sample n	Degrees of freedom for F	F_{crit}	$F' = \dfrac{F_{crit}}{1 + n\theta}$	Power
2	4, 5	5.19	1.04	0.47
3	4, 10	3.48	0.50	0.74
4	4, 15	3.06	0.34	0.85
5	4, 20	2.87	0.26	0.90
10	4, 45	2.58	0.12	0.97

8.4.2 *Influences of* α, n, σ_e^2, σ_A^2 *and* a

The effects of choosing different probabilities of Type I error are shown in Figure 8.5. Where α is larger, power is greater because the critical value is then smaller and the probability of exceeding it is greater. Similarly, as n increases, the power of the test must increase because n is a divisor of the critical value in Equation 8.12. This is illustrated in Figure 8.6.

Any increase in σ_e^2, the variation among replicates in the populations sampled, will decrease power (Figure 8.7) because θ will be reduced (see Equation 8.9), thus reducing the divisor of the critical value in Equation 8.12.

In contrast, where the differences among the levels of a factor are larger, σ_A^2 is larger and therefore θ and the power are greater (Figure 8.7). Greater effect sizes (i.e. larger departures from the null hypothesis) are more likely to be detected.

Finally, in contrast to the power of a fixed factor, the number of levels of the factor sampled in the experiment (a) is not determined by the logic of the hypothesis. It is usually set small by logistical constraints (there isn't enough money, time or personnel to sample more sites, tanks, plots, etc.). a does not feature in the evaluation of θ. Its only influence on power is to increase the precision of the estimate obtained of the variance (σ_A^2) of the population of levels being examined. As a increases, the precision of the

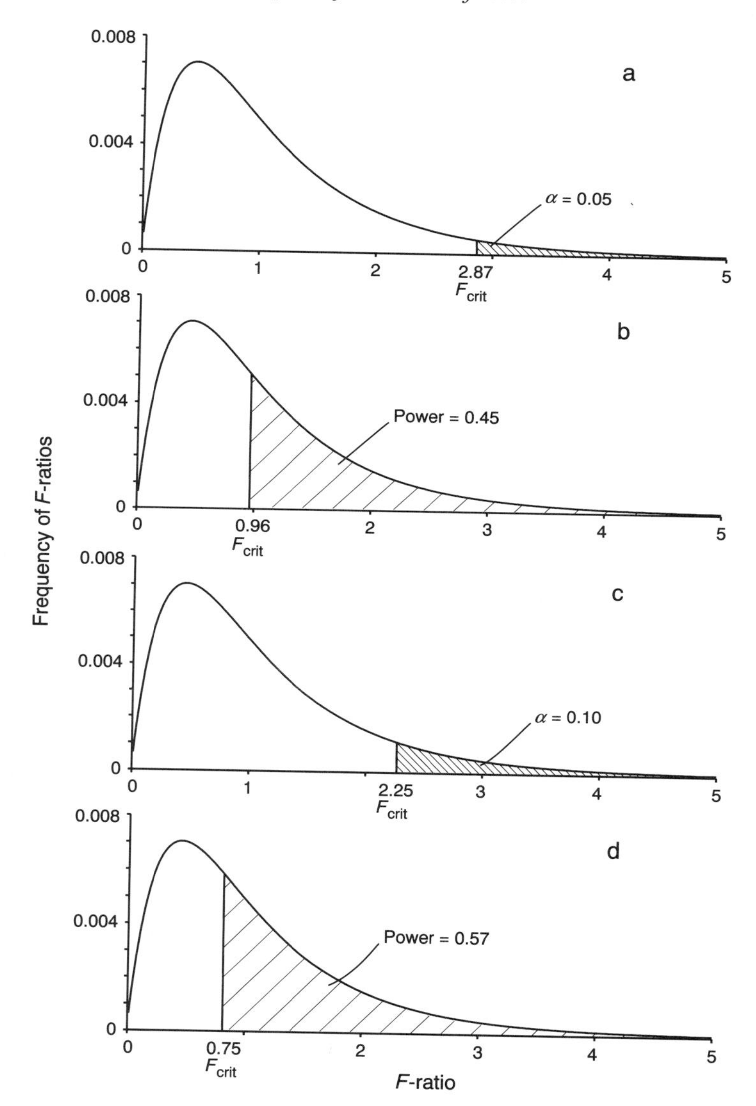

Figure 8.5. Power of an analysis of variance of a random factor. $a = 5$ randomly chosen levels of a factor, each sampled with $n = 5$ replicates. All populations have $\sigma_e^2 = 200$. (a) Null hypothesis true, $\sigma_A^2 = 0$. Probability of Type I error, $\alpha = 0.05$, $F_{crit} = 2.87$ with 4 and 20 degrees of freedom. (b) Null hypothesis false, $\sigma_A^2 = 80$. $F_{crit}/(1 + n\theta) = 0.96$. Power is indicated by the shaded region. (c), (d) as (a), (b), respectively, except that $\alpha = 0.10$; $F_{crit} = 2.25$ in (c), $F_{crit}/(1 + n\theta) = 0.75$ in (d). Power when $\alpha = 0.10$ is greater than when $\alpha = 0.05$ (compare (d) with (b)).

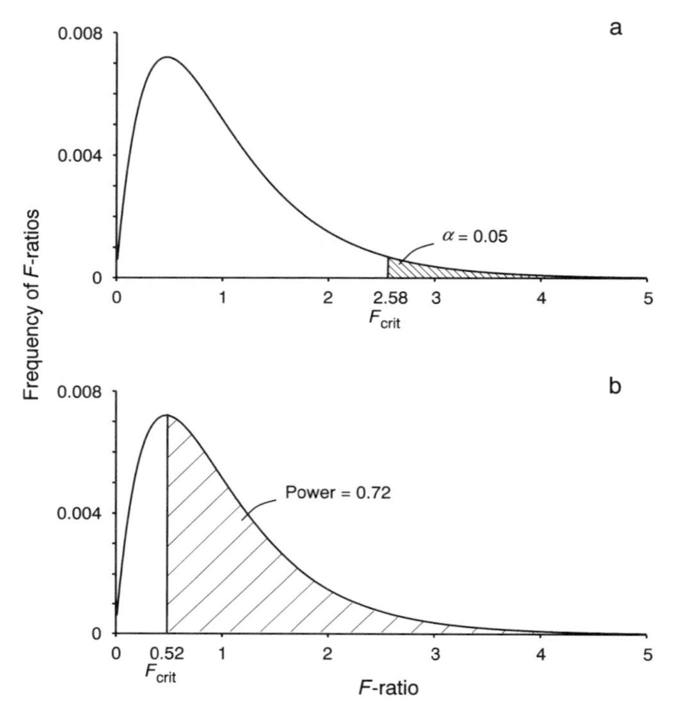

Figure 8.6. Effect of increasing size of sample on power of an analysis of variance of a random factor, each sampled with $n = 10$ replicates, $\sigma_e^2 = 200$. (a) Null hypothesis true, $\sigma_A^2 = 0$, $\alpha = 0.05$; $F_{crit} = 2.58$ with 4 and 45 degrees of freedom. (b) Null hypothesis false, $\sigma_A^2 = 80$; $F_{crit}/(1 + n\theta) = 0.52$. Power is greater with increased n; compare (b) with Figure 8.5b.

estimate of σ_A^2 increases, making it more likely to be detected. So, power increases with increased intensity of sampling the random factor. This is illustrated in Figure 8.8.

The influence on power of an experiment of various aspects of the design or other features of the experiment is summarized in Table 8.4.

8.4.3 Construction of an alternative hypothesis

This is much more difficult for a random than for a fixed factor. There is usually no simple way to predict, from an explanatory model, what might be the *variance* among a population of differences, representing the levels of a factor.

Sometimes, there is prior information from other studies that can be used to estimate the likely variation among levels of a factor. For example,

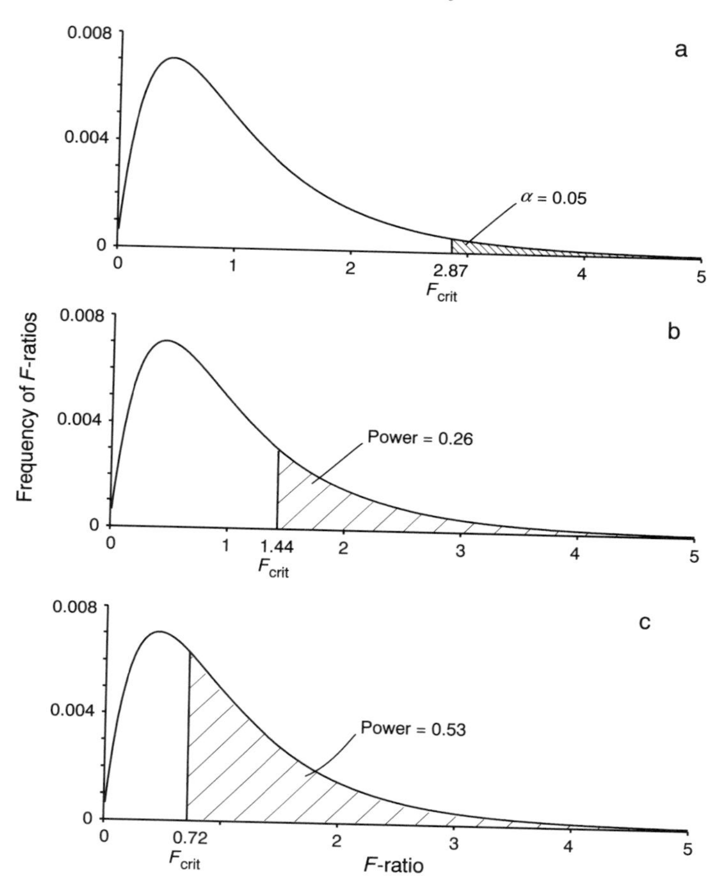

Figure 8.7. Effect of increased within sample variation or effect size on power of analysis of variance of a random factor, each sampled with $n = 5$ replicates. All populations have $\sigma_e^2 = 400$. (a) Null hypothesis true, $\sigma_A^2 = 0$; $F_{crit} = 2.87$, with 4 and 20 degrees of freedom. (b) Null hypothesis false, $\sigma_A^2 = 80$, $\alpha = 0.05$, $F_{crit}/(1 + n\theta) = 1.44$. Power is smaller than with $\sigma_e^2 = 200$ (compare (b) with Figure 8.5b). (c) Null hypothesis false, $\sigma_A^2 = 120$; $\alpha = 0.05$; $F_{crit}/(1 + n\theta) = 0.72$. Power is greater with increased effect size; compare (c) with (b).

consider variation among sites (say, sites of 100 m × 100 m at kilometre intervals through woodland) in the average rate of mortality of small mammals. Studies of possible causes of mortality will include propositions that there will be variation among places used in experiments in the mean rates of mortality due to a postulated mechanism (e.g. predation or disease). The null hypothesis of no difference will be then contrasted against the alternative that variance among places is some value calculated from the observed,

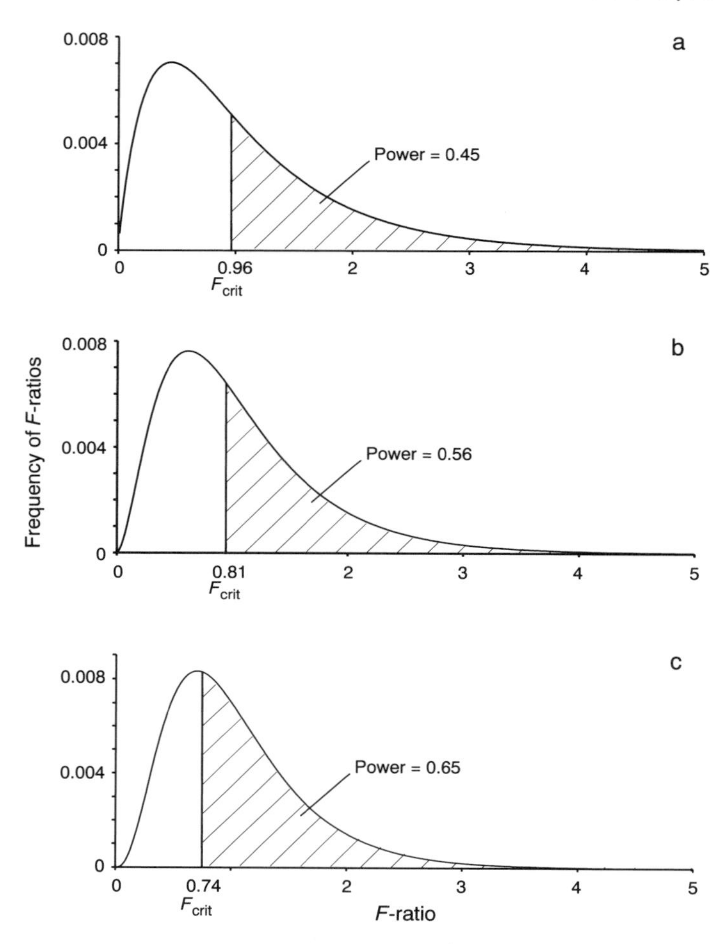

Figure 8.8. Effect of increased number of levels, a, on power of analysis of a random factor. In all cases, $n = 5$ replicates in each sample and all populations have $\sigma_e^2 = 200$. (a) Null hypothesis false, $\sigma_A^2 = 80$; $a = 5$ populations; $F_{crit}/(1 + n\theta) = 0.96$; (b) as (a), but $a = 7$ populations sampled, $F_{crit}/(1 + n\theta) = 0.81$; (c) as (a), but $a = 9$, $F_{crit}/(1 + n\theta) = 0.74$. Power increases with increasing number of levels of the factor sampled.

natural rates from site to site. It is then possible to calculate the power of a given planned experiment to detect differences among sites. Alternatively, the appropriate number of replicated sites or samples within sites to achieve a requisite or adequate power can be calculated.

In other situations, it is sometimes possible to model variances of a population of levels of a random factor. An example is given, by

Table 8.4. Summary of influences on power of analysis of variance

	Effect on power to detect differences in	
Effect of	Fixed factor	Random factor
Increased probability of Type I error (α)	Increased	Increased
Larger size of sample (n)	Increased	Increased
Greater variance within populations (σ_e^2)	Decreased	Decreased
Larger effect size (k_A^2 or σ_A^2)	Increased	Increased
Larger number of samples (a)	Not relevant	Increased

Underwood (1993) for a more complex case in the detection of environmental impacts. Consider a study of dispersion of insects. Large numbers of marked insects are released from a set point and as many as possible are trapped at chosen distances from the point of release. Under a model of random dispersal in all directions, the average numbers expected at any distance from the point of release can be hypothesized as conforming to a negative exponential distribution. There will be reasonably uniform densities in any direction at a given distance from the origin.

Alternatively, if there are vagaries in rates of dispersal due, for example, to wind, there will be considerable variation among sites in different directions at a chosen distance from the point of release. If the speeds, directions and variances in these characteristics are known, it is possible to model the predicted variance among sites in the number of insects. Such models provide a basis for an alternative to the null hypothesis of no difference in mean number of insects per site at each distance from the point of release.

A worked example is given in Table 8.3 of how to determine the appropriate number of samples (n) to achieve adequate power in an analysis of a random factor. In the example, it is again assumed that the consequences of a Type I or a Type II error are about the same. Consequently, because it is decided to set the former (α) to 0.05, β is set to be equal to α. The requisite power is therefore 0.95 and the sample size should be $n = 10$ (see Table 8.3).

8.5 Alternative analysis of ranked data

For a single-factor analysis of variance, there is an alternative, so-called non-parametric analysis – the Kruskal–Wallis test (Kruskal & Wallis,

1952). This is often used by ecologists who are concerned that the assumptions underlying the use of analysis of variance are too restrictive. As discussed in Chapter 7, in general these assumptions are not. Nevertheless, sometimes data are far better examined by their comparative or relative magnitudes, rather than their absolute values. So, ranks, rather than absolute values, might be an appropriate basis for analysis, regardless of the assumptions.

The procedure is straightforward. The data are ranked from smallest to largest for all n replicates in all a treatments. The ranks are then used to calculate sums of squares, using the same formulae as used with ordinary data (Table 7.6). Then, the Kruskal–Wallis statistic, H, is calculated as described in Siegel (1953) and Hollander & Wolfe (1973). Multiple comparisons (see Section 8.6) require pair-wise comparisons of sets of ranks with appropriate adjustments of probabilities of Type I error (Colquhoun, 1971; Hollander & Wolfe, 1973; Conover, 1980; Winer *et al.*, 1991).

Note, however, that the Kruskal–Wallis procedure is not free from restrictive assumptions. In ecological studies, it is often used as an alternative to an analysis of variance because the data have heterogeneous variances. Homogeneity of variances and independence of data are, however, also assumptions for the Kruskal–Wallis test, as is the requirement that the distributions are continuous and of the same general shape. It is not appropriate to use a rank-order procedure to solve a problem that it does not solve.

8.6 Multiple comparisons to identify the alternative hypothesis

8.6.1 Introduction

Once an analysis of variance has rejected the null hypothesis that a set of a means are the same, we face a problem. If the analysis is of a random factor, this is often the end of the analysis. The means differ, so the null hypothesis is false. Put another way, there is now evidence that there is variance among means of the sampled populations; σ^2_A is *not* zero. That is usually all that is necessary in the analysis of a random factor because that is all that is specified by the hypothesis.

In the case of a fixed factor, however, that is not the usual end-point. Usually, an analysis of a fixed factor is done because mechanisms have been proposed to explain certain patterns of observations. From the model proposing the mechanisms, certain predictions are hypothesized, which lead to specific predictions about what is going to happen in the experiment. Where there are only two treatments, as in the cases in

Chapter 6, once a significant difference has been demonstrated, which of the two is larger and which smaller is obvious by inspection.

Where there are more than two treatments, the pattern of difference cannot usually be identified by inspection. There are too many possibilities. We need a procedure to identify the precise alternative to the null hypothesis. There should be an orderly procedure for comparing all the means, in a logical and ordered way, to define which alternative to the null hypothesis is most likely to apply. There are many such procedures and there is much potential confusion about how to choose among them.

8.6.2 *Problems of excessive Type I error*

The major difficulty comes about because any attempt to identify, unambiguously, a specific alternative to the null hypothesis must involve several comparisons among the means of the treatments. If each comparison is done with a pre-set probability of Type I error, α, the set of tests has a much greater probability of error, as explained in Section 6.9.2.

Definitions of probability of Type I error were nicely separated by Ryan (1959), following work by Tukey (1949a) and Hartley (1955). The nominal level of Type I error is that chosen for each comparison – the error rate per comparison. In any comparison of two means after an analysis of variance, this is traditionally set at $P = 0.05$. Then there is the error rate per experiment, i.e. the probability that out of all the comparisons made in a particular analysis (those in a single experiment) there will be an error. If there are $a = 3$ means and three pair-wise comparisons, this is a rate of $P = 0.14$, as explained in Section 6.9.2, if $\alpha = 0.05$ for each comparison.

Finally, if numerous experiments are done to test the same null hypothesis, there is a rate of error measuring the probability that a given *experiment* will contain an error. This has been called, clumsily, the 'experiment-wise' rate of error (Ryan, 1959).

Following Ryan (1959), these differences can be illustrated as follows. Suppose there have been, over the years, 30 experiments on a particular hypothesis. In each experiment, there were three treatments and therefore three pair-wise comparisons, making a total of 90 tests. Of these (of course, we could never know this in the real world), four gave significant differences due to Type I errors and these occurred in three of the experiments.

The error rate per comparison is then 4/90 (=0.044); there were four errors in 90 tests. The error rate per experiment was 4/30 (=0.133);

there were four errors in 30 experiments. Finally, the 'experiment-wise' rate of error is 3/30 (= 0.10); three of the 30 experiments resulted in errors. In more complicated experiments, with more than one factor, there are also what are called 'family-wise' rates of error, for sets of comparisons on each experimental factor (Huitema, 1980).

The issue of concern is to be clear what rate of error is actually appropriate for a given experiment. For the above example, it would be quite fraudulent to report rates of Type I error as 0.05 (the rate chosen for each comparison). In a single experiment it is 0.13. So, it is incumbent on you to report the results objectively – even if you choose to have a larger probability of error than 0.05. You could do experiments and comparisons with rates of error much larger, but it is inappropriate to describe them as if the rate was, in fact, 0.05.

What follows is an attempt to consider what to do under different circumstances. The problem, the issues and the different options have been bashed about over the years by numerous authors. Prominently among these are the studies of simulated sets of data by Einot & Gabriel (1975) and Ramsey (1978). A very fine summary, with yet more simulations, has been provided for ecologists by Day & Quinn (1989). They made specific recommendations for many types of circumstances. By and large, the following is consistent with their recommendations. They did, however, overemphasize the problems of Type I error at the expense of the power of tests (i.e. they played down the inevitable problems of Type II error). So, the following discussion cannot concur with all of their conclusions.

8.6.3 **A priori** *versus* **a posteriori** *comparisons*

An essential issue to resolve before embarking on multiple comparisons is whether an alternative to the null hypothesis has been identified before the experiment. In one sort of experiment, there is a model and hypothesis constructed so that a specific pattern of differences among treatments has been identified. This is the defined or specified alternative to the null hypothesis and is identified before the experiment (i.e. it is defined in advance, or *a priori*). Examples will be given below.

In contrast, in other types of experiments, the hypothesis does not provide clear expected patterns of differences among treatments. The hypotheses are much more exploratory. For example, the prediction is that there will be seasonal differences in mean abundance of a population – but which seasons will differ cannot be deduced from existing observations and models. The experiment will test whether or not there are seasonal differences that can be used, after the data are available, to determine which seasons differ. Thus, all comparisons are made in an effort to determine *a posteriori* (after the event, with the 20–20 vision of hindsight) what is the specific alternative to the null hypothesis. Again, some examples will be considered below.

In an *a priori* determined set of comparisons, two features matter. First, the number of comparisons to be made is smaller than all possible comparisons among treatments. So, the rate of Type I error per experiment will not be as great as in *a posteriori* comparisons. This almost always guarantees that the power of the set of comparisons will be greater than in *a posteriori* comparisons. Second, the *order* or *sequence* in which tests are done is also specified by the *a priori* defined hypothesis, which guarantees a more orderly, or more logical interpretation of the outcome than is sometimes the case in *a posteriori* comparisons.

Both considerations (better power and logic) should encourage better structure of hypotheses. The more completely the hypothesis is specified in advance, the more useful the experiment and its interpretation will be. Thus, when the alternative hypothesis has been spelled out sufficiently so that the anticipated differences and similarities among means are identified in advance, much better experiments and interpretations can be made. If the null hypothesis is then rejected by the analysis of variance, the experimenter immediately knows where to start looking to see whether the proposed alternative has occurred.

8.6.4 **A priori** ***procedures***

8.6.4.1 Two examples

Here is a simple example (see Chapter 6 for more discussion of this type of experiment). Differences from place to place in the mean abundance of a species are explained by the model that predators in some areas cause declines in abundance of the observed species. The hypothesis is proposed that removal of predators from experimental plots will cause increased density of prey to match the larger abundances seen in some parts of the area. Procedural control areas will be needed to identify any artefact due to the procedures used to remove predators.

A priori, we can specify that the following would be an interpretable outcome of the experiment: mean abundance where predators are removed will be greater than either type of control (and will, of course, match the larger abundances where predators are naturally sparse – see Chapter 6). In contrast, if the procedures used cause a difference in density compared with the untouched controls, we have created an artefact that prevents *any* sensible conclusion about predation. Thus, the null and alternative hypotheses are:

$$H_0: \quad \mu_C = \mu_{PC} = \mu_E \qquad H_A: \quad \mu_C = \mu_{PC} < \mu_E \tag{8.13}$$

where C is control, PC is procedural control and E is experimental treatment. *A priori* to the experiment we have specified what pattern of differences matches the prediction about predation. Other patterns do not support the proposed model – even if they lead to significant differences. So, finding evidence that $\mu_C < \mu_{PC} = \mu_E$ would provide no evidence for predators being important. Rather, it would suggest that removal of predators caused an increase in densities for some other reason (maybe disturbance created more useable habitat, or cages enhanced the shelter from unfavourable weather).

Alternatively, $\mu_C < \mu_{PC} < \mu_E$ would imply an effect of predators, but in areas where some procedural artefact has already changed abundances. This is not a logically interpretable result given the stated hypothesis (see Chapter 6).

So, to distinguish between the null hypothesis and the specified alternative we can define *a priori* a series of tests and a sequence in which to do them.

1. There must be a difference among means in the analysis of variance, otherwise H_0 is retained.
2. $\bar{X}_C$ must be compared with $\bar{X}_{PC}$ and found to be not different, otherwise some difference has occurred other than that specified as H_A.
3. If $\bar{X}_C$ does not *differ* from $\bar{X}_{PC}$, $\bar{X}_C$ and $\bar{X}_{PC}$ must be compared with $\bar{X}_E$ and must be shown to be different, in the specified direction ($\bar{X}_E$ must be greater).

Two comparisons between treatments and the order to do them have been completely specified *a priori*.

As another example, suppose that it has been observed that there are more insects about at some times than at others. The model proposed to explain this is that abundances of annoying insects are greatest on warm, dry days than at any other times. Data have been collected on many days as part of some routine, hypothesis-free monitoring programme, but they may now be used to test this model. The data have been classified as: (1) warm/damp, (2) cool/damp, (3) warm/dry and

(4) cool/dry days. You predict (as the hypothesis) that mean abundance will be greater in a sample of the third category than any other. You have no interest in differences among the other three conditions – your hypothesis requires abundances on warm and dry days to be larger. Your null and alternative hypotheses are:

$$\begin{aligned} &H_0: \quad \mu_1 = \mu_2 = \mu_3 = \mu_4 \\ &H_A: \quad \mu_3 > \mu_1; \quad \mu_3 > \mu_2; \quad \mu_3 > \mu_4 \end{aligned} \qquad (8.14)$$

where μ_3 represents mean abundance on warm, dry days. You have no interest in differences that may occur among the other weather conditions. So, *a priori* you have determined that, if the null hypothesis is rejected by the analysis of variance, you will make the three specified comparisons ($\bar{X}_3-\bar{X}_1$; $\bar{X}_3-\bar{X}_2$; $\bar{X}_3-\bar{X}_4$). You will not need to compare $\bar{X}_1$ with $\bar{X}_2$ or $\bar{X}_4$, nor $\bar{X}_2$ with $\bar{X}_4$. Of course, if any of the three comparisons were not significant, you would stop because the hypothesis requires $\bar{X}_3$ to be different from each of the other means.

8.6.4.2 A priori *tests*

These may involve pair-wise *comparisons* as in step (2) of the first example above, or any of the comparisons in the second example. Alternatively, they may be *contrasts*, where some averaging of means occurs. An example was the final part (step (3)) of the first example described above.

The procedures used depend on information obtained from the analysis of variance (e.g. that variances were equal). Here, the only procedure described is when sample sizes are the same (reflecting my bias against unbalanced analyses; see Section 7.10). The comparisons or contrasts are all constructed as the following linear equation (see details in any textbook or in the detailed review by Rosenthal & Rosnow, 1985):

$$L = c_1\bar{X}_1 + c_2\bar{X}_2 + \ldots + c_i\bar{X}_i + \ldots + c_a\bar{X}_a \qquad (8.15)$$

where $\bar{X}_i$ represents the mean of treatment i, c_i represents the *coefficient* given to $\bar{X}_i$ for this contrast and L represents the linear sum of the products of means and their coefficients.

Any linear combination of coefficients will lead to a valid comparison or contrast, with one degree of freedom, if:

$$\sum_{i=1}^{a} c_i = 0 \qquad (8.16)$$

In the first case discussed in the previous section, there are two contrasts:

$$\left.\begin{aligned} L_1 &= 1\bar{X}_{\mathrm{C}} - 1\bar{X}_{\mathrm{PC}} + 0\bar{X}_{\mathrm{E}} \\ L_2 &= 0.5\bar{X}_{\mathrm{C}} + 0.5\bar{X}_{\mathrm{PC}} - 1\bar{X}_{\mathrm{E}} \end{aligned}\right\} \tag{8.17}$$

where L_1 is the comparison of $\bar{X}_{\mathrm{C}}$ and $\bar{X}_{\mathrm{PC}}$, which must be done first. The coefficients (c_i values) are 1 and -1 for these two treatments because their difference is being tested. The experimental group ($\bar{X}_{\mathrm{E}}$) has a coefficient of zero because it is not involved in the test. This is a valid comparison ($\sum_{i=1}^{a} c_1 = 0$).

The test is simple: you construct a sum of squares (which is also a mean square because it has one degree of freedom) and compare it with the residual or within sample mean square in the analysis of variance. The general formula is:

$$\mathrm{MS}(L_k) = \frac{L_k^2 n}{\left(\sum_{i=1}^{a} c_i^2\right)} \tag{8.18}$$

In the case of the two contrasts identified above (Equations 8.17):

$$\mathrm{MS}(L_1) = \frac{L_1^2 n}{(1+1+0)} = \frac{(\bar{X}_{\mathrm{C}} - \bar{X}_{\mathrm{PC}})^2 n}{2} \tag{8.19}$$

$$\mathrm{MS}(L_2) = \frac{L_2^2 n}{(0.25+0.25+1)} = \frac{\left(\dfrac{\bar{X}_{\mathrm{C}} + \bar{X}_{\mathrm{PC}}}{2} - \bar{X}_{\mathrm{E}}\right)^2 n}{1.5} \tag{8.20}$$

For each of these, the appropriate test is the F-ratio of:

$$\frac{\mathrm{MS}(L_k)}{\mathrm{MS}_{\text{within samples}}}$$

with 1 and $a(n-1)$ degrees of freedom. Note that, in this example, if the first comparison is significant, the testing stops. If the control and procedural controls differ, the specified alternative to the null hypothesis cannot apply. A worked example is given in Table 8.5.

In theory, these comparisons or contrasts can all be done using a predetermined probability of Type I error, on the grounds that each comparison or contrast is an important, independent step in determining whether the specified alternative hypothesis is correct. Therefore, an error rate per comparison should be appropriate, because each comparison is an important step; see the discussion by Day & Quinn (1989).

Table 8.5. Examples of *a priori* multiple comparisons following analysis of variance (details of experiments are in Section 8.6.4)

(a) Experimental removal of predators ($n = 5$ replicates)

Treatment	Control (C)	Procedural control (PC)	Experimental (E)
Mean	27.8	33.6	55.8
s^2	58.8	77.2	64.0

Source of variation	Sum of squares	Degrees of freedom	Mean square	F-ratio	
Among treatments	2184.1	2	1092.1	16.4	$P < 0.001$
Within treatments	800.0	12	66.7		
Total	2984.1	14			

Comparison:

$L_1 = (\bar{X}_C - \bar{X}_{PC})^2 n/2 = 84.1 \quad F = 84.1/66.7 = 1.26, P > 0.05$

$L_2 = \left[\frac{(\bar{X}_C + \bar{X}_{PC})}{2} - \bar{X}_E\right]^2 n \Big/ 1.5 = 2100.3 \quad F = 2100.3/66.7 = 36.48, P < 0.001$

The hypothesis and model are supported and there is no evidence of artefacts due to the manipulation.

(b) Abundances of insects under different conditions of weather ($n = 6$ samples)

	1 Warm/Damp	2 Cool/Damp	3 Warm/Dry	4 Cool/Dry
Mean	9.50	13.16	19.33	10.00
s^2	0.56	0.64	1.01	0.91

Source of variation	Sum of squares	Degrees of freedom	Mean square	F-ratio	
Among weathers	368.07	3	122.69	157.29	$P < 0.001$
Within weathers	15.64	20	0.78		
Total	383.71	23			

Comparison: $L_1 = (19.33 - 9.50)^2/12 = 8.05 \quad F = 8.05/0.78 = 10.32, P < 0.017$

$L_2 = (19.33 - 13.16)^2/12 = 3.17 \quad F = 3.17/0.78 = 4.06, P > 0.017$

$L_3 = (19.33 - 10.00)^2/12 = 7.25 \quad F = 7.25/0.78 = 9.29, P < 0.017$

The hypothesis and model are wrong. There are differences among conditions, but mean abundance in cool, dry conditions is not greater than under other conditions.

This argument is not always the only issue. Often, means will have to be used several times to complete the pre-determined set of tests. Under these circumstances, several means may be used more than once, leading to non-independent and *non-orthogonal* tests. Two tests (L_k and L_l) are orthogonal in a balanced design if (using the terminology of Equations 8.15 to 8.17 above):

$$\sum_{i=1}^{a} c_{ik} \times c_{il} = 0 \tag{8.21}$$

where c_{ik} are the coefficients in contrast L_k and c_{il} are the coefficients for contrast L_l. In the case considered so far, the two contrasts *are* orthogonal:

$$\sum_{i=1}^{a} c_{i1} \times c_{i2} = (1 \times 0.5) + (-1 \times 0.5) + (0 \times -1) = 0 \tag{8.22}$$

In the second example in Section 8.6.4.1, the contrasts are:

$$\left.\begin{aligned} L_1 &= (-1 \times \bar{X}_1) + (0 \times \bar{X}_2) + (1 \times \bar{X}_3) + (0 \times \bar{X}_4) \\ L_2 &= (0 \times \bar{X}_1) - (1 \times \bar{X}_2) + (1 \times \bar{X}_3) + (0 \times \bar{X}_4) \\ L_3 &= (0 \times \bar{X}_1) + (0 \times \bar{X}_2) + (1 \times \bar{X}_3) + (-1 \times \bar{X}_4) \end{aligned}\right\} \tag{8.23}$$

from which:

$$\left.\begin{aligned} L_1, L_2 \sum_{i=1}^{a} c_{i1} \times c_{i2} &= (-1 \times 0) + (0 \times -1) + (1 \times 1) + (0 \times 0) = 1 \\ L_1, L_3 \sum_{i=1}^{a} c_{i1} \times c_{i3} &= (-1 \times 0) + (0 \times 0) + (1 \times 1) + (0 \times -1) = 1 \\ L_2, L_3 \sum_{i=1}^{a} c_{i2} \times c_{i3} &= (0 \times 0) + (-1 \times 0) + (1 \times 1) + (0 \times -1) = 1 \end{aligned}\right\} \tag{8.24}$$

and the tests are not orthogonal because $\bar{X}_3$ (representing warm, dry conditions) must be used in every test.

Summarizing the previous findings of several others, Day & Quinn (1989) recommended that the Dunn–Sidák procedure be used when a series of non-orthogonal comparisons is to be done. This procedure is described in more detail by Ury (1976) and in the textbook by Sokal & Rohlf (1981). What is needed is a procedure to adjust the probabilities of Type I error in each test, so that the rate of error for the experiment is maintained at the pre-set level (usually $P = 0.05$). If α is the chosen probability of Type I error, the Dunn–Sidák procedure uses:

$$\alpha' = 1 - (1 - \alpha)^{1/r} \tag{8.25}$$

where r is the number of non-orthogonal comparisons to be made.

So, in the two examples considered above, the first one only has two, orthogonal comparisons, each as an F-ratio with 1 and $a(n - 1)$ degrees of freedom. If $\alpha = 0.05$, each test is done using the tabulated value of F for $\alpha = 0.05$. This is illustrated in Table 8.5.

In the second example, there would be $r = 3$ tests (unless one happened to be non-significant). So, if α is chosen to be 0.05 for the experiment:

$$\alpha' = 1 - 0.95^{1/3} = 0.017 \tag{8.26}$$

Each test is done as an F-ratio with 1 and $a(n - 1)$ degrees of freedom, using $\alpha' = 0.017$ as the appropriate level of significance. This is also illustrated in Table 8.5.

Thus, *a priori* planned comparisons are straightforward. They have an excellent logical structure because they require careful thought about the appropriate hypotheses. They provide a consistent framework for evaluating the results of the analysis, including the order in which to proceed with the tests.

For further information about these procedures, see Day & Quinn's (1989) summary and the references therein, or consult the text by Rosenthal & Rosnow (1985). Don't get too obsessed about problems of excessive Type I error. In the first example, there were only two contrasts. In the second, there were three, but there is already evidence that at least two of the means differ (because the analysis of variance was significant, which is why you are making these comparisons). Excessive Type I error in this example would cause you erroneously to find your hypothesis to be true (errors would cause you to find the expected differences when they are not present). This would lead you back to your model to refine it (see Chapter 2) thus, perhaps, wasting some time and energy but at least causing you to probe further. The problems of excessive Type I error will not, under these circumstances, be too great.

In contrast, in the second case, each comparison now has a reduced power compared with what it would be were $\alpha = 0.05$ as originally chosen. So, staving off excessive Type I error for the whole experiment may lead to excessive Type II error for any single comparison (and therefore for the whole experiment, which requires all three tests to be significant for the alternative hypothesis, as defined, to be considered correct). Commonsense dictates that being careful requires thought and not just slavish acceptance of some adjustment to α.

8.6.5 **A posteriori** *comparisons*

8.6.5.1 What do they do?

A posteriori comparisons are much more common than *a priori* ones in biological studies, because hypotheses are usually preliminary. For example, some model leads to the hypothesis that mean abundance of an organism differs seasonally, but does not specify which seasons may differ. To examine this, all seasons are compared once it has been shown that there are differences (i.e. the analysis of variance was significant). *A posteriori* comparisons are tests of all possible pairs of means, in a chosen order, so that, if it can be, the alternative to the null hypothesis will be defined. For example, if there are mean abundances in four seasons and the null hypothesis that they are all equal has been rejected, each mean must now be compared with every other one. Suppose the six possible comparisons are made and the outcome is interpreted as: $\bar{X}_{\text{spring}} > \bar{X}_{\text{summer}}$, $\bar{X}_{\text{spring}} > \bar{X}_{\text{autumn}}$, $\bar{X}_{\text{spring}} > \bar{X}_{\text{winter}}$, $\bar{X}_{\text{summer}} > \bar{X}_{\text{autumn}}$, $\bar{X}_{\text{summer}} > \bar{X}_{\text{winter}}$; $\bar{X}_{\text{autumn}}$, $\bar{X}_{\text{winter}}$ do not differ. This series of paired comparisons leads, uniquely, to the following alternative to the null hypothesis:

$$\bar{X}_{\text{spring}} > \bar{X}_{\text{summer}} > \bar{X}_{\text{autumn}} = \bar{X}_{\text{winter}} \tag{8.27}$$

It is precisely this sort of alternative that *a posteriori* multiple comparison procedures are designed to determine. There are many possibilities, but the procedures attempt to identify which actually applies.

The problem of excessive Type I errors is much more acute where all means are compared, rather than a pre-determined subset, as an *a priori* series of comparisons. This is because, usually, more tests are done than in *a priori* defined comparisons. So, all procedures described have been attempts to control the rate of error per experiment, by reducing the error rate per comparison. Also, there has to be some procedure to decide in what order to make comparisons. This has resulted in *sequential* procedures being favoured, whereby a series of tests is done in an orderly manner, proceeding to the next test in the light of what has been concluded so far.

It turns out that an unambiguous definition of an alternative to the null hypothesis can be identified only if the following conditions apply to the outcome of the comparisons. Means of all treatments are in groups (including groups containing only one mean) such that there are:

1. No differences among the means within a group.
2. Every mean in one group differs from all means in any other group.

Otherwise, no clear-cut result has been obtained and no alternative to the null hypothesis has been identified. This point will recur later.

8.6.5.2 Student–Newman–Keuls (SNK) procedure

To illustrate a sequential procedure, the Student–Newman–Keuls (SNK) test is used here. This is a sequential extension of *t*-tests (Newman, 1939; Keuls, 1952). It has, under some circumstances, excessive Type I error per experiment (Einot & Gabriel, 1975; Snedecor & Cochran, 1989). How to deal with this is discussed later (see Ryan, 1960; Ramsey, 1978; Day & Quinn, 1989).

The example is about observations that densities of a particular species of plant vary across shaded and open woodland, open fields, hedgerows and marsh. The model to explain this is that observed variations are due to different availabilities of resources from one habitat to another, leading to different carrying capacities. The hypothesis being examined is that there are consistent differences in densities among the five habitats where the species is found (habitats are therefore a fixed factor). In each habitat, the density of plants was sampled with $n = 6$ representative quadrats scattered across the countryside. The first hypothesis to be examined is that there are different abundances among the different habitats. The means and analysis are in Table 8.6.

The first step is to arrange the means in ascending order. This provides the sequence in which to test them in pairs. The logic is that the most likely two means to be different are those most apart in ascending order. We can already conclude that these two differ because the analysis of variance tells us that at least two means differ and these are the most different. The next most likely differences must be in those means most different, except for the two most extreme ones (i.e. means 2 to 5 and 1 to 4 in the ranked, ascending order). Thus, the sequence for testing is 5 to 1, then 5 to 2, 4 to 1; then 5 to 3, 4 to 2 and 3 to 1. Finally, adjacent pairs are tested (5 to 4, 4 to 3, 3 to 2 and 2 to 1). This is the *sequence* of the test.

In each set of comparisons, what is needed is a standard error for the difference between two means. Here, because only balanced analyses (with n the same for all treatments) are considered, this is simply calculated. The standard error for each sample mean is the square-root of the sample variance divided by the size of sample ($SE = \sqrt{(s^2/n)}$, see

Table 8.6. Numerical calculation of Student–Newman–Keuls' and Ryan's multiple comparisons. Data are densities of plants in five habitats ($n = 7$ quadrats in each habitat). Note that, in this example, Ryan's precautions are unnecessary (there are only two groups of means); the test is provided solely for illustration

(a) Analysis of variance

Habitat	Shaded woodland	Open woodland	Open fields	Hedgerows	Marsh
Mean	6.4	7.1	3.8	2.6	4.1
s^2	2.18	0.67	1.77	1.35	2.06

Source of variation	Sum of squares	Degrees of freedom	Mean square	F-ratio	
Among habitats	99.26	4	24.82	15.38	$P < 0.01$
Within habitats	48.18	30	1.60		
Total	147.44	34			

Standard error for means = $\sqrt{(1.60/7)} = 0.48$

(b) Multiple comparisons. For SNK test, P is nominally 0.05. For Ryan's test, P is as calculated in Equation 8.30. Q values for Ryan's test were calculated from tabulated values of $Q_{0.05}$ and $Q_{0.01}$ (with g and 30 degrees of freedom). Any difference between two means larger than the relevant product $D = Q \times \text{SE}$ is significant at the specified probability (nominally 0.05 for SNK and 0.05 for Ryan's test), see Equation 8.29 and Section 8.6.5.4. Significance is denoted by an asterisk

						SNK test			Ryan's test		
Rank order	1	2	3	4	5						
Ranked means	2.6	3.8	4.1	6.4	7.1	g	Q	$D = Q \times \text{SE}$	p	Q_p	$D = Q_p \times \text{SE}$
Comparisons	$^{5-1}4.5^*$					5	2.89	1.97	0.05	4.63	2.22
	$^{4-1}3.8^*$	$^{5-2}3.3^*$				4	3.49	1.84	0.04	4.14	1.99
	$^{3-1}1.5$	$^{4-2}2.6^*$	$^{5-3}3.0^*$			3	3.84	1.68	0.03	3.62	1.74
	$2-1^a$	$3-2^a$	$^{4-3}2.3^*$	$^{5-4}0.7$		2	4.10	1.39	0.02	2.89	1.39

[a] Tests not done because 3–1 was not significant.

Conclusion: Mean density in shaded and open woodland is similar, but greater than in other habitats, which are also different from each other.

Section 5.2). Here, however, we have already concluded that the variances of the sampled populations are the same (it is one of the tested assumptions before we did the analysis of variance; see Section 7.16). So, we have a common estimate of variance, pooled from all samples, in the mean square among treatments. Thus:

$$\text{Standard error} = \sqrt{\left(\frac{\text{Mean square within samples}}{n}\right)} \tag{8.28}$$

with $a(n-1)$ degrees of freedom, i.e. those associated with this mean square (Table 8.7).

For each pair to be compared, we need to calculate a test statistic, analogous to that used for t-tests to compare means of two samples (as in Chapter 6). The test statistic is:

$$Q_{ij} = \frac{\bar{X}_i - \bar{X}_j}{\text{SE}} \tag{8.29}$$

where $\bar{X}_i$ and $\bar{X}_j$ are the two means being compared and SE is the standard error defined above (Equation 8.28). The distribution of Q is tabulated when the null hypothesis is true and there is no difference between two means (see e.g. Snedecor & Cochran, 1989; Winer *et al.*, 1991) and its distribution depends on the range of number of means across which any two are compared and the degrees of freedom for the standard error (i.e. $a(n-1)$, see above). The range of number of means is five ($g = 5$ in Table 8.6b) for the first comparison (smallest to largest means), four ($g = 4$ in Table 8.6b) for the next step (second smallest to largest; smallest to second largest) and so on, as in Table 8.6b. It is, however, more convenient to use the observed differences between the means $(\bar{X}_i - \bar{X}_j)$ directly. So, the difference between each pair of means $(\bar{X}_i - \bar{X}_j)$ is compared to $D = Q \times \text{SE}$, as in Table 8.6. Now, if $(\bar{X}_i - \bar{X}_j) > D$, the difference is considered to be significant.

So, the tests proceed sequentially from the furthest apart to the closest. At each step, the calculated differences between pairs of means are compared with D to evaluate them for significance. Where means do not differ, testing stops. So, in Table 8.6b, because mean 3 is not significantly different from mean 1, the comparisons of means 1 and 2, and means 2 and 3 are not done. There is already evidence that a single group of means exists that includes the smallest three (1 and 3 do not differ). Nevertheless, further tests are needed to determine whether the largest two means (4 and 5) differ, so the final row of tests is completed for this comparison.

Table 8.7. Illustration of Student–Newman–Keuls' multiple comparison producing an illogical result. Means are from six experimental groups, with $n = 6$ replicates

(a) Data and analysis of variance

Treatment	1	2	3	4	5	6
Mean	59.8	52.7	66.7	73.4	48.2	50.3

Source of variation	Sum of squares	Degrees of freedom	Mean square	F-ratio	
Among treatments	2987.45	5	597.49	29.42	$P < 0.001$
Within treatments	609.30	30	20.31		
Total	3596.75	35			

Standard error for means = $\sqrt{(20.31/6)} = 1.84$

(b) SNK test (note that no correction for excessive Type I error is needed – no differences are considered different!). Apparently significant differences at nominal $P = 0.05$ are shown as asterisks. Q values are with g and 30 degrees of freedom

Rank order	1	2	3	4	5	6			
Ranked means	48.2	50.3	52.7	59.8	66.7	73.4	g	Q	$D = Q \times \text{SE}$
Comparisons	$^{6-1}$25.2*						6	4.30	11.27
	$^{5-1}$18.5*	$^{6-2}$23.1*					5	4.10	10.75
	$^{4-1}$11.6*	$^{5-2}$16.4*	$^{6-3}$20.7*				4	3.84	10.06
	$^{3-1}$4.5	$^{4-2}$9.5*	$^{5-3}$14.0*	$^{6-4}$13.6*			3	3.49	9.15
			$^{4-3}$7.1	$^{5-4}$6.9	$^{6-5}$6.7		2	2.89	7.57

Conclusion: Although treatment 4 (rank 6) has a larger mean than treatment 5 (rank 1), it is impossible to determine consistent groupings. Horizontal lines underline treatments that do not differ:

Rank	1	2	3	4	5	6
Treatment	5	6	2	1	3	4

Although commonly described, this *cannot* present a logical alternative to the null hypothesis!

8.6.5.3 *Interpretation of the tests*

The procedure is straightforward; the rules are simple. In this case, there is a clear interpretation and an obvious alternative to the null hypothesis has been identified. This will not always happen. The multiple comparisons are not as powerful as the original F-test in the analysis of variance. So, there are going to be cases where the analysis of variance causes rejection of the null hypothesis, but the multiple comparisons do not allow identification of a valid alternative to it. This is not at all unreasonable. It is a far simpler matter, requiring far less information, to determine that a null hypothesis is false and therefore one of a set of many possible alternatives applies than it is to identify which particular alternative applies. Thus, sometimes the multiple comparisons do not show any differences – even between the most extremely different pair of means. Commonsense and logic tell us that these two means *must* differ, but that is already known from the analysis of variance. The multiple comparison (SNK) procedure is needed for a much more precise result than that.

The second failure to identify an alternative to the null hypothesis occurs when contradictions are caused by the multiple comparisons. This has led to all sorts of irrational conclusions because the purpose of the procedure described in detail in Section 8.6.5.1 has not been remembered.

Suppose that instead of a clear-cut pattern (as in the example in Table 8.6), you get the results in Table 8.7. There is no doubt that means 1 to 3 are smaller than means 5 and 6. Nor is there doubt that means 1 to 3 are similar and means 5 and 6 are similar. The problem is mean 4. It is different from the group containing 1, 2 and 3 because it differs from 1 and 2. It does not, however, differ from 3. At the same time, it cannot be included in a group with 6, but is not different from 5. The result is not interpretable (as in Table 8.7). The procedure has *failed* to provide an alternative to the null hypothesis.

What can be concluded is that there are differences, but the precise alternative hypothesis has not been identified. More experimentation is needed with increased sizes of samples because increasing n decreases the standard errors and decreases the sizes of critical values of Q because degrees of freedom are $a(n-1)$ for the standard error. Or, the results provide clues as to what may be happening, so that more refined or precise hypotheses can be proposed for subsequent experiments. The pattern of rank order of the means will provide the new observations for new hypotheses and experiments. In particular, these might now have *a priori* determined statements about the alternative hypothesis and therefore more powerful multiple comparisons to detect it (as in Section 8.6.4).

8.6.5.4 Controlling Type I errors: Ryan's procedure

One very popular multiple comparison is Duncan's (1955) multiple-range test. This test is almost certainly the worst procedure to use and it has received severe criticism since its first appearance (Scheffé, 1959), particularly on its mathematical basis. In simulations, its control of Type I error is the least satisfactory of many procedures. There is no convincing case for its favoured status amongst biologists.

The other most widely used procedure in ecology is the Student–Newman–Keuls (SNK) test described above. As pointed out not long after it was first described, the sequential studentized range test, or SNK procedure, has excessive probability of Type I error under some circumstances (Ryan, 1960; Einot & Gabriel, 1975; Ramsey, 1978; Snedecor & Cochran, 1989).

The situation for SNK tests is worst when means of treatments fall into groups such that the groups are spaced widely apart. Then, differences between pairs of means within the groups will tend to be detected as significantly different when they are not much more often than should be the case at the specified probability of Type I error (Day & Quinn, 1989; Snedecor & Cochran, 1989).

Ryan (1960) proposed to remedy this by adjusting the probability of Type I error at each step down in the range of means. The suggestion was pursued further by Einot & Gabriel (1975) and Ramsey (1978). The adjustment recommended by Day & Quinn (1989) is Einot & Gabriel's (1975) version, rather than Ryan's (1960) or Ramsey's (1978). The procedure is to modify α according to how many means are in the group being compared. So, if there are a means in the whole experiment, the modified probability of Type I error is:

$$\alpha' = 1 - (1 - \alpha)^{g/a} \tag{8.30}$$

where g is the number of means across which two are being compared. When the largest and smallest means are compared $g = a$ (i.e. the two means are compared across the whole set of a means in the experiment). When the second largest mean is compared to the smallest and the largest is compared to the second smallest, $g = a - 1$. This contains the rate of error for the whole experiment, as shown in Table 8.7.

There are several ways to obtain the critical values of Q for these modified probabilities (the tabled values are at $P = 0.05$ or 0.01 and do not provide critical values for other probabilities). You can use the algorithm given by Lund & Lund (1983) or, for some sizes of experiment, the formula for interpolation given by Day & Quinn (1989).

What must also be taken into account is the effect of reducing the probability of Type I errors on the power of the tests. Practical experience with thousands of analyses of variance compels me to the view that excessive Type I error is not the biggest problem in multiple comparisons of biological data. The larger issue is the inability of tests to identify the appropriate alternative to a null hypothesis. The tests often fail – which is entirely attributable to lack of power. In the first 100 analyses I saw or did after I first read a draft of the article by Day & Quinn (1989) (in the period mid-1987 to late 1988), I compared the outcome of SNK and Ryan's tests. In only one analysis was there a difference that could easily be identified as due to Type I error being more likely in SNK tests. In 80 of the analyses, there was no identifiable alternative hypothesis.

So, the first consideration for a general procedure should be to maximize power. Ryan's test, as modified by Einot & Gabriel (1975), correspondingly has smaller power than SNK tests. The first choice should be SNK tests. If the means fall into groups that are widely spaced, results of SNK tests start to become suspect. There have to be at least five means and at least three groups before Type I errors in comparisons within groups can increase. If this seems to be the pattern (widely spaced groups), you should use Ryan's test (see Day & Quinn, 1989, and as described above). If there is a difference between the two procedures, so that more differences are encountered among pairs of means within larger groups when using SNK tests, make your judgement based on the Ryan's rather than the SNK tests.

As a final point, multiple comparisons are often done when they are irrelevant. If, for example, the experiment is to test an hypothesis that rate of predation on mites differs at different distances from the edge of a field, then rate is measured in replicate plots at specified distances. The relationship should then best be analysed as a regression of rate of predation (Y) against distance (X) (or by polynomial regressions as in Section 13.7). There is no need for means at adjacent distances to be different (as is the pattern sought by multiple comparisons). The relationship between predation and distance is what needs to be examined. It is not at all uncommon for multiple comparisons to be used under such inappropriate circumstances (see Dawkins, 1981). Adjacent distances or any adjacent values of X in any regression do not have to differ, even if the relationship of some variable with X is significant. The logical basis of multiple comparisons must not be lost in the rush to use them.

9

Nested analyses of variance

9.1 Introduction and need

As made abundantly clear earlier, replication of experimental and sampling units is mandatory to maintain any logical basis for making inferences from experiments. Without replication, there is no way to demonstrate that statistically significant differences among experimental treatments are due to the applied experimental treatments and not simply due to chance variation among the units measured. Replication provides the material from which statistical estimates can be made to provide measures of the intrinsic variation among units.

To obtain replication it is necessary that the appropriate choice and scale of experimental units are used. Underwood (1981) and Hurlbert (1984) found a distressingly large proportion of published studies had no or the wrong replicates, even though inferences were made from quantitative data and statistical tests.

One of the major uses of nested or hierarchical designs of experiments is to ensure appropriate replication. A simple example will illustrate the point. Suppose plants in open fields have been observed to vary in their production of fruit. A model to explain this has, so far, successfully withstood various experimental tests of its predictions. It is that some areas are visited by more insect pollinators and therefore the plants are more fertile. Such a pattern is illustrated in Figure 9.6, below. From this, the hypothesis is proposed that preventing insects from visiting plants in experimental plots will reduce the reproductive success of plants in those plots. A mesh cage is placed over an experimental plot. A similar cage with a few openings to admit insects is established somewhere else as a procedural control and a third area is marked out, but otherwise untouched. After the following season, n individual plants (say $n = 10$) in each plot are examined and the number or proportion of fertile flowers recorded for each plot. Thus, there are numerous replicate plants. The data can be examined using a one-factor analysis of variance ($a = 3$ treatments, $n = 10$ replicates in each treatment) to test the null hypothesis of no difference among mean fertility in each plot.

Suppose the outcome is a significantly large value of F and rejection of the null hypothesis. Is this evidence to support the hypothesis and model? Is this a demonstration that the differences are due to the experimental treatments? Of course not. There are likely to be differences from one place to another in the average fertility of the plants – this was the original observation! There are numerous differences between areas of a study-site that might affect fertility of the plants. There are differences in water, nutrients, past history of disturbance of the soil, spatially variable arrivals of pests and parasites or accidents. There is no logical basis for concluding that the differences found are due to the pollinators. The differences among areas due to experimental treatments are completely *confounded* by any other differences (see also Section 6.2).

Alternatively, the outcome of the experiment could be the opposite one: no differences among the treatments. Again, this is not useful evidence to conclude that the pollinators are irrelevant to the variations in fertility and thus to reject the hypothesis and model. Intrinsic differences among plots, not due to the number of insects visiting each area may cause fertility to be greater in the plot that happened to be chosen for excluding the insects. So, there could be an apparent effect, even the predicted effect, of the pollinators. As a result, they increase the fertility of plants in the two plots that actually had smaller fertility of plants, so that the intrinsic differences are cancelled out.

What is required is replication of the experimental plots. There must be adequate replication of the plots to allow an estimate of the intrinsic variability among areas that has nothing to do with the differences that might be due to the experimental treatments. The *plots*, not the *plants*, are the experimental units to which the treatments have been applied. The *plots*, not just the *plants*, must be replicated to solve the problem of confounding.

A nested experimental design is one that uses replication of experimental units in at least two levels of a hierarchy. In the above example, these are plots and plants. The plots belong uniquely to one of the treatments (which contrasts these designs with those that are orthogonal or factorial as described in Chapter 10). This defines nesting – one set of experimental treatments has different levels or representations in each of the levels of another treatment. The replicated plots are levels of the factor 'plots' and are nested in the different levels of the manipulation of insects. The plants are also nested, because each individual plant can belong to only one of the plots. These designs are widespread in biology and are extensions of the one factor experiment considered earlier.

9.2 Hurlbert's 'pseudoreplication'

Hurlbert (1984) reviewed a range of different types of ecological experiment to discuss their shortcomings in terms of inadequate (or no) replication. Anyone interested in recognizing mistakes in the design and interpretation of experiments should read his paper.

The neologism 'pseudoreplication' was coined by Hurlbert (1984) to highlight the problem that investigators *had* replicates, but not of the appropriate units (as is the case of the plants and plots in the previous section). The older term 'confounding' is still a good one. It is probably a better term because it draws attention to what is needed. It is not replication as such that is the problem. The difficulty is to separate out the differences among treatments that are due to the experimental factor from any differences due to other factors. The logic of the experiment requires this, so that any differences found can be unambiguously attributed to the process proposed in the model. Other differences confound such conclusions, making logical interpretation impossible.

9.3 Partitioning of the data

Data collected in a nested or hierarchical design are replicated samples (each of size n) from the replicated plots or aquaria, or whatever, sample units (with b replicated plots for each experimental treatment). In the case of the experiment on pollination, there are three treatments (exclusion, procedural control, control). Suppose each is set up in five separate, randomly scattered, independent plots ($b = 5$). In each unit, there are numerous plants and six of them are randomly sampled to determine percentage fertility ($n = 6$).

To ensure proper allocation of replicates to experimental treatments, the 15 plots are scattered at random throughout the study-area (as in Figure 9.1) so that they are independently intermingled (or interspersed to use a term often used by statisticians). Each set of five plots then should equally and unbiasedly estimate the variation in fertility that occurs from one spot to another across the site.

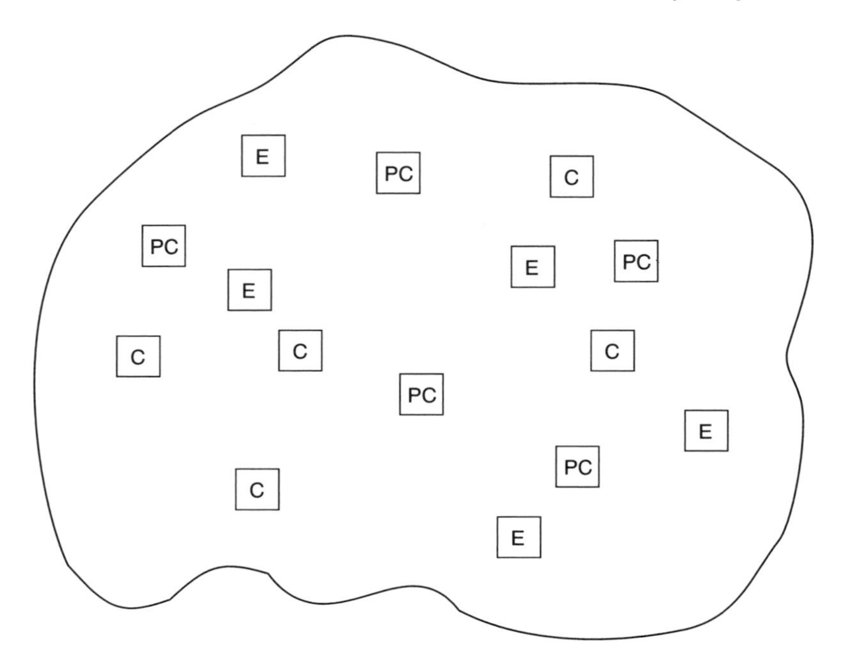

Figure 9.1. A nested experimental design. Five plots of each of three experimental treatments are established. In each plot $n = 6$ plants are sampled. Plots are scattered at random over the study-site: E, experimental; PC, procedural control; C, untouched control.

For example, suppose that every now and then a fungus attacks a patch of plants, altering their fertility. Suppose about one in three plots is attacked by fungus. The experiment could be expected to have about one in three plots affected by fungus. If the experiment examined only one plot in each treatment it would therefore, on average, demonstrate a difference among treatments, even if the actual proposed mechanism (pollination by insects) was irrelevant. By having replicated plots, it is a reasonable proposition that fungal attacks (or any other unplanned or unknown process) will, on average, affect a similar number of plots in each treatment. Therefore, no difference due to fungal attacks would show up as a difference among treatments.

Data from the experiment would then be as in Table 9.1. In general terms, the data conform to a matrix, as in Table 9.2. This has a total amount of variability, measured exactly as in the one-factor case (Equation 7.12) by the sum of squared deviations of all data from the overall mean of the entire set. The only difference from the earlier case

Table 9.1. Data from an experimental analysis of the effects of insects on the fecundity of plants. There are three treatments (control, cages with openings, cages that exclude insects; $a = 3$), each replicated in $b = 5$ plots. In each plot, $n = 6$ plants are sampled and the percentage of fertile flowers recorded. Data are X_{ijk} where i is the treatment, $j(i)$ is the plot in treatment i and k is the plant in plot j in treatment i. Means are described in Section 9.3. The data are analysed in Table 9.4

	Treatment (i)														
	1					2					3				
	Control					Cages with openings					Cages				
Plot (j)	1	2	3	4	5	1	2	3	4	5	1	2	3	4	5
Replicate (k)															
1	82	79	90	75	38	92	62	67	95	70	74	47	60	43	47
2	67	84	100	93	64	80	97	64	93	62	76	71	88	53	44
3	73	70	65	99	80	83	63	85	100	77	72	54	86	48	16
4	70	71	99	95	74	77	77	83	80	80	71	56	84	79	43
5	83	67	84	92	87	52	88	79	83	71	60	77	45	70	49
6	95	80	63	95	79	73	77	88	76	87	74	66	48	45	55
Mean($\bar{X}_{j(i)}$)	78.3	75.2	83.5	91.5	70.3	76.2	77.3	77.7	87.8	74.5	71.2	61.8	68.5	56.3	42.3
Variance	108.0	45.4	264.0	71.1	309.0	181.0	188.0	98.3	90.2	76.3	33.0	129.0	394.0	217.0	185.0
Mean($\bar{X}_i$)			79.8					78.7					60.0		
Mean ($\bar{X}$)								72.8							

Table 9.2. Matrix of data from a nested analysis of variance with a experimental treatments $(1 \ldots i \ldots a)$ each replicated in b experimental units $(1 \ldots j \ldots b$ in each $i)$ and sampled with n replicates $(1 \ldots k \ldots n$ in each $j(i))$

Treatment			1			…			i			…			a		
Unit	1	…	j	…	b		1	…	j	…	b		1	…	j	…	b
	X_{111}		X_{1j1}		X_{1b1}				X_{ij1}								X_{ab1}
	X_{112}		X_{1j2}		X_{1b2}				X_{ij2}								X_{ab2}
	⋮		⋮		⋮				⋮								⋮
	X_{11k}		X_{1jk}		X_{1bk}				X_{ijk}								X_{abk}
	⋮		⋮		⋮				⋮								⋮
	X_{11n}		X_{1jn}		X_{1bn}				X_{ijn}								X_{abn}
Sample mean	$\bar{X}_{1(1)}$		$\bar{X}_{j(1)}$		$\bar{X}_{b(1)}$			…	$\bar{X}_{j(i)}$	…			…				$\bar{X}_{b(a)}$
Treatment mean			$\bar{X}_1$						$\bar{X}_i$						$\bar{X}_a$		

$$\bar{X}_{j(i)} = \sum_{k=1}^{n} X_{ijk}/n$$

$$\bar{X}_i = \sum_{j=1}^{b}\sum_{k=1}^{n} X_{ijk}/bn = \sum_{j=1}^{b} \bar{X}_{j(i)}/b$$

$$\bar{X} = \sum_{i=1}^{a}\sum_{j=1}^{b}\sum_{k=1}^{n} X_{ijk}/abn = \sum_{i=1}^{a} \bar{X}_i/a = \sum_{i=1}^{a}\sum_{j=1}^{b} \bar{X}_{j(i)}/ab$$

(Table 7.3) is that variability must be added up over all n replicates, all b plots and all a treatments, i.e. over three levels of variability.

The total variability can be partitioned into three components:

$$\text{Total sum of squares} = \underbrace{\sum_{i=1}^{a}\sum_{j=1}^{b}\sum_{k=1}^{n}(X_{ijk} - \bar{X}_{j(i)})^2}_{\text{(i)}} + \underbrace{\sum_{i=1}^{a}\sum_{j=1}^{b}\sum_{k=1}^{n}(\bar{X}_{j(i)} - \bar{X}_i)^2}_{\text{(ii)}} + \underbrace{\sum_{i=1}^{a}\sum_{j=1}^{b}\sum_{k=1}^{n}(\bar{X}_i - \bar{X})^2}_{\text{(iii)}} \quad (9.1)$$

These three terms are, without much difficulty, identifiable in terms of the data, as follows:

(i) Variation among replicates (X_{ijk}) in a sample, as deviations from the sample mean ($\bar{X}_{j(i)}$) for a given unit (j) in a particular experimental treatment (i). This is variability within samples and is exactly equivalent to the measure of variation within samples in the one-factor case, except there are now b samples (one for each unit) in each of the a treatments.

(ii) Variation among units within each treatment. The mean of each of the b units in a given treatment (i) is $\bar{X}_{j(i)}$. This measures how different they are from the mean of the treatment $\bar{X}_i$. If units do not differ from one another, the mean of any variable measured in them should be the same, except for sampling error. So, in each treatment, this term measures the variability among units within experimental treatments.

(iii) Variation among experimental treatments, exactly as in the one-factor case, except that the mean value in each treatment is estimated from n replicates in each of b units, rather than from a single sample. This is very similar to the outcome in the one-factor case, except that the middle term has intruded to measure the variability among experimental units imposed on the design to avoid confounding among the experimental treatments.

The data in the example in Table 9.1 are analysed using these formulae (Tables 9.3 and 9.4). The actual calculations are according to the 'machine formulae' – simpler, but algebraically identical forms that do not introduce rounding errors.

9.4 The linear model

As with the single-factor case (Equation 7.7), a linear model can be used to describe the data.

$$X_{ijk} = \mu + A_i + B_{j(i)} + e_{ijk} \tag{9.2}$$

where X_{ijk} is any replicate (k) in any replicate plot (j) in a given experimental treatment (i).

As before, μ is the overall mean of all populations being sampled. In the case here, μ is the average fertility of all populations of plants in all the plots and is estimated from all 15 plots in the field. As before, A_i measures how different the mean of one experimental treatment is from the overall mean, if the null hypothesis is false. If the null hypothesis is true and there are no differences among the means of the experimental treatments, all the A_i terms are zero.

Thus:

$$\text{H}_0\text{:} \quad \mu_1 = \mu_2 \ldots = \mu_i \ldots = \mu_a (= \mu) \tag{9.3}$$

is equivalent to:

$$\text{H}_0\text{:} \quad A_1 = A_2 \ldots = A_i \ldots = A_a (= 0) \tag{9.4}$$

The $B_{j(i)}$ terms represent the differences in means of populations in each of the j replicate units nested in the i experimental treatments. Thus, if the plants in the experimental plots *do* have different mean fertilities from plot to plot, because of genetic, historical, geographical and any other differences in their ecology, the $B_{j(i)}$ terms indicate these differences. In contrast, if there are no differences among the means in the replicate units, the $B_{j(i)}$ terms are zero.

The final, e_{ijk} term is exactly as before. It represents the individual variability in fertility from one plant (k) to another in a particular plot.

The relationships between A_i, $B_{j(i)}$ and e_{ijk} are illustrated in Figures 9.2 and 9.3. This model is an obvious and non-complicating extension of the one used for the single-factor analysis. The analysis is a *nested* one because the b replicate units are nested (or completely contained) in a particular treatment (i). This is, of course, exactly what occurred in the single-factor case, where the n replicates are nested in a particular treatment.

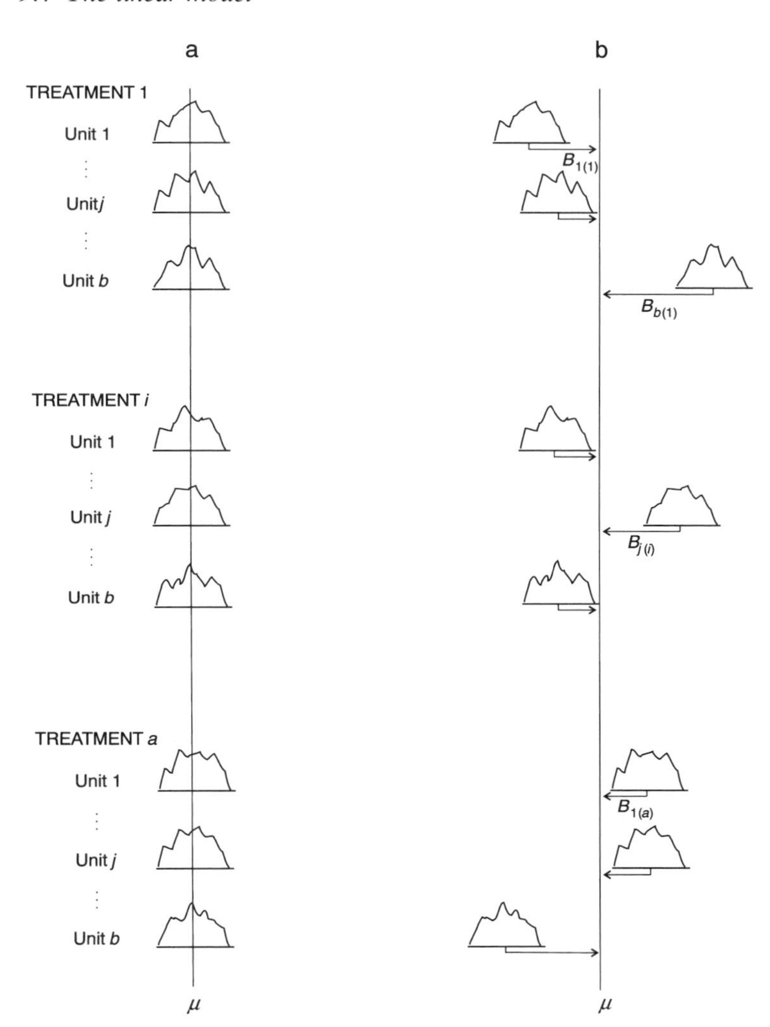

Figure 9.2. Graphical illustration of the terms in the linear model for a nested analysis of variance. Two possible outcomes of an experiment are identified (see Figure 9.3 for the others). There are a treatments, each replicated in b experimental units (plots, tanks, patches, etc.). In each unit, there is a population of possible measures with mean $\mu_{j(i)}$, which will be sampled by n replicates in the experiment. (a), (b) The null hypothesis is true, the means of all populations in each experimental treatment (μ_i values) are equal to μ. In (a), there are no differences among the means of the units in each treatment ($\mu_{j(i)}$ values all equal μ_i). In (b), the units in each treatment differ. See the discussion in the text.

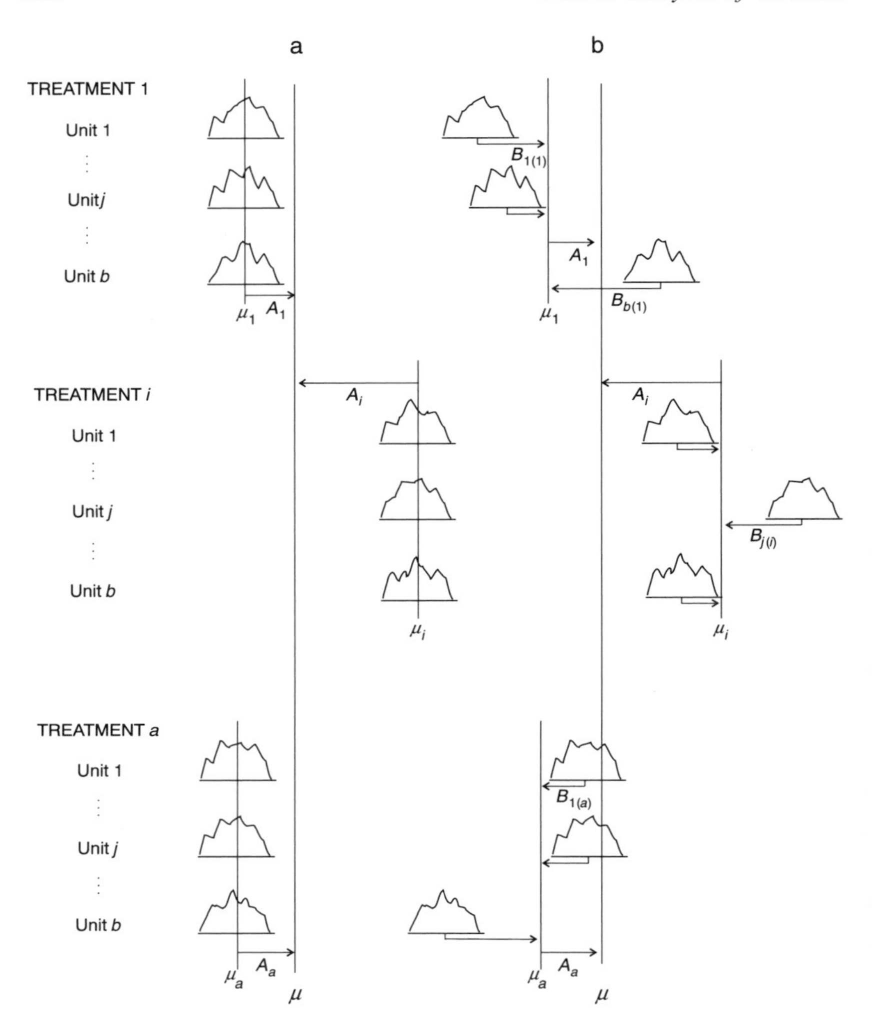

Figure 9.3. Graphical illustration of the terms in the linear model for a nested analysis of variance. Two possible outcomes of an experiment are identified (see Figure 9.2 for the others). There are a treatments, each replicated in b experimental units (plots, tanks, patches, etc.). In each unit, there is a population of possible measures with mean $\mu_{j(i)}$, which will be sampled by n replicates in the experiment. (a), (b) The null hypothesis is false, the means of populations in each experimental treatment (μ_i values) are not all equal to μ. In (a), there are no differences among the means of the units in each treatment ($\mu_{j(i)}$ values all equal μ_i). In (b), the units in each treatment differ. See the discussion in the text.

To be independent of one another (Section 7.14), the replicates must be properly nested. There must be no possibility that a specific replicate occurs again elsewhere in the experiment (in another treatment).

The experimental design is hierarchical because it has a hierarchy of two tiers of replication. The entities investigated (about which the hypotheses are proposed) are the replicates (in this case, the individual plants). Variability among them must be estimated to allow valid comparisons among the treatments. Unfortunately, the arrangement of experimental units causes the plants to be in geographically separated groups. There is no way of avoiding confounding of separate units (i.e. the experimental units), unless they are replicated in each treatment. Thus, the *experiment* forces the inclusion of a second scale of replication.

As indicated in Figures 9.2 and 9.3, there is a logical basis for proceeding with an analysis to test for differences among the means of the experimental treatments. It doesn't matter whether or not there are differences among the replicated units. In either case, there is a separate, independent, *unconfounded* measure of the magnitude of the A_i terms.

In any experimental treatment (the ith treatment), if there is variation among the replicate units, the means of whatever is being measured will not be identical in the different units. As shown in Figure 9.2b, the means differ by amounts easily identified as $B_{j(i)}$ terms, representing the difference of unit j (in treatment i) from the overall mean in treatment i.

As in the single-factor case (Figure 7.2), the experimental treatments will have different means if the null hypothesis is false. The differences are the amounts specified by A_i terms in the linear model.

There are therefore four possible scenarios:

1. The null hypothesis is true and all the treatments have the same mean (μ) and all A_i values are zero. There is no variability among the means in the replicated units; all units in each treatment have the same mean (and all $B_{j(i)}$ terms are zero for every treatment). This is illustrated in Figure 9.2a.
2. The null hypothesis is true (as in (i) above), but there is variation among the means of the experimental units ($B_{j(i)}$ terms are *not* all zero). This is illustrated in Figure 9.2b.
3. There is no variation among the means of the replicated units in each treatment (as in (1) above) but the null hypothesis is not true and the means of the treatments (averaged over all units in each treatment) are not equal. This is illustrated in Figure 9.3a.

4. Finally, there may be variability among the units in each treatment (as in (2) above), but the null hypothesis is also not true (as in (3) above). This is illustrated in Figure 9.3b.

Provided that the A_i values can be estimated independently of the $B_{j(i)}$ values, as shown diagrammatically in Figure 9.2, it makes no difference to testing the null hypothesis. The situation where the null hypothesis is false can be distinguished from that where the null hypothesis is true, regardless of whether there is or is not variation among the replicate units. Thus, in principle, Figure 9.3a can be distinguished from Figure 9.2a and Figure 9.3b can be distinguished from Figure 9.2b. The nested design successfully separates, or unconfounds, A_i terms from $B_{j(i)}$ terms.

9.5 Degrees of freedom and mean squares

The analysis of variance of the nested model is completed, as for the previous case (Sections 7.6 and 7.7), by determining the degrees of freedom associated with each sum of squares and then calculating the mean squares (Tables 9.3 and 9.4). The degrees of freedom are (absolutely as before) calculated using the operational rule in Section 3.10.

So, in Table 9.3, the degrees of freedom for the total sum of squares is $(abn - 1)$. There are *abn* data and one thing – their mean ($\bar{X}$) – must be known to calculate the sum of squares. For the within samples source of variation, there are *abn* data, but the mean for each replicate unit (each $\bar{X}_{j(i)}$) must be known. There are *ab* of these, *b* units in each of *a* treatments. So, there are $abn - ab = ab(n - 1)$ degrees of freedom. For the variability among replicate units in each treatment, there are *b* units in a given treatment. The mean ($\bar{X}_i$, averaged over all units in that treatment) must be known. Hence, there are $(b - 1)$ degrees of freedom in each treatment and $a(b - 1)$ over all *a* treatments. The degrees of freedom for variation among treatments are self-evident (and identical with the single-factor case; see Section 7.6). The sum of the degrees of freedom for all other sources of variation must total those for the total sum of squares. It is a useful skill to be able to calculate degrees of freedom properly, so that you can check the sum over the whole analysis, to be sure that all are present and accounted for.

Having determined the degrees of freedom, the mean squares are then calculated. What remains is to determine what has been measured by the various mean squares. We will use the same procedure as for the single-factor case and first substitute the various terms in the calculation

Table 9.3. Nested analysis of variance of an experiment with a treatments $(1 \ldots i \ldots a)$ replicated in b units in each treatment $(1 \ldots j \ldots b$ in each $i)$ and each sampled with n replicates $(1 \ldots k \ldots n$ in each $j(i))$

Source of variation	Sum of squares	Degrees of freedom	Mean square	Mean square estimates
Among treatments = A	$\sum_{i=1}^{a}\sum_{j=1}^{b}\sum_{k=1}^{n}(\bar{X}_i - \bar{X})^2 = \mathrm{SS_A}$	$a - 1$	$\mathrm{MS_A} = \mathrm{SS_A}/(a-1)$	$\sigma_e^2 + n\sigma_{\mathrm{B(A)}}^2 + \frac{bn\sum_{i=1}^{a}(A_i - \bar{A})^2}{(a-1)}$
Among units within each treatment = B(A)	$\sum_{i=1}^{a}\sum_{j=1}^{b}\sum_{k=1}^{n}(\bar{X}_{j(i)} - \bar{X}_i)^2 = \mathrm{SS_{B(A)}}$	$a(b-1)$	$\mathrm{MS_{B(A)}} = \mathrm{SS_{B(A)}}/a(b-1)$	$\sigma_e^2 + n\sigma_{\mathrm{B(A)}}^2$
Within samples	$\sum_{i=1}^{a}\sum_{j=1}^{b}\sum_{k=1}^{n}(X_{ijk} - \bar{X}_{j(i)})^2 = \mathrm{SS_W}$	$ab(n-1)$	$\mathrm{MS_W} = \mathrm{SS_W}/ab(n-1)$	σ_e^2
Total	$\sum_{i=1}^{a}\sum_{j=1}^{b}\sum_{k=1}^{n}(X_{ijk} - \bar{X})^2 = \mathrm{SS_T}$	$abn - 1$		–

Table 9.4. Nested analysis of variance of data (in Table 9.1) from experimental exclusion of insects to investigate influences on fecundity of plants

Cochran's test for heterogeneity of variances, $C = \frac{394}{2311} = 0.17$, $P > 0.05$

Analysis of variance

Source of variation	Sum of squares	Degrees of freedom	Mean square	F-ratio	
Among treatments	6865.39	2	3432.70	8.32	$P < 0.01$
Among plots (within treatments)	4951.39	12	412.62		
Within samples	11552.64	75	154.04		
Total	23369.42	89			

of sums of squares by their equivalents from the linear model. This produces (see Section 7.4):

$$\bar{X}_{j(i)} = \mu + A_i + B_{j(i)} + \bar{e}_{j(i)} \tag{9.5}$$

where $\bar{e}_{j(i)}$ is the average of all 'error' terms from the sample replicates of the jth unit in the ith experimental treatment. It is entirely equivalent to the term $\bar{e}_i$ in the simpler analysis in Section 7.5

$$\bar{X}_i = \mu + A_i + \bar{B}_i + \bar{e}_i \tag{9.6}$$

where $\bar{B}_i$ is the average of $B_{j(i)}$ terms in the ith experimental treatment. This requires that we examine closely the meaning of these terms. The best way to envisage them is to imagine that each unit in the experiment varies from the others in some specific manner. For example, the rate of successful germination by the plants considered earlier varies from place to place because of chance variations in the degree to which insects arrive, fungi attack, the soil is excessively dry, the mixture of male and female plants – a whole host of causes. So, each unit has its unique combination of influences on the fecundity of the plants. These combined influences are represented by $B_{j(i)}$, its 'unitness', the unique properties that influence the fecundity of plants in each patch of the world.

If you consider all the possible pieces of the world on which the experimental treatment might have been placed, the experiment consists of b randomly chosen patches in each of the a treatments. In the example here, there are five replicates of each of three treatments. Provided that the units have been sampled at random – i.e. randomly chosen pieces of countryside containing plants were used in the experiment – each

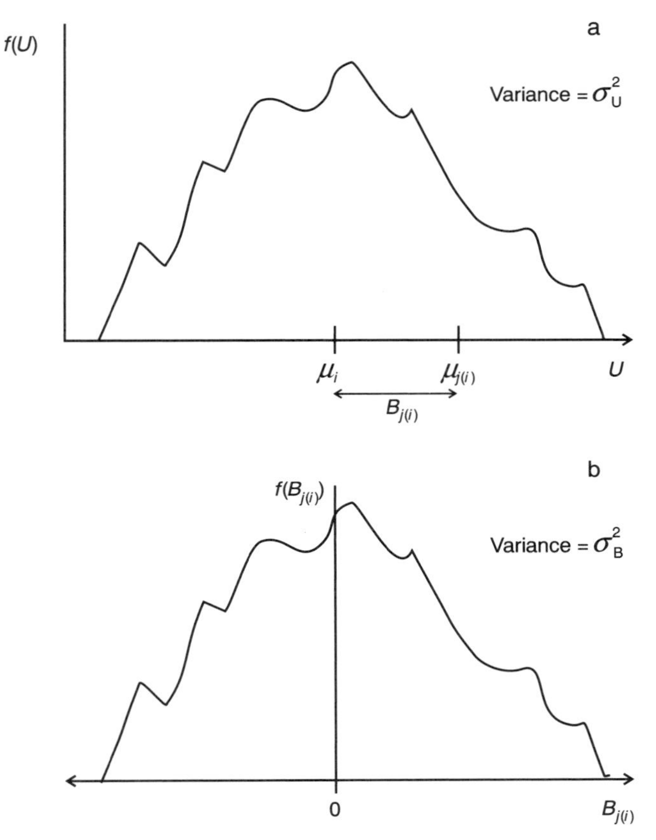

Figure 9.4. Illustration of the variability among experimental units in a nested analysis of variance. The units differ because of history, geography, disturbances, initial conditions, numerous factors. In (a), the variability among units is shown as a frequency distribution of a variable called U. The mean in any experimental treatment (μ_i) is the mean of U, which has variance σ^2_U. The mean in any unit j in experimental treatment i is $\mu_{j(i)}$ and differs from μ_i by an amount $B_{j(i)}$. (b) The distribution of B values is thus the same distribution as U, centred on zero, with variance σ^2_B $(= \sigma^2_U)$.

treatment contains a representative sample of the population of units that might have been included in the experiment. Each experimental treatment is applied to a representative sample of replicate units taken from the same distribution of places where the fecundity could have been measured.

A graphical method for illustrating this is provided in Figure 9.4, which shows a frequency distribution representing that of the 'unitness' of a large number of possible plots in the study-area. Some plots will have plants with a greater than average fecundity. Some will have less than average fecundity. It is therefore reasonable to assume a roughly symmetrical distribution of influences on fecundity, centred on the average fecundity. Any experimental unit picked from this distribution will differ in the fecundity of plants from the mean over the whole study-site by an amount unique to it. This difference – the influence of that specific plot – is the amount called $B_{j(i)}$ in Equation 9.2. The mean of the population of B values is zero and its variance is σ^2_B, as in Figure 9.4b.

In each experimental treatment, a sample of b values of B is used (i.e. a sample of b units is chosen). Each is a value of B_j for that treatment and $j = 1, 2 \ldots b$. For the ith experimental treatment, there is a sample of $B_{j(i)}$ values, read as B_j nested in experimental treatment i. This sample has a mean of $\bar{B}_i$, which is an estimate of zero (see Figure 9.4b), but differs from zero by an amount due to the sample of values not being an absolutely exact representation of the population of possible values. So, in experimental treatment i, there is a sample of $B_{1(i)}, B_{2(i)}, \ldots B_{b(i)}$ with a sample mean $\bar{B}_i$, estimating zero. The variance of this sample of values is:

$$s^2_B = \frac{\sum_{j=1}^{b} (B_{j(i)} - \bar{B}_i)^2}{(b-1)} \qquad (9.7)$$

which estimates σ^2_B in Figure 9.4.

Each of the experimental treatments has a similar sample of values of $B_{j(i)}$ assigned to it, provided that the experimental units are assigned fairly to each treatment. Except for sampling error, the $\bar{B}_i$ terms are equal for all experimental treatments. So, provided that samples of units are *independently and representatively* allocated to each treatment, the experimental treatments start out with no differences in mean fecundity, except for sampling error of the samples of units in each treatment.

Note that, as in Section 9.1, there is no need to try to pretend that experimental units are equal in the mean fecundity of their plants. That would require that all areas have an identical fecundity (thus σ^2_B in the distribution in Figure 9.4b would have to be zero). That is not possible to achieve; if there is variation, it is a property of the study site.

What matters instead is to ensure that the variability among units is equally represented in all experimental treatments, by having equivalent, comparable samples of units in every treatment. If replicate units are all chosen in the same way and assigned at random to the treatments, this should be achieved.

The preceding explanation accounts for $\bar{B}_i$ in Equation 9.6. $\bar{e}_i$ is the average of all e_{ijk} terms over all n replicates in all b units in that experimental treatment. It is, of course, the average of the b values of $\bar{e}_{j(i)}$ (from Equation 9.5) in that particular treatment. Again, like the $\bar{e}_{j(i)}$ values, $\bar{e}_i$ values will be zero except for sampling error.

The overall mean of all the data is:

$$\bar{X} = \mu + \bar{A} + \bar{B} + \bar{e} \tag{9.8}$$

where $\bar{A}$ represents the average of all treatment effects (see also Section 9.4). $\bar{B}$ represents the average of all the $\bar{B}_i$ terms from the different treatments. $\bar{e}$ represents the average of all error terms over the entire experiment. Except for sampling error, $\bar{B}$ and $\bar{e}$ should be zero.

By replacing the means in Equation 9.1 with their equivalents in Equations 9.5, 9.6 and 9.8, it is possible to determine what the sums of squares measure (for details, see, for example, Underwood, 1981; Winer *et al.*, 1991).

Each component of the equations can be expanded and simplified. If the assumption of *independence among replicate units* and *among treatments* is made (as described in Section 7.5 and discussed in detail in Section 7.14), the results for differences among levels of factor A are:

$$\sum_{i=1}^{a}\sum_{j=1}^{b}\sum_{k=1}^{n}(\bar{X}_i - \bar{X})^2 = bn\sum_{i=1}^{a}(\bar{e}_i - \bar{e})^2 + bn\sum_{i=1}^{a}(\bar{B}_i - \bar{B})^2 + bn\sum_{i=1}^{a}(A_i - \bar{A})^2 \tag{9.9}$$

If data are not independent, the equation would also include the cross-products, such as $n\sum_{i=1}^{a}\sum_{j=1}^{b}(\bar{e}_i - \bar{e})(\bar{B}_i - \bar{B})$, which would make the following analyses invalid, because these terms are assumed to be zero. The results of the algebra are shown as mean squares in Table 9.3. Most of the estimates provided by mean squares in Table 9.3 are familiar from the one-factor case in Chapter 7. Note that σ_{B}^2 is the variance of the population of $B_{j(i)}$ values (see earlier and Figure 9.4).

9.6 Tests and interpretation: what do the nested bits mean?

9.6.1 **F*-ratio of appropriate mean squares***

The result of all these algebraic machinations is indicated in Table 9.3. The mean square among experimental treatments includes a component of variance for the replicates (σ_e^2) and another component for variation among replicated units (σ_B^2). If the null hypothesis is true, there are no differences among treatments (all $\mu_i = \mu$) and all the A_i terms are zero (Equation 9.4). Therefore the term $\sum_{i=1}^{a} (A_i - \bar{A})^2$is zero. Under any other circumstances (i.e. any alternative to the null hypothesis), this term is greater than zero. Thus, from Table 9.3:

$$\text{H}_0: \quad \text{Mean square A} = \text{Mean square B(A)}$$

$$\text{therefore} \quad \frac{\text{Mean square A}}{\text{Mean square B(A)}} = 1 \tag{9.10}$$

$$\text{H}_\text{A}: \quad \frac{\text{Mean square A}}{\text{Mean square B(A)}} > 1 \tag{9.11}$$

and this can be tested by an F-ratio with $(a - 1)$ and $a(b - 1)$ degrees of freedom. The rationale for this was fully explained for the single-factor case in Section 7.7.

9.6.2 Solution to confounding

Interpretation of a significantly large value of F depends upon multiple comparisons (as described in Sections 7.12 and 8.6) see below. What is, however, clear is that the structure of the test demonstrates *why* this sort of experimental design is the solution to the problem of confounding originally raised. The comparison used in the test (A versus B(A)) is unconfounded. It is valid whether or not there is variation among replicate units in the experimental treatments. If σ_B^2 is zero, the F-ratio calculated in the test is still valid; if there are no differences among treatments, mean square A will still be the same as the mean square B(A).

If, however, there had been no replicated units in each treatment, σ_B^2 could not have been estimated. If, for example, a group of n replicate plants were sampled in one patch of each treatment (as at the beginning of Section 9.1), there would still potentially be variation among patches due solely to their intrinsic differences ($B_{j(i)}$ terms). As a result, the estimate of differences among treatments (the mean square among treatments) still includes three terms, one of which is σ_B^2. The problem is that there could then be no estimate of variation among replicate units – there is none. So, the mean square among treatments could only be compared with that within samples.

Obviously, under such circumstances, F could be significantly larger than 1 because of differences among treatments or because of intrinsic variance among experimental units (i.e. variations *not* due to the experimental manipulations). Without provision of replicate units, no separate estimate of σ_B^2 is possible. The significance of any statistical test would thus be compromised. Without replicate units in each treatment, it is impossible to interpret the result to mean that there are differences among treatments. This would be rational only if there were some independent evidence that there is no significant variation among the individual units assigned to each experimental treatment. It is difficult to see how such evidence could be gathered unless some nested experimental procedure has been used before. Even then, the assertion that lack of variability among units (plots, patches, tanks, etc.) in previous experiments means a lack of such variability in the present experiment is itself an inductive guess. Hurlbert (1984) usefully discussed the problems of detecting variations among replicate plots in ecological experiments.

The nested analysis of variance solves the problem because it provides an independent estimate of the variability among replicated units (σ_B^2) and therefore a valid test whether or not such variation exists.

9.6.3 Multiple comparisons

When the null hypothesis is rejected, multiple comparisons are needed on the means of the experimental treatments. These are exactly as described for the single-factor case (as in Section 8.6), except that care must be taken to ensure that the correct standard errors and degrees of freedom are used. Instead of the within sample mean square being used to construct standard errors, the appropriate mean square is that among units within each treatment (i.e. B(A) in Table 9.3). This is used to construct standard errors because it includes all sources of variation within each treatment.

An example using the Student–Newman–Keuls test is provided in Table 9.5. Nothing further needs to be discussed here (but the general rules for construction of correct multiple comparisons are presented in Section 11.9).

9.6.4 Variability among replicated units

One other aspect of nested designs that can usefully be analysed to determine its significance is the variation among the replicate sampled units (B(A) in Table 9.3). There are three major reasons for wanting to do this.

First, if it could be demonstrated that such variability is non-existent despite the original possibility (i.e. σ_B^2 is indeed zero for your experiment),

Table 9.5. Student–Newman–Keuls multiple comparison of means after a significant nested analysis of variance. Data are in Table 9.1 and analysed in Table 9.4. The standard error for the three means is calculated from the mean square among plots within each treatment. All other procedures are as explained in Section 8.6.5 (Table 8.6). Q values are for probability of Type I error = 0.05, with g and 12 degrees of freedom (i.e. degrees of freedom for the mean square used to calculate the standard error)

Treatment:	Control	Cages with openings	Cages
Mean:	79.8	78.7	60.0

$$\text{Standard error} = \sqrt{\left[\frac{\text{Mean square plots (treatments)}}{\text{No. of data in each mean}}\right]} = \sqrt{\left(\frac{412.62}{30}\right)} = 3.71$$

Rank order			SNK test		
1	2	3	g	Q	$D = Q \times SE$
60.0	78.7	79.8			
3–1 19.8*			3	3.41	12.65
2–1 18.7*	3–2 1.1		2	2.84	10.54

Conclusion: Cages that exclude insects significantly reduce fecundity of plants and there are no artefacts due to the experimental manipulations.

this component of the analysis could be eliminated, making more powerful tests available. The issues involved and a recommended procedure are discussed below in Section 9.7.

Second, if variability among replicate units in each experimental treatment does occur, it is often informative to check whether it is similar for all treatments. If not, something may have gone wrong in the experiment. For example, in the experimental exclusion of insects described previously, it is always possible that some mishap befell one of the cages, so that insects got in. In that treatment, the variance among units would probably be very large, compared to the intrinsic variability naturally found among units, as measured among units in the other two treatments. It would be very helpful to ascertain whether variation among units turned out to be similar. This can be examined by partitioning the sum of squares measuring variation among units into its component parts, each representing variation among units in one of the experimental treatments. Variation among units is measured by the sum of squares as in Table 9.3.

The variation among units in a particular experimental treatment (i) is:

$$\sum_{j=1}^{b}\sum_{k=1}^{n}(\bar{X}_{j(i)}-\bar{X}_i)^2=\frac{\sum_{j=1}^{b}\left(\sum_{k=1}^{n}X_{ijk}\right)^2}{n}-\frac{\left(\sum_{j=1}^{b}\sum_{k=1}^{n}X_{ijk}\right)^2}{bn} \tag{9.12}$$

which has (b – 1) degrees of freedom. The relevant calculations have been done in Table 9.6 for the experimental removal of pollinating insects discussed earlier (in Section 9.1) and below.

The third and final reason why the variation among replicate units might be of particular interest is in analyses of models explaining spatial variability itself. For example, the original observations on variability in fecundity of plants in relation to number of pollinating insects could have been something like the data in Figure 9.5. There is more variability among plots where there are more insects. Two quite different interpretative models could now be considered to explain the pattern.

First, as before, the insects are responsible for the increased mean fecundity of the plants and are responsible for the variability of fecundity. For example, where insects are numerous, they are sometimes concentrated

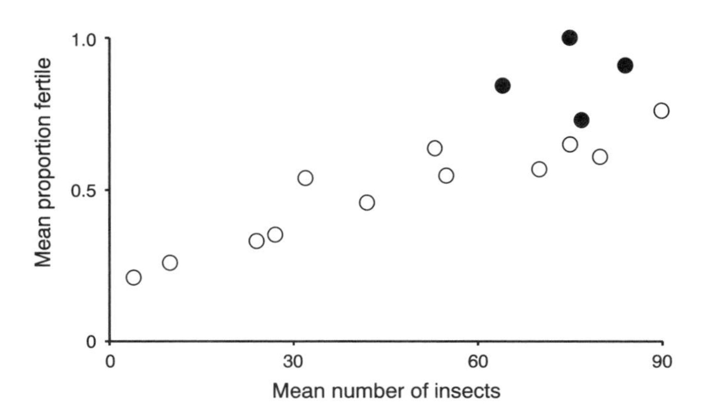

Figure 9.5. Relationship between average number of potentially pollinating insects and the fecundity of plants. Data are mean proportion of fertile flowers per plant in ten replicate plants and mean number of insects caught in five replicate traps from each of a random set of plots in a field. Note the greater variability among plots where there are larger numbers of insects. Filled circles are the 'high spots' discussed in the text.

on relatively few plants, causing an increase in fecundity of those individuals, but not increasing mean fecundity over the whole patch by very much. Alternatively, in some plots there are large numbers of insects spread relatively evenly over the plants, raising mean fecundity in the patch to much greater levels. This would explain the pattern in the data. This model leads to the hypothesis that removal of insects, as before, leads to a decrease in mean fecundity and to a decrease in variability among patches. The caged plots should have smaller mean fecundity and smaller variance in fecundity among plots than do the plots with insects in (i.e. mesh controls and open controls). It would, *en passant*, also lead to the prediction that variability in fecundity among plants in a plot should increase with increasing numbers of insects – but we shall leave that intriguing possibility (note, though, the comment at the end of Section 2.4).

A second interpretation of the data is that there is, indeed, a general relationship between fecundity and number of insects, but that is not the only process operating. Another pollinator, for example nectar-eating birds, also influences fecundity, but are much more patchy in their visits to the plants. They also tend to visit the same plots that have numerous insects (why this should be so would be the subject of a different experimental study). Where birds visit plots with large numbers of insects, they raise fecundity. This causes some plots (shown as filled circles in Figure 9.5) to have greater fecundity than generally found where there are numerous insects. Hence, the *birds* cause the increased variation in mean fecundity.

Thus, the hypothesis is now that there will be greater fecundity where insects are present (open and mesh controls) than where they are excluded (cages). There will also be greater variation among plots where birds are present (open areas) than where they are excluded (mesh controls and cages).

The different hypotheses are illustrated in Figure 9.6. Note that there may or may not be a significant difference detected between mean fecundity in the presence or absence of birds (as required by the second model) because of the variability among plots. If the second model were supported by the analysis, in terms of the variation among plots, there would be considerable incentive to design better experiments to separate the effects of insects from those of birds.

The two models can be contrasted in terms of their predictions about variation among plots (or units in the experiment) as:

Hypothesis from model 1: $\sigma^2_{\mathrm{B(Cage)}} < \sigma^2_{\mathrm{B(Mesh\ control)}} < \sigma^2_{\mathrm{B(Open)}}$

Hypothesis from model 2: $\sigma^2_{\mathrm{B(Cage)}} = \sigma^2_{\mathrm{B(Mesh\ control)}} < \sigma^2_{\mathrm{B(Open)}}$

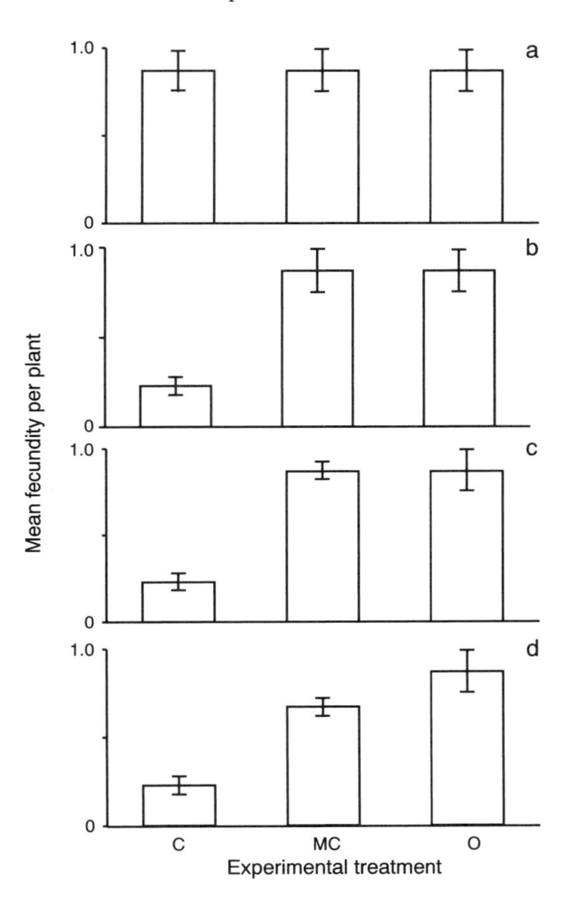

Figure 9.6. Possible results from an experiment to test models for differences among patches in fecundity of plants. Experimental treatments are: C, cage to exclude insects and birds; MC, mesh control (i.e. cages with holes to let in insects) to exclude birds; O, open control to which birds and insects have access. Standard errors are from mean square among units, i.e. include variability among units. Two models are proposed in the text, where more details are provided. (a) Neither mean nor variation among units is affected: models 1 and 2 are wrong. (b) Insects cause increased fecundity, but access to birds makes no difference to variation among units: model 2 is wrong. (c), (d) Insects cause increased fecundity and access to birds causes increased variability: model 2 is supported, model 1 is wrong. In (d), there is also increased fecundity due to birds, which may or may not be detected, depending on the power of the experiment.

where $\sigma^2_{B(...)}$ indicates variance among replicate units nested in the specified experimental treatment. The null hypothesis is that differences in variation among patches are not due either to insects or to birds and therefore any other pattern will occur in the relative magnitude of the three variances.

This could be examined by, first, partitioning the variation among units within treatments into its components (as in Table 9.6a) and then proceeding to test the component mean squares (MS) against each other, as shown in Table 9.6b. The sequence of events must be:

1. If $MS_{B(1)}$ is larger than $MS_{B(2)}$, model 1 must be wrong (note the inequality in predictions from model 1). For model 2, test for differences, using the two-tailed F-ratio:

   $F = MS_{B(1)}$ versus $MS_{B(2)}$, with 4 and 4 degrees of freedom

   (a) If significant (at $P = 0.05$), model 2 is also wrong. Both models are wrong (although the results would be very difficult to explain given the original observations).
   (b) If not significant, proceed to (3) to examine model 2.

2. If $MS_{B(2)}$ is larger than $MS_{B(1)}$, test for Model 1, using a 1-tailed F-ratio (noting the direction of difference specified by the hypothesis).

   $F = MS_{B(2)}/MS_{B(1)}$, with 4 and 4 degrees of freedom

   (a) If significant (at $P = 0.05$), model 2 is wrong. Proceed to 4 to examine model 1.
   (b) If not significant, model 1 is wrong, proceed to 3 to examine model 2.

3. If $MS_{B(3)} < MS_{B(2)}$, model 2 is wrong (note the inequality in predictions from model 2). Otherwise, test as a one-tailed F-ratio:

   $F = MS_{B(3)}/MS_{B(2)}$, with 4 and 4 degrees of freedom

   (a) If significant, model 2 is supported. Proceed triumphantly to publish, but examine again with more rigour in the model, by hypotheses and tests (see Section 2.8).
   (b) If not significant, model 2 is wrong. Both models are wrong. Start thinking again.

4. Test as a two-tailed F-ratio:

   $F =$ (larger of $MS_{B(2)}$ or $MS_{B(3)}$)/(smaller of $MS_{B(2)}$ or $MS_{B(3)}$) with 4 and 4 degrees of freedom

   (a) If significant, model 1 is wrong. Find a new model.
   (b) If not significant, model 1 is supported. Publish, but examine again with more structure to the model and more powerful tests (Section 2.8).

Note, there may be some problems of excessive Type I error in a series of such tests, so caution is necessary (or the probabilities of Type I error in each test can be reduced to compensate).

Table 9.6. Repartitioning of variation among replicated units nested in experimental treatments

(a) Analysis of variation among units in an experiment on pollination of plants by insects; experimental treatments are as in Table 9.1. Results here are from a different set of data

Source of variation		Sum of squares		Degrees of freedom		Mean square		F-ratio	
Among treatments	= A	3062.8		2		1531.4		4.01	$P < 0.05$
Among plots(treatments)	= B(A)	4582.8		12		381.9		2.04	$P < 0.05$
Plots(cages)	= B(1)		1057.3		4		264.3		
Plots(mesh controls)	= B(2)		376.2		4		94.1		
Plots(controls)	= B(3)		3149.2		4		787.3		
Within plots		14040.1		75		187.2			
Total		21685.7		89					

(b) Analysis of experimental variation to test hypotheses about different causes of variability (see Section 9.6.4 for details)

1. $F_{\text{2-tailed; 4, 4 df}} = \dfrac{MS_{B(1)}}{MS_{B(2)}} = \dfrac{264.3}{94.1} = 2.81,\ P > 0.05$

Model 1 is wrong, go to (3)

3. $F_{\text{1-tailed; 4, 4 df}} = \dfrac{MS_{B(3)}}{MS_{B(2)}} = \dfrac{787.3}{94.1} = 8.37,\ P < 0.05$

Model 2 is supported; insects do not influence variation, birds apparently do.

Note also, that differences in variability among units from one experimental treatment to another may cause excessive Type I error in a test for differences among means of treatments (A in Tables 9.3 and 9.6), as discussed in Section 7.16. It is not clear whether any transformation of the data would solve the problem.

What is clear, however, is that examination of variability among replicate units in each treatment is an informative procedure. Either no differences should be expected, in which case their discovery must cause concern and caution about the experiment; or there are specified possible differences because of the underlying hypotheses tested by the experiment. In this case, tests for specific patterns of difference in variability within experimental treatments are important.

9.7 Pooling of nested components

9.7.1 Rationale and procedure

The test for differences among experimental treatments is not a particularly powerful one unless there are large effect sizes or large numbers of replicated units in each treatment (see Sections 8.3 and 8.4). The number of replicates within units (n) does not contribute to the degrees of freedom in the test and increases both the numerator and the denominator of the F-ratio (Table 9.3).

If it happens to turn out, despite earlier caution, that there is no variability among replicated units, it is tempting to try to improve on the capacity of the analysis to detect differences among treatments. This can be done using the following *post hoc* procedure for pooling terms in a model for analysis of variance. Suppose that a preliminary test of variation among experimental units in each treatment reveals *no* significant variation:

$$F = \text{MS}_{\text{B(A)}}/\text{MS}_{\text{within samples}}$$
with $a(b-1)$ and $ab(n-1)$ degrees of freedom

is not significant (at, say, $P = 0.05$ probability of Type I error).

One of two conclusions is now possible – either there really is no variability among units ($\sigma^2_{\text{B(A)}} = 0$) *or* there is such variability, but of insufficient magnitude to be detected. In the latter case, the experiment is not powerful enough to detect the alternative to the null hypothesis (i.e. $\sigma^2_{\text{B(A)}} > 0$).

If we *knew* that $\sigma^2_{\text{B(A)}}$ was zero (i.e. we drew the first conclusion), we could eliminate that component of mean squares from the analysis. We could then examine the model that:

$$X_{ijk} = \mu + A_i + e_{ijk} \tag{9.13}$$

without taking any notice of the experimental units. If $\sigma^2_{\text{B(A)}} = 0$, all $B_{j(i)}$ terms in Equation 9.2 are zero (see Figures 9.2 and 9.3 and Section 9.4). In other words, we could do the analysis as a single-factor analysis with bn replicates in each treatment. There is no possibility of confounding differences among experimental treatments with differences among experimental units – the latter have been *shown* not to exist. Thus, we would *pool* together the two estimates of variance within samples, as shown in Table 9.7.

The rationale behind this is that, if $\sigma^2_{\text{B(A)}}$ is zero, the mean square$_{\text{B(A)}}$ estimates σ^2_e, which is also estimated by the mean square within samples. These can only differ because of sampling error, so we can average them together to provide a combined estimate. The averaging must take into account the precision of the two estimates (a more precise estimate should be given more influence, or more weight) when they are combined. So, we take the weighted average – weighting according to their degrees of freedom.

$$\text{Weighted average} = \frac{a(b-1)\text{MS}_{\text{B(A)}} + ab(n-1)\text{MS}_{\text{W}}}{a(b-1) + ab(n-1)} \tag{9.14}$$

This is of course identical with:

$$\text{Weighted average} = \frac{\text{SS}_{\text{B(A)}} + \text{SS}_{\text{W}}}{a(b-1) + ab(n-1)} \tag{9.15}$$

SS being sum of squares (see Table 9.7). This combined estimate of σ^2_e can be used in a test for differences among treatments.

Having concluded that $\sigma^2_{\text{B(A)}}$ is zero, it can be eliminated from the mean square estimate MS_{A} (as in Table 9.7). Thus, a test for differences among treatments can now be done as:

$F = \text{MS}_{\text{A}}/\text{MS}_{\text{P}}$, with $(a-1)$ and $a(bn-1)$ degrees of freedom

This will nearly always be a more powerful test than the one of A versus B(A) before pooling was done. So, pooling the estimates of variation within samples is a useful and helpful procedure.

9.7.2 Pooling, Type II and Type I errors

It is, however, crucial that thought and caution be applied in equal measure to the consequences of pooling when it is inappropriate. There

Table 9.7. *Post hoc* pooling of estimates of variation. The analysis of variance of a nested experiment, with a treatments, b replicated units in each treatment and n replicates in each unit is shown in Table 9.3

Pooled estimate of σ_e^2; $\sigma_{B(A)}^2$ assumed zero after $F = MS_B/MS_W$ is not significant.

Source of variation	Sum of squares	Degrees of freedom	Mean square	Mean square estimates
A	SS_A	$a - 1$	$MS_A = SS_A/(a-1)$	$\sigma_e^2 + \frac{bn\sum_{i=1}^{a}(A_i - \bar{A})^2}{(a-1)}$
Pooled = B(A) + W	$SS_P = SS_{B(A)} + SS_W$	$a(bn - 1)$	$MS_P = SS_P/a(bn-1)$	σ_e^2
Total	SS_T	$abn - 1$	–	–

are serious problems for an analysis if pooling is done when it should not be. It was pointed out in the previous section that pooling is useful when there is evidence that the variation among units is zero. The test for the null hypothesis that $\sigma^2_{B(A)} = 0$ is used to examine this issue. If the test is not significant, there is no evidence to refute the null hypothesis.

This is, however, not at all the same conclusion as the inference that variation among units *is* zero. All we know is that the evidence to hand does not indicate the opposite. In other words, there is always the risk of making a Type II error (retaining the null hypothesis when it is false) because the available evidence is inadequate or insufficient to identify the existing variation among units. The null hypothesis may be wrong, but the test has failed to reject it (see Sections 5.11, 5.13, 8.3, 8.4).

The consequence of unwittingly making a Type II error here would be to proceed to pool the two sources of variation (B(A) and W in Table 9.7) and to test for differences among treatments using the pooled mean square as the divisor of the test. If, however, there is undetected variability among units, the resulting pooled mean square now estimates:

$$\text{Pooled mean square} = \frac{a(b-1)\sigma^2_e + a(b-1)n\sigma^2_{B(A)} + ab(n-1)\sigma^2_e}{a(b-1) + ab(n-1)}$$

$$= \sigma^2_e + \frac{(b-1)n\sigma^2_{B(A)}}{(bn-1)} \qquad (9.16)$$

This is derived from the mean square estimates (as in Table 9.3) multiplied by their degrees of freedom to calculate what the sums of squares estimate and then pooling as in Equations 9.14 and 9.15.

As a result, if the null hypothesis of no differences among the means of the treatments is true, the mean square among treatments would still be larger than the pooled mean square. Thus, if pooling were done when $\sigma^2_{B(A)} > 0$, the F-ratio for testing the null hypothesis of no differences among means of treatments is expected to be greater than 1. Hence, using such a test leads to excessive Type I error, because such biased F values are more likely to reject the null hypothesis than is calculated in the tabled distribution of the F-ratio. More values than anticipated will exceed a chosen critical value of F.

To protect against this potential for error, it has been recommended that pooling be done only when there is a great likelihood of detecting variation among units in the preliminary test of $\sigma^2_{B(A)}$ over σ^2_e. Winer *et al.* (1991) recommended that the preliminary test be done with a very large probability of Type I error, so that the probability of Type II

Table 9.8. Analysis showing *post hoc* pooling. The data come from an experiment like that shown in Table 9.1 (and analysed in Table 9.4). There are $a = 3$ treatments, $b = 5$ units in each treatment and $n = 6$ replicates in each unit

(a) Analysis of variance

Source of variation	Degrees of freedom	Mean square
Among treatments = A	2	357.1
B(A)	12	108.5
Within samples = W	75	104.3
Total	89	

Testing for variation among units ($\sigma^2_{B(A)}$):

$$F = \frac{MS_{B(A)}}{MS_W} = 1.04$$

which has $P > 0.05$ with 12 and 75 degrees of freedom. It is also smaller than 1.89, the critical value of F at $P = 0.25$. Therefore pooling is permissible.

(b) Pooled analysis

Source of variation	Degrees of freedom	Mean square
A	2	357.1
Pooled	87	104.9
Total	89	

Testing for differences among treatments:

$$F = \frac{MS_A}{MS_P} = 3.40,\ P < 0.05;\ \text{2 and 87 degrees of freedom}$$

Note that, without pooling:

$$F = \frac{MS_A}{MS_{B(A)}} = 3.29,\ P > 0.05;\ \text{2 and 12 degrees of freedom}$$

error is kept small (see Section 8.4). They proposed that the F-ratio used to test for non-zero $\sigma^2_{B(A)}$ be done with $\alpha = 0.25$, i.e. a one in four chance of rejecting the null hypothesis when it is true. By this device, it is assumed that it is extremely unlikely to retain the null hypothesis erroneously. Even small values of $\sigma^2_{B(A)}$ will cause rejection of the null hypothesis.

Thus, the recommended pooling procedure is as follows:

1. Test the null hypothesis $\sigma^2_{B(A)} = 0$, with $P = 0.05$, using

   $F = MS_{B(A)}/MS_W$

   If significant, no pooling is warranted. Therefore test for differences among treatments using:

   $F = MS_A/MS_{B(A)}$ with $(a-1)$ and $a(b-1)$ degrees of freedom

2. If the first test is *not* significant at $P = 0.05$, use the critical value of F for probability of Type I error $P = 0.25$. If significant, no pooling is possible, proceed as in (1) above. If not significant, pool $MS_{B(A)}$ and MS_W. Test for differences among means of treatments using:

   $F = MS_A/MS_P$, with $(a-1)$ and $a(bn-1)$ degrees of freedom

The procedure is illustrated in Table 9.8, where pooling sufficiently increased the power of the test for differences among experimental treatments to be detected.

Finally, note that the pooled mean square as an estimate of σ^2_e will generally cause multiple comparisons of the means of treatments to be more sensitive (powerful) because of the larger number of degrees of freedom associated with the pooled mean square. This may make pooling a desirable procedure for multiple comparisons, even though the analysis without pooling was already powerful enough to detect differences among treatments.

So, provided caution is the guiding principle, pooling is a valuable procedure. There appears to have been no work or modelling on the degree to which using F with a probability of Type I error of 0.25 may prevent Type II error in the decision to pool. It is, as yet, an arbitrary recommendation, but one that is based on common sense.

9.8 Balanced sampling

One point worth considering in a nested analysis is that the sizes of samples (n) should be the same in all b units in each treatment and the number of units (b) should be equal in all a treatments. This balanced sampling is necessary to avoid problems with the use of F-ratios. The problems arise because different numbers of replicates from unit to unit and from treatment to treatment cause differences in the multipliers of $\sigma^2_{B(A)}$ in the mean squares for A and B(A). The algebra to demonstrate this is complicated, but the consequence is clear. If the $\sigma^2_{B(A)}$ terms have different multipliers in the two mean squares, an F-ratio constructed from the ratio of the two mean squares *cannot* be expected to be 1 under the null hypothesis of no differences among the means of treatments. Suppose this null hypothesis is true and the two mean square estimates are written as:

MS_A estimates $\sigma_e^2 + n_1\sigma_{B(A)}^2$
$MS_{B(A)}$ estimates $\sigma_e^2 + n_2\sigma_{B(A)}^2$

then $F = MS_A/MS_{B(A)} = 1$ only if n_1 and n_2 are identical. Therefore, the probability associated with any observed value of F will match the tabulated values only where n_1 and n_2 are identical.

Numerous computer packages (and formulae in such texts as that by Sokal & Rohlf (1981)) provide the wherewithal to do an unbalanced analysis despite this problem. There are also procedures using residual maximal likelihood estimators (Robinson, 1987). Earlier, I have given reasons for preferring balanced sampling and this topic is discussed more in Chapter 10. Some of the available procedures provide approximations to allow adjustment of F-ratios in an attempt to make them conform to the tabled distribution of F. These are, often, very approximate approximations. Similar problems arise when b varies from one experimental treatment to another. There is no difficulty programming general formulae to allow calculation of sums of squares and mean squares. Many such programs exist. The problem is being able to provide a valid distribution of F-ratios as a test statistic.

In addition, sampling with varying numbers of replicates also creates some heterogeneity of variances, among units in each treatment and among treatments. The variances around means are affected by the size of sample (see Section 5.2) and therefore means from larger samples are estimated more precisely (i.e. with smaller standard errors). This is not solved by the procedures used to approximate F-ratios. The problems such heterogeneity causes were discussed in Section 7.16. There are procedures to sort out problems of analyses with unbalanced data (e.g. Weerahandi, 1995), but only for relatively simple nested experimental designs.

Finally, as discussed in Section 7.10, there is usually no possible justification for deciding that the means of some experimental treatments or means of some replicated units should be estimated with greater precision than others – which is inevitable when sizes of samples differ.

It is always better to plan the experiments with balanced sampling. Take extra replicates and establish extra experimental units if you think some will be lost or impossible to sample at the end of the experiment. Use the procedures discussed in Section 7.10 to balance the samples if they turn out, through mishap, to have become unequal.

9.9 Nested analyses and spatial pattern

One of the problems commonly encountered at the start of a research programme is the difficulty of knowing at what scale a process (or a set of interactive processes) may be operating. Consider the simple example of the pattern of abundance of a population. There may be ecological processes influencing abundances at several spatial scales. Dispersal of seeds or larvae may create large-scale gradients in densities. At a smaller scale, patchiness of microhabitats or other resources may create patchiness in the local density of the organism. At a different larger scale, there may be gradients of physical environmental features, leading to gradients in density. Predators may operate in patchy ways, creating gaps and localized differences in the abundance of the species. All sorts of behavioural responses to the habitat and to other individuals may influence the dispersion of a species. And so on.

As a result, it is often confusing to know how to start with any investigation of abundance. Nested sampling designs are one tool (of which there are others, such as the range of procedures reviewed by Pielou (1969)) to help to sort out spatial patterns. The philosophical underpinning is to use the nested allocation of sampling units – quadrats – to gather data to test the general hypothesis that there is variation in abundance or spatial scale a where $a = 1, 2 \ldots s$ different scales. The model from which this is derived is the series of statements in the preceding paragraph. So, you start with the observation that the numbers of an organism seem to vary from one place to another across a habitat. The model to explain this is very non-specific and in two parts. First, there are many processes creating variations in density of organisms and these operate at different spatial scales. Second, the species being observed conforms to these generalizations. Hence the hypothesis. The null hypothesis is that there is no variability in density at each of the specified scales. To test this, quadrats or traps or nets, or whatever unit, are sampled at each of a series of inclusive scales. The data are collected by balanced, independent samples at each scale and can be examined by a nested analysis of variance.

Sometimes, there are numerous clues from the habitat itself and from previous studies to specify what might constitute appropriate scales to sample. As an example, consider the distribution of an insect pest that lives on leaves of fruit trees on farms in some areas of countryside.

Farms with fruit trees are separated by several kilometres of other habitat. Orchards on a farm are not homogeneous, single plantations. Without knowing anything about the ecology of insects, it is at least arguable that a preliminary description of abundances of the insects should include any variability from farm to farm, orchard to orchard, tree to tree, branch to branch, twig to twig and leaf to leaf. An appropriate sampling unit would be a chosen area of leaf; insects would be counted in such an area on each of the appropriate number of leaves.

Without specifying them here, it is easy to imagine the sorts of processes that could potentially affect numbers of insects per leaf at all the spatial scales measured. Stratification of sampling into these scales is therefore a sensible way to start. Note that it is not proposed that tests of null hypotheses (there is no difference in density from farm to farm) are tests of hypotheses about *processes* causing patterns. The procedure is entirely a method of determining at what scales there are variations in densities and how large these variations are. The patterns identified then indicate at what scales processes need to be invoked in models to explain the observed patterns. Without knowledge of the patterns (Andrew & Mapstone, 1987) a considerable amount of energy can be misdirected into studies of irrelevant processes (see examples in Underwood, 1990).

So, a set of farms (with orchards; other farms are irrelevant, as in Section 3.4) is identified in the study-area. Several, say three as an arbitrary number, are chosen at random to sample. On each farm, there are four to six patches of orchard. Of these, two are chosen at random. In each orchard, there are numerous trees, of which, say, six are chosen at random. And so on – three branches are picked to represent each tree; on each branch four twigs are chosen and four leaves are sampled on each twig. Thus, there are totals of 3 farms, 6 orchards, 36 trees, 108 branches, 432 twigs and 1728 leaves to be sampled. The various levels of sampling are randomly nested inside the next higher level. Thus, a particular tree can belong only to a specific orchard and farm. The sampling design is fully nested and the analysis would be as in Table 9.9.

At each level sampled, the null hypothesis is that there are no differences among the means of units sampled within the next higher level (e.g. there are no differences in mean number of insects per unit area of leaf on twigs within each branch). These null hypotheses can be tested by an F-ratio for each level, as:

$$F_{\text{level}\,x} = \text{MS}_{\text{level}\,x}/\text{MS}_{\text{level}\,x+1}$$

Table 9.9. Nested analysis of variance of spatial variations in densities of insects. Three farms are sampled. On each farm, two orchards, six trees in each orchard, three branches on each tree, four twigs on each branch and four leaves on each twig are sampled

Source of variation		Degrees of freedom	Mean square estimates
Among farms	= F	2	$\sigma^2_W + 4\sigma^2_{T(B(I(O(F))))} + 16\sigma^2_{B(I(O(F)))} + 48\sigma^2_{I(O(F))} + 288\sigma^2_{O(F)} + 566\sigma^2_F$
Among orchards (farms)	= O(F)	3	$\sigma^2_W + 4\sigma^2_{T(B(I(O(F))))} + 16\sigma^2_{B(I(O(F)))} + 48\sigma^2_{I(O(F))} + 288\sigma^2_{O(F)}$
Among individual trees within orchards	= I(O(F))	30	$\sigma^2_W + 4\sigma^2_{T(B(I(O(F))))} + 16\sigma^2_{B(I(O(F)))} + 48\sigma^2_{I(O(F))}$
Among branches within trees	= B(I(O(F)))	72	$\sigma^2_W + 4\sigma^2_{T(B(I(O(F))))} + 16\sigma^2_{B(I(O(F)))}$
Among twigs within branches	= T(B(I(O(F))))	324	$\sigma^2_W + 4\sigma^2_{T(B(I(O(F))))}$
Within samples (= W)		1296	σ^2_W
= Among leaves within twigs	= L(T(B(I(O(F)))))		
Total		1727	

with degrees of freedom x and degrees of freedom $(x + 1)$, where x indicates a source of variation in the analysis and $x + 1$ indicates the source immediately below it. So, to test the null hypothesis of no differences in mean number of insects per leaf on trees within each orchard:

$$F_{\text{trees}} = \text{MS}_{\text{trees(orchards)}}/\text{MS}_{\text{branches(trees)}},$$

with 30 and 72 degrees of freedom

The rationale behind these tests is obvious from the mean square estimates in Table 9.8b. For any level in the analysis (level x), the null hypothesis of no differences among means is equal to the null hypothesis that σ_x^2 is zero. Consequently, the mean square for that level will equal that of the level next down in the hierarchy, because they contain exactly the same components except σ_x^2. If the null hypothesis is false, $\sigma_x^2 > 0$ and the ratio of the xth mean square to the $(x + 1)$th mean square will be >1.

Using an analysis like this allows detection of spatial variation at scales that relate to natural, identifiable components of the habitat (twigs, trees, farms, etc.). Sampling can be optimized using cost–benefit procedures, as described later (Section 9.11).

Similar analyses can, of course, be done when there is no obvious structure in the habitat that might automatically be considered to be an important influence or to be correlated with important influences on patterns of distribution. Examples are analyses of densities of plants on an open grassland, barnacles on a stretch of rocky shore, bivalves in a muddy habitat. It is also impossible to perceive any clues as to what might constitute patchiness or gradients in abundance of organisms that are cryptic or buried in the substratum. In all of these cases, more arbitrary choices are made about scales. So, you might arbitrarily divide the habitat into several 100 m × 100 m plots and pick two or three to sample (say three). In each, you might randomly choose several (say three) 10 m × 10 m areas. In each of these, you choose, perhaps four, 1 m × 1 m plots and then you take five small core-samples from each plot to count small arthropods in the soil (an example from marine environmental work is in Morrisey *et al.*, 1992). A nested analysis of variance of variability among 100 m plots, 10 m plots in 100 m plots, 1 m plots in 10 m plots and among replicated cores within each sample would provide evidence of spatial scales at which variations in density occurred. This would obviously provide a thoughtful and/or experienced ecologist with observations on which to base models about processes influencing local dispersion and density (such as in Green & Hobson, 1970).

It would also provide the pilot estimates of costs (of laying out and moving among the plots and of analysing the cores) and variances to allow cost–benefit procedures to optimize sampling designs for any future work (see Section 9.11).

Nested designs of sampling programmes are, of course, not the only way to determine spatial patterns (Pielou, 1969) in density of an organism. Two commonly used alternatives are spatial autocorrelation (Cliff & Ord, 1973; Legendre & Fortin, 1989) and so-called block sampling (Kershaw, 1957; Greig-Smith, 1964; Pielou, 1974). Both use regularly spaced sampling units and identify scales of variability in a far less arbitrary manner than is described above. In this sense, they are far superior to having to make judgements about how to divide the area into scales for sampling. They do, however, both require very large efforts for sampling an area, if sufficient data points are to be accumulated to use the procedures. Spatial autocorrelation requires very long transects of replicated sample units for any analysis that includes more than small spatial scales (Pielou, 1974). The procedure re-uses data non-independently to examine correlations at increasing spatial intervals. Block sampling also involves non-independent repeated use of data in quadrats to examine increasing sizes of quadrats (Greig-Smith, 1964). Neither technique provides estimates of the magnitude of variability at different spatial scales that can be extracted independently from one another from the data.

A judicious mix of various techniques will usually provide the best way to examine spatial patterns. Only a nested sampling scheme is possible when there are real, microhabitat differences causing spatial scales to be sampled, as is the case with the insects on leaves. Nested analyses are also necessary (i.e. not simply an option) where different spatial scales must be incorporated into an experimental design to unconfound interpretation of differences at larger spatial scales (see the previous sections).

9.10 Nested analysis and temporal pattern

The arguments and discussion about spatial patterns are also relevant for temporal sampling. As described by Underwood (1993, 1994), there are quite common irrational interpretations of so-called seasonal data. It is, for example, not uncommon for samples to be taken at one time in each of four seasons in a year. At each time of sampling, several independent, representative units are sampled and the data are analysed for seasonal trends. The replicates are, however, spatial ones and there is no temporal replication to unconfound the possible seasonal differences from any faster fluctuations. For example, the densities of animals per unit area sampled may vary rapidly, so that there is considerable variation in the number present from one week to the next. There might be as much variability from one week to another as from one season to another. The consequences of this for the interpretation of larger temporal patterns are disastrous. Any difference among samples three months apart may not be a seasonal pattern at all (Figure 9.7a–c). It may simply reflect unreplicated sampling of weekly variability. Alternatively, if fluctuations at shorter time-scales occur, they may cancel out a real seasonal trend, so that there are no apparent differences among seasons. Such problems are illustrated in Figure 9.7d–f (and see Underwood, 1993, 1994).

The solution is to sample more often – either to detect the temporal trend by some regression procedure, or to create replication in the times of sampling. The latter is achieved by a temporally nested sampling scheme. In each season, several samples are taken at different times, randomly and independently scattered in the season. This is shown in Figure 9.7 for two different cases. At each time of sampling, several replicates are taken, scattered over the whole area to be sampled. This is important because otherwise the replicate samples will be in a small area of the study-site. If a different small area is sampled at each time, there is no way to separate temporal differences from spatial ones – any difference from one time to another could simply be due to differences from one area to another. But, if the same restricted area is sampled at each time, this is now the study-site and it is much *smaller* than had been described (and presumably than is required by the hypothesis causing the data to be collected).

Now, because differences from time to time *within* a season can be separated from differences from season to season, there is no confounding of seasonal and smaller time-scales (Figure 9.7 and Table 9.10). As with spatial sampling, there can be several temporal scales that may be of interest (annual, seasonal, in response to events such as floods or rainfall, etc.). The choice of scales to sample may be determined by theory about what sorts of temporal variation there are, fixed by observations about apparent variations in the population or arbitrarily chosen to represent a series of possible temporal patterns (for example, variability among days, weeks, months, seasons and years). Alternatively, temporal auto-correlations can be used (see e.g. Box & Jenkins, 1976;

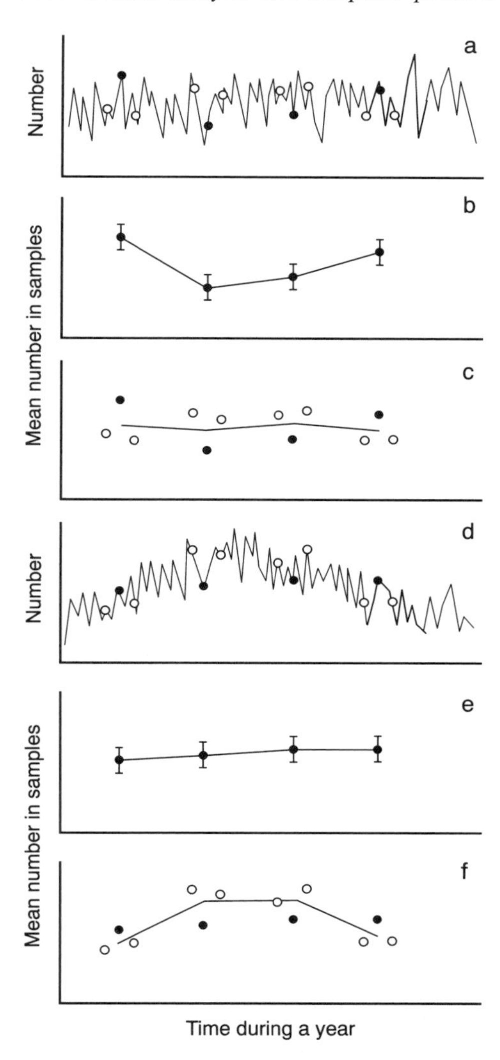

Figure 9.7. Confounding and nested sampling to examine temporal patterns. In (a), a noisy, fluctuating population with no seasonal trends in abundance is sampled. (b) The apparent pattern in data from single times of sampling in each season (filled circles in (a)). The pattern is caused by confounded variation at shorter time-scales. (c) The real lack of pattern shown by data averaged from three times of sampling in each season (filled and empty circles, in (a)). In (d), a noisy, fluctuating population has a marked seasonal cycle of abundance. (e) The apparent lack of pattern in data from single times of sampling in each season (filled circles in (d)). (f) The real pattern shown by data from three times of sampling in each season (filled and empty circles, in (d)).

Table 9.10. Nested analysis of temporal patterns. Several (t) times of sampling are scattered in each of the four seasons. At each time, n replicate sample units are examined. An illustration of the sampling design is in Figure 11.7. Note that if there is no variability among times of sampling in each season, *post hoc* pooling can be done

Source of variation	Degrees of freedom	Mean square estimates	F-ratio
Among seasons	3	$\mathrm{MS_S} = \sigma_e^2 + n\sigma_{\mathrm{T(S)}}^2 + \dfrac{tn\sum_{i=1}^{4}(S_i - \bar{S})^2}{3}$	$\mathrm{MS_S}/\mathrm{MS_{T(S)}}$
Among times (seasons)	$4(t-1)$	$\mathrm{MS_{T(S)}} = \sigma_e^2 + n\sigma_{\mathrm{T(S)}}^2$	$\mathrm{MS_{T(S)}}/\mathrm{MS_W}$
Within samples	$4t(n-1)$	$\mathrm{MS_W} = \sigma_e^2$	
Total	$4tn - 1$		

S_i indicates a seasonal difference if there are differences among seasons (equivalent to A_i in previous analyses).
$\sigma_{\mathrm{T(S)}}^2$ represents variance at time-scales shorter than seasonal.
σ_e^2 represents variance among replicates at all times of sampling.

Gottman, 1981; Connor, 1986; and see the discussion in the previous section).

If nested temporal scales are sampled, considerable care must be taken to ensure independence in the data from one time to the next (Section 7.14). This may be very difficult if a single site is revisited repeatedly.

One solution to the problem of non-independence in time would be to divide the study-site into numerous small patches. At each time of sampling, several independent patches are picked and replicate sample units examined in each one. The analysis of such a sampling design is indicated in Table 9.11. Now, there is no repeated sampling on the same patches. Of course, patches must be sufficiently scattered to ensure that sampling one of them does not influence adjacent ones. For example, sampling one area of scrub with nets to catch insects may cause sufficient disturbance to reduce the densities of insects in surrounding areas. A subsequent time of sampling will not give data on densities that are independent of this first time.

Despite prolonged pre-occupation with spatial patterns, few ecologists have examined temporal patterns in such a manner that allows clearly unconfounded interpretations.

Table 9.11. Nested analysis of temporal patterns. Several (t) times of sampling are scattered in each of the four seasons. At each time, p randomly chosen plots are sampled in the study area. These are different at (and therefore nested in) each time of sampling. At each time, n replicates are taken in each plot. If there is no variation among times within seasons, *post hoc* pooling can be done. More pooling can be done if there are no differences among plots

Source of variation	Degrees of freedom	Mean square estimates
Among seasons	3	$MS_S = \sigma_e^2 + n\sigma_{P(T(S))}^2 + pn\sigma_{T(S)}^2 + \frac{tpn\sum_{i=1}^{4}(S_i - \bar{S})^2}{3}$
Among times (seasons)	$4(t-1)$	$MS_{T(S)} = \sigma_e^2 + n\sigma_{P(T(S))}^2 + pn\sigma_{T(S)}^2$
Among plots (times(seasons))	$4t(p-1)$	$MS_{P(T(S))} = \sigma_e^2 + n\sigma_{P(T(S))}^2$
Within samples	$4tp(n-1)$	$MS_W = \sigma_e^2$
Total	$4tpn-1$	

S_i indicates a seasonal difference if there are differences among seasons (equivalent to A_i in previous analyses).
$\sigma_{T(S)}^2$ represents variance at time-scales shorter than seasonal.
$\sigma_{P(T(S))}^2$ represents variance among plots at any time of sampling in any season.
σ_e^2 represents variance among replicates at all times of sampling.

9.11 Cost–benefit optimization

One useful property of nested sampling and experimental designs is that they provide exact solutions for optimizing the use of replication at the various levels included in the experiment. Consider the simplest case of an experimental manipulation of grazers to determine their effects on the concentration of unpalatable chemicals in their food plants. Variations in concentrations of such chemicals have been shown to be positively correlated with the abundances of grazing insects. The model to explain this observed correlation is that insects cause the response by the plants; where there are more grazers, the plants respond by production of more allelochemicals. Where grazing is less intense, the plants reduce their concentrations of defensive chemicals.

To test the likelihood of this being correct, it is proposed that removing insects from areas where plants have a large concentration of the chemicals will eventually cause a decrease, relative to plants that continue to have intense grazing. Insects are excluded by mesh cages, which each cover several plants. The cages cast considerable shade, so mesh controls with adequate holes in the sides to ensure insects freely enter are needed in addition to unmanipulated patches of plants. Thus, the three treatments ($a = 3$) are replicated in b experimental units (cages, mesh controls or marked plots). In each unit, some number of plants (n) is to be assayed to get a representative measure of the concentration of allelochemicals.

Optimization of replication is appropriate to ensure that the number of replicates per unit (n) and the number of units per treatment (b) are chosen to maximize the power of the test (see Sections 8.3 and 8.4) for the minimal possible cost in terms of money per replicate or per unit, or the time taken to get the data or maintain the experiment. The principles of this cost–benefit procedure are straightforward. The notion is to reduce the cost to a minimum while maximizing the 'benefit' in terms of power to detect differences among treatments. As demonstrated in Sections 8.3 and 8.4, the power of the experiment depends on the probability of Type I error, the effect size, the intrinsic variability in the system studied and the size of samples. For any probability of Type I error and effect size, the ideal is to maximize the precision with which means are estimated for the minimal possible size of samples (i.e. cost of experiment). Thus, the best designs have the smallest standard errors (as a measure of precision – see Section 5.2) for the smallest possible cost. Examples of calculations for real experiments in marine ecology are given by Kennelly & Underwood (1984, 1985).

In the case of a nested design, there are two sources of variation in the standard errors (σ_e^2 and $\sigma_{\mathrm{B(A)}}^2$ both contribute) and two costs of replication (the cost per replicate X_{ijk} and the cost of the b replicated units). The standard error associated with the estimation of the means of the treatments is the square-root of the variance within a treatment divided by the number of values used to calculate the mean (see Sections 5.2 and 8.6). It is convenient not to use the square-root, so the variance associated with estimated means of treatments is:

$$\begin{aligned}\text{Variance} &= (\text{Standard error})^2 \\ &= \frac{\text{Variation within treatments}}{\text{Size of sample used to estimate means of treatments}} \\ &= \frac{\text{Mean square among units within each treatment}}{\text{No. units} \times \text{No. replicates in each unit}} \qquad (9.17)\end{aligned}$$

The mean square among units (B(A) in Table 9.3) is the variance within the treatments – it includes the variance among replicates in each unit and the variance among units. The mean of each treatment is averaged over n replicates in each of b units. Therefore the variance of estimated means of treatments is as in Equation 9.18. According to the results in Table 9.3:

$$\text{Variance} = V \text{ estimates } \frac{\sigma_e^2 + n\sigma_{\mathrm{B(A)}}^2}{bn} \tag{9.18}$$

The cost of the experiment can be estimated or measured in accordance with the following:

$$\text{Cost of each treatment} = C = bnc_k + bc_j \tag{9.19}$$

where c_k is the cost of each of the n replicates per experimental unit (i.e. of the bn replicates in all b units in a treatment) and c_j is the cost of each of the b experimental units. To design the most efficient experiment, the product of these must be kept as small as possible, so:

$$VC = c_k\sigma_e^2 + \frac{c_j\sigma_e^2}{n} + nc_k\sigma_{\mathrm{B(A)}}^2 + c_j\sigma_{\mathrm{B(A)}}^2 \tag{9.20}$$

The minimal value of this product can be found by first differentiating with respect to n and setting the differential to zero to identify its minimum. Yes – we have all forgotten differentiation – a fact that should cause its inventor to turn in his grave (which allows us now to dig up Newton's and Leibnitz's remains and determine which of them really did invent calculus; see the controversy in Hawking, 1988).

$$\frac{\mathrm{d}(VC)}{\mathrm{d}n} = \frac{-c_j\sigma_e^2}{n^2} + c_k\sigma_{\mathrm{B(A)}}^2 = 0 \tag{9.21}$$

$$c_k\sigma_{\mathrm{B(A)}}^2 = \frac{c_j\sigma_e^2}{n^2} \quad \text{and} \quad n = \sqrt{\left(\frac{c_j\sigma_e^2}{c_k\sigma_{\mathrm{B(A)}}^2}\right)} \tag{9.22}$$

If we obtain estimates of the costs and the variances, we can calculate the optimal number of replicates per experimental unit (n). The costs are obtained from the literature, previous similar experimental and quantitative work and, as discussed below, pilot studies. The estimates of variance can also be trawled from the literature (occasionally), from previous experience and estimates or from pilot studies. Assume some preliminary work has already been done to allow the calculation of mean squares as in a nested analysis. Specifically, assume a small experiment has been done with $n = 3$ replicate plants in $b = 2$ replicate units for the three experimental treatments. The analysis of variance is in Table 9.12, where estimates of costs and variance are calculated.

Table 9.12. Example calculation of costs and variances per treatment in a nested analysis of variance with $a = 3$ treatments (cages, mesh controls, open areas), $b = 2$ replicated units per treatment and $n = 3$ replicate plants per unit

(a) Analysis of variance of data from pilot study

Source of variation	Degrees of freedom	Mean square	Mean square estimates
Among treatments = A	2	138.3	
Among units(treatment) = B(A)	3	80.6	$\sigma_e^2 + 3\sigma_{B(A)}^2$ (see Table 9.3)
Within samples	12	48.7	σ_e^2
Total	17	–	–

Costs: per replicate = C_k = \$11.00; per unit = C_j = \$108.

(b) Determining n

Variance within samples = σ_e^2 estimated as 48.7

Variance among units = $\sigma_{B(A)}^2$ estimated as

$$\frac{\text{Mean square B(A)} - \text{Mean square within}}{n} \quad \text{(from MS estimates)}$$

$$= (80.6 - 48.7)/3 = 10.6$$

Therefore, from equation 9.22:

$$n = \sqrt{\left[\frac{(108)(48.7)}{(11)(10.6)}\right]} = 6.7 \approx 7$$

(c) Determining b

Using cost equation: $C = \$600$ available per treatment $= bnC_k + bC_j$
$= (b \times 7 \times 11) + (b \times 108)$

Therefore: $b = 600/185 = 3.2 \approx 3$ to keep it within the available cost.
Or:
Using variance of treatment means:

$$V = (\sigma_e^2 + n\sigma_{B(A)}^2)/nb \quad \text{(see Equation 9.18)}$$

$$= (48.7 + 7 \times 18.1)/7b = 175.4/7b$$

SE is to be no more than 10% of mean; mean = 20.5, SE = 2.05, $V = 4.20$

Therefore: $b = 175.4/(7)(4.20) = 5.9 \approx 6$

Once the above formula (Equation 9.22) has been used to determine n, it is necessary to calculate b. There are two options for doing this, depending on whether the crucial limiting factor in the design of the experiment is its total cost or the precision which must be achieved.

Of the two, being limited by the total time or money (or other limiting constraint) is more usual in ecology and is also easier to understand. The total resources available for the experiment must be divided among the a experimental treatments. Usually, the division will be reasonably equal. It is, however, often cheaper to set up a control for some field experiment because it requires no expensive installation or maintenance of fences or enclosures or watering machines and so forth. Some adjustment must be made to the following description of the procedure to deal with situations where costs are very different from one treatment to another. The total resources allocated or available to the experiment divided by the number of treatments gives C in Equation 9.19. The only unknown in the equation (once the costs per replicate and per unit have been ascertained) is b, because n comes from Equation 9.22. Solving for b is straightforward, as in Table 9.12c. The cost–benefit procedure thus provides the optimal allocation of effort to b and to n, according to their relative variances and costs. In this case, there should be seven replicates in each of three replicated units in each treatment (i.e. a total of 9 units and 63 replicate plants). The cost of the experiment is now $C = \$555$ per treatment (from Equation 9.19) and a total cost of \$1665.

As a result of this procedure, it is possible using Equation 9.18 to estimate the standard errors that would result from such an experiment. Once the resources to be allocated have been determined, the precision of the estimates of means of the treatments and therefore the power of the experiment to detect differences are set. The alternative approach to an optimal sampling design uses a pre-determined level of precision to optimize n and b. The result then determines the cost of the experiment. Suppose that it has been decided, in advance of the experiment, that the standard errors must be no greater than 10% of the means. A pilot study has suggested that the treatment where grazers are experimentally excluded has the greatest mean concentration of allelochemicals, at about 20.5 g per 100 g of plant tissue. Thus, the standard errors should be no larger than 2.05 g per 100 g. This sets V, the variance of the sample means (as in Equation 9.18) to be 4.20. The only unknown in Equation 9.19 is b, given n and the two costs c_k and c_j. Re-arranging to solve for b is easy, as in Table 9.12c. Thus, there should be seven replicates in each of six replicated units per treatment (i.e. a total of 18 units and 126 plants). The total cost of the experiment will now be $C = \$879$ per treatment (Equation 9.19) and the whole experiment will cost \$2637.

Sometimes, of course, the optimization of n and b according to a set standard error will result in fewer experimental units being needed than was originally guessed – so that an experiment would be cheaper than is found by arbitrarily allocating resources to it and solving for b using the cost equation. All this suggests that *planning* the experiment with some quantitative goals (in the form of precision needed or power to detect specified alternative hypotheses) will always be a more realistic way of finding out about nature.

It is surprising, given the lack of cost–benefit analyses in the literature, that so many experiments get funded by grant agencies. Suppose it has been determined that an experiment should be done with the standard errors pre-determined to be, say, S, in order to achieve adequate power for detecting differences among treatments. The cost of the experiment to achieve this is \$2637 as in the example above. The grant agency supplies only \$1800 (due to the mysterious and arbitrary cuts imposed by never-identified policies and procedures). You now should not do the experiment at all – the resources are inadequate.

Alternatively, supposing you request an arbitrary sum of \$1800 to do the experiment. It is an arbitrary calculation based on some arbitrary choice of number of replicates (n) and units (b). You obviously have no opinion about the power or precision of your experiment, so to do it requires funding only for the minimal n and minimal b to allow an analysis. The cost of the previously described experiment, with n and b equal to 2 (their minimal possible value) is (from Equation 9.19) \$780, which is all the agency should give you (if it bothered to fund an experiment for which power and precision don't matter). For such an experiment, the standard error of means will be (from Equation 9.18) 4.61 or about 22% of the largest mean. The power of such an experiment to detect differences among experimental treatments can be shown to be small.

Optimization using cost–benefit procedures is a valuable tool in experimental design. Its use is, however, limited by the experimenter's capacity to determine, in advance, the appropriate precision. If this determination were possible, one could also calculate the minimal alternative to the null hypothesis which the experiment should detect (see Sections 8.3 and 8.4). Under these circumstances, n would be chosen using the cost–benefit procedure and b should be chosen to keep the probability of Type II error small (i.e. to create adequate power to detect differences if the null hypothesis were not true).

Finally, where there are more levels of nesting in the experiment, because replication is required at a number of scales in the experiment, the procedures extend to allow an exact determination of the number of replicates at all scales except the largest one. The final level of replication is again determined from the cost equation or by a choice of maximal standard error. A worked example was given in Underwood (1981).

9.12 Calculation of power

The procedures for calculation of the power of a nested analysis of variance are straightforward extensions of the procedures described for a single factor experiment (Section 8.4) when the experimental (or top) source of variation is a random factor (Section 8.2). So, in the examples of sampling at different spatial scales described in Section 9.9, the largest spatial scales (farms in one case and 100 m × 100 m plots in the second example) were random factors.

For any level of the analysis below the highest one, the source of variation is a random factor – the units sampled are, by definition, representatives of a large number that could equally well have been chosen as samples. Thus, orchards, trees, branches, twigs were all randomly chosen and other orchards, trees, etc., could have been used instead. The 10 m × 10 m and 1 m × 1 m plots in the second example in Section 9.9 were randomly chosen. For these, as with a random highest level, the determination of power is based on the variance component procedure in Section 8.4. Power depends on the sample size, the intrinsic variability, the effect size and the probability of Type I error (α). For any given α (say, $P = 0.05$ of Type I error), what we need to know is the variability so that we can calculate the power for chosen effect sizes and sizes of samples.

Any level (w) of a nested analysis of variance has a mean square:

$$\mathrm{MS_W} \text{ estimates } \sigma_e^2 + l\sigma^2_{u(v(w\ldots))} + lm\sigma^2_{v(w\ldots)} + lmn\sigma^2_{w(\ldots)} \tag{9.23}$$

where l, m, n, etc., represent the number of replicate units at each successively higher level in the hierarchy and u, v, $w \ldots$, etc., represent the successively higher levels of the hierarchy. σ_e^2 represents variability among replicate units at the lower level. To test for significant differences among the units at some level (w) requires an F-ratio versus the mean square of the level below (v):

$$\mathrm{MS}_v \text{ estimates } \sigma_e^2 + l\sigma^2_{u(v(w\ldots))} + lm\sigma^2_{v(w(\ldots))} \tag{9.24}$$

Thus, to test the null hypothesis of no differences among units at level w requires an F-ratio:

$$F = \frac{\mathrm{MS}_{w(...)}}{\mathrm{MS}_{v(w(...))}} \tag{9.25}$$

with the degrees of freedom of these two mean squares. In terms of the procedure described in Section 8.4:

$$\theta = \frac{\sigma^2_{w(...)}}{\sigma^2_{v(w(...))}} \tag{9.26}$$

where z is the total number of replicates in that level of the analysis. This is best illustrated by an example. Consider a simplified version of the situation described in Table 9.9, where differences in mean density of insects are being examined at several nested spatial scales. Insects are to be counted in n replicate quadrats in each of c replicate plots in each of b replicate fields in each of a randomly chosen farms. Thus, the mean square for differences among farms estimates:

$$\sigma^2_e + n\sigma^2_{\mathrm{C(B(A))}} + cn\sigma^2_{\mathrm{B(A)}} + bcn\sigma^2_{\mathrm{A}} \tag{9.27}$$

where σ^2_{A}, $\sigma^2_{\mathrm{B(A)}}$, $\sigma^2_{\mathrm{C(B(A))}}$ and σ^2_e represent variation among farms, fields, plots and quadrats, respectively. The mean square among fields estimates:

$$\sigma^2_e + n\sigma^2_{\mathrm{C(B(A))}} + cn\sigma^2_{\mathrm{B(A)}} \tag{9.28}$$

Using the above formulation, the power to detect differences among farms is based on θ (as defined in Chapter 8, where all other details can be found):

$$\theta = \frac{\sigma^2_{\mathrm{A}}}{\sigma^2_e + n\sigma^2_{\mathrm{C(B(A))}} + cn\sigma^2_{\mathrm{B(A)}}} \tag{9.29}$$

The crucial issue is to have a sensible and realistic estimate of the appropriate magnitude of σ^2_{A} that would represent an appropriate alternative to the null hypothesis. This was discussed in Section 8.4. There are no new issues to consider just because there are more components of variances in the relevant mean squares. Note, however, that, in this example, the highest level of the analysis (the farms) were themselves levels of a random factor. The entire analysis conforms to the requirement that every level is a random factor. There are no fixed factors. If the highest level in the analysis is a fixed factor, this procedure for determining power is not applicable (see the discussion of mixed models in Chapter 10).

9.13 Residual variance and an 'error' term

Throughout this discussion of nested analyses, the 'lowest' component of the sources of variation has always been referred to as the 'within samples' variability. This measures the variance among replicates within the samples, σ_e^2. It will be convenient to use a specific and simpler term for this. It is often called the *residual* or *error* variation. Either term is acceptable provided that it is a genuine measure of variation among replicates within samples completely nested inside every other source of variation. In the case of nested analyses, every term below the highest level is completely nested in the higher levels. Any term (or level) of replication could therefore serve as a residual source of variation for all higher levels.

In many analytical models, considerable care must be taken to ensure that a term used as an error or residual term really represents proper independent replication. This is discussed in full in Chapter 12. It is, however, convenient to introduce the terminology now.

In the linear model, for a nested analysis (Equation 9.2), the residual component of variation was represented by e_{ijk}. Technically, following the convention used for the nested replicated units (the B term in Equation 9.2), this should be written differently. The replicated units were denoted by $B_{j(i)}$, meaning unit j nested in treatment i. The brackets identify the nesting, or specific inclusion within a particular level of another factor. Later, in non-nested models, it will be important to distinguish between those sources of variation that are nested and those that are not. Using brackets, as in this convention, will enable this distinction. Thus, a source of variation C nested in levels of a source of variation B will be denoted as C(B). For example, in Table 9.9, the individual trees (I) nested in each orchard (O) nested in each farm (F) are listed as I(O(F)). The brackets keep track of the nesting. So it is with the relevant subscripts. The kth tree in the jth orchard on the ith farm is $I_{k(j(i))}$.

In the light of this convention, the residual term should be rewritten, so that Equation 9.2 becomes:

$$X_{ijk} = \mu + A_i + B_{j(i)} + e_{k(j(i))} \tag{9.30}$$

This terminology will be used from here on. Note that the linear equation for the single-factor model (Equation 7.7) should be:

$$X_{ij} = \mu + A_i + e_{j(i)} \tag{9.31}$$

The final point to make about residual variances is that mentioned above. Given the definition of residual, in the case of the insects on the

Table 9.13. The Residual term in an analysis of variance. The experiment has $i = 1 \dots a$ treatments, $j = 1 \dots b$ replicate units in each treatment and $k = 1 \dots n$ replicates in each unit. SS_A, $SS_{B(A)}$ and SS_T are as in Table 9.3

(a) Using the mean of each unit ($\bar{X}_{j(i)}$) as the replicates; there are now b replicates in each of the a treatments.

Source of variation	Sum of squares	Degrees of freedom	Mean square estimates
A′	$\frac{\sum_{i=1}^{a}\left(\sum_{j=1}^{b}\bar{X}_{j(i)}\right)^2}{b} - \frac{\left(\sum_{i=1}^{a}\sum_{j=1}^{b}\bar{X}_{j(i)}\right)^2}{ab}$	$a-1$	$\frac{\sigma_e^2}{n} + \sigma_{B(A)}^2 + \frac{b\sum_{i=1}^{a}(A_i - \bar{A})^2}{(a-1)}$
	$= \frac{\sum_{i=1}^{a}\left(\sum_{j=1}^{b}\sum_{k=1}^{n}X_{ijk}\right)^2}{bn^2} - \frac{\left(\sum_{i=1}^{a}\sum_{j=1}^{b}\sum_{k=1}^{n}X_{ijk}\right)^2}{abn^2}$		$= \frac{MS_A}{n} = MS_{A'}$
	$= \frac{SS_A}{n}$		
B(A)′	$\sum_{i=1}^{a}\sum_{j=1}^{b}\bar{X}_{j(i)}^2 - \frac{\sum_{i=1}^{a}\left(\sum_{j=1}^{b}\bar{X}_{j(i)}^2\right)^2}{b} = \frac{SS_{B(A)}}{n}$	$a(b-1)$	$\frac{\sigma_e^2}{n} + \sigma_{B(A)}^2 = MS_{R'}$
Total	$\sum_{i=1}^{a}\sum_{j=1}^{b}\bar{X}_{j(i)}^2 - \frac{\left(\sum_{i=1}^{a}\sum_{j=1}^{b}\bar{X}_{j(i)}\right)^2}{ab} = \frac{SS_T}{n}$	$ab-1$	–

The test for A is $F = MS_{A'}/MS_{R'} = MS_A/MS_{B(A)}$ with $(a-1)$ and $a(b-1)$ degrees of freedom.

(b) The analysis in (a) is a single-factor experiment. X_{ij} is the mean of n plants in each of the $j = 1 \ldots b$ replicates in each of the $i = 1 \ldots a$ treatments.

Source of variation	Sum of squares	Degrees of freedom	Mean square estimates
A	$\dfrac{\sum_{i=1}^{a}\left(\sum_{j=1}^{b} X_{ij}\right)^2}{b} - \dfrac{\left(\sum_{i=1}^{a}\sum_{j=1}^{b} X_{ij}\right)^2}{ab}$	$a - 1$	$\sigma_e^{2'} + \dfrac{b\sum_{i=1}^{a}(A_i - \bar{A})^2}{n(a-1)}$
Residual	$\sum_{i=1}^{a}\sum_{j=1}^{b} X_{ij}^2 - \dfrac{\sum_{i=1}^{a}\left(\sum_{j=1}^{b} X_{ij}\right)^2}{b}$	$a(b - 1)$	$\sigma_e^{2'}$
Total	$\sum_{i=1}^{a}\sum_{j=1}^{b} X_{ij}^2 - \dfrac{\left(\sum_{i=1}^{a}\sum_{j=1}^{b} X_{ij}\right)^2}{ab}$	$ab - 1$	

$\sigma_e^{2'}$ in this design includes σ_e^2 and $\sigma_{B(A)}^2$ components and any other unknown component that is smaller and nested in the replicate plants.

trees (Table 9.9), the leaves, twigs, branches, trees, orchards could each have been considered as a residual. It makes no sense to do this, in this case, because of the possibilities of *post hoc* pooling. It is, however, quite sensible to imagine that, instead of the number of insects per unit area of leaf, the total number of insects on a set of n leaves could have been counted. This measure is replicated on twigs on each branch. In this case, the twigs would form the replication at the lowest level in the analysis, but are still fully nested in everything else and can serve as a properly defined residual. In such an analysis, hypotheses about the variability among twigs could not be tested because, by summing the insects over a set of leaves, there is no measure of variability within twigs (i.e. among leaves on each twig).

The same argument holds for branches (the number of insects could have been summed over n leaves and m twigs on each branch), making the branches the replicates used in the residual term – and so on for trees and orchards. In this example, the whole purpose was to estimate variability in abundances at every scale in the experiment, so lumping things together would defeat the purpose.

In some nested designs, it may be convenient to eliminate terms from the model by adding or averaging data over some level of replication. Because the terms are completely nested in all higher levels, this preserves the structure of the test. Consider the example in Table 9.3. There are $1 \ldots i \ldots a$ experimental treatments, each replicated in $1 \ldots j \ldots b$ units in each treatment. In each unit, there are $1 \ldots k \ldots n$ replicates. The model is as in Equation 9.2 and the analysis is in Table 9.3. Now, suppose instead the replicates in each unit are sampled, but the data are averaged. There is no intrinsic interest in the variability among such replicates, nor among the units. The replicates are taken only to ensure that the sample in each unit is properly representative of the unit.

As a specific example, consider the experiment on grazers discussed in Section 9.11 and in Table 9.12. There is no particular need to be interested in the variability among the plants in each unit. There is, however, a real need to sample several plants in each unit. If only one plant were taken, it may not be at all representative of the population of plants in the units. So, several randomly chosen plants are examined to provide a representation of the unit. These were kept separate in the analysis in Table 9.12. The alternative would have been to average the data over all plants in each unit, providing a single, but representative measure of allelochemicals in each unit. The analysis is now the single-factor one in Table 9.13b.

Note that, because averages over the n replicates are used, the sums of squares in Table 9.13b are those in Table 9.3 divided by n.

The mean square estimates are now as shown in Table 9.13b. A valid test for differences among A is still provided by $F =$ means $\text{square}_A/\text{mean square}_{B(A)}$ and the fact that these mean squares are those from Table 9.3 divided by n is immaterial – the n cancels. So, in this case, a valid test for A was possible using B(A) as the appropriate residual. This was possible because all the terms were nested. In this experiment, it makes no sense to do the analysis as in Table 9.13b, because, if there is no variability among levels of B(A), pooling would give a better test for A in Table 9.12.

The point is, however, that the residual in a properly designed experiment always includes all smaller, but unidentifiable nested components of variation. In the case of the nested analysis of spatial pattern, the ecologically interesting issues might have concerned only orchards and farms, with trees as replicates. The fact that branches, twigs and leaves can be identified as smaller, nested scales of potential variation is unimportant. They could be ignored (provided the sampling of each tree is representative of the tree, e.g. by counting insects on random leaves from several twigs and branches scattered over the tree). Their contributions to variation will still exist, but unseparated, in the mean squares for trees, orchards and farms. In a nested analysis of variance, the residual term always contains numerous unidentifiable components of variation. Because it is a properly sampled residual term, completely nested in every higher term, these unknown components are present in all higher mean squares (as shown in Table 9.13) and cannot cause confounding in tests about the higher levels.

It is therefore crucial to ensure that the residual term has this property and the design of the experiment or sampling programme *must* be carefully done to ensure there is a proper residual. Examples where this is not likely are given in Chapter 12.

10
Factorial experiments

10.1 Introduction

Factorial experiments are investigations of more than one experimental treatment examined simultaneously. They are widespread in ecology, but probably not widespread enough. Examples are experiments on, say, competition done simultaneously in several places. Observations are available that densities of animals of one species (A) are smaller where another species (B) that feeds on the same seeds is numerous. The model proposed to explain this is that there is competition for food between the two species and, where B are numerous, they decrease the abundance of A. The hypothesis is proposed that removal of B from some areas will cause an increase in abundance of A. The null hypothesis is that there will be no difference in density of A, or A will decrease in abundance where B are removed, compared with plots where B are left intact. The experimental treatments are plots where B are removed by trapping and then continued trapping is done to keep removing any immigrants during the experiment. Procedural control plots are established in which traps are set, but then triggered so that no animals can be caught. This creates the same disturbance to the habitat (e.g. due to people walking around) as in the experimental plots. Finally, there are undisturbed control plots. At the end of the experiment, traps are used in all three sets of plots to estimate the densities of species A.

The original observations also indicated that densities of species A are different in different places and, *in all areas*, there is a tendency for density of A to be smaller where B are numerous. Thus, there is a general model of competitive interactions that applies to lots of places. To examine this proposition, the hypothesis is that removal of species B *in several different places* will have the same effect on numbers of A. The experiment is therefore done simultaneously in several sites, chosen simply to represent the sorts of place where the two species occur.

In this sort of experiment, there is a fixed factor – the experimental treatments – with three levels. There is also a random factor: the places chosen to do the experiment. This has a number of levels chosen by the experimenter (or set by logistical constraints). The sites are chosen solely to represent different parts of the habitat. Any set of such sites would be appropriate to include in the experiment – the ones chosen are a representative set of places. Any set would do, so this is a random factor (see Section 8.2).

This is a factorial experiment. It includes a combination of two different experimental factors.

A second example is the situation where it has been observed that, in each of three species of mice, individuals with parasitic ticks are smaller than those without. The model proposed to explain this is that, in each species, young individuals that have ticks grow more slowly than those that are uninfested. The faster-growing individuals remain uninfested. From this, the hypothesis is proposed that removal of ticks from young mice will cause them to grow more quickly than those that are left with ticks on them and they will grow at the same rate as individuals that had no ticks at the start. To test this, three samples of mice of a given species are chosen. One has no ticks and the other two have ticks. The ticks are removed from one of these latter two sets, leaving the others as controls. This procedure is done with each of the three species.

There must be an experimental removal of ticks from one treatment. If the so-called 'natural experiment' were done by comparing growth of mice in areas naturally with and without ticks, they should be different. These were the original observations! The difference in growth could be, as an example, due to the competing model that genetically slower-growing mice, for whatever reason, are more likely to have ticks than are mice that naturally grow more quickly. The experimental removal of ticks would unconfound this model from the one being tested above.

There are two experimental factors (species and treatments). Both factors in the experiment are fixed. They are both in the experiment because they are required by the hypotheses.

Factorial experiments are those that seek to investigate more than one experimental treatment, simultaneously and in combination. In many cases, there really are two or more factors included because hypotheses require them. Such experiments include studies on effects of diet on different genders of animals, effects of different fertilizers on plants at a range of densities, etc. In other cases, one treatment or factor is examined in a range of places or times, to ascertain the generality of findings, or to ensure that appropriate replication is examined. Again, this is a random factor.

In general, factorial experiments are designed to be *orthogonal*. Orthogonality is the property that every level of one factor is present in the experiment in combination with every level of the other factor (or combinations of levels of other factors if there are more than two others). Ensuring orthogonality is important in order to analyse interactions, as discussed below (Sections 10.6 and 10.11).

Where two factors are not orthogonal to each other in an experiment, they should, to retain a valid structure and a valid analysis, be organized so that one is independently sampled as a nested factor. Thus, if two experimental factors are not orthogonal, some levels of one factor are not present with all levels of the other and differences are confounded. Unless the second factor is made properly nested (Chapter 9), the analysis will be invalid.

A simple example is illustrated in Table 10.1. In the first case, there is a fully orthogonal arrangement of experimental treatments and places (Table 10.1a). Every treatment is set up, with n randomly placed, independent replicates, in each of four places. Such a design is factorial and most useful for determining first whether there are differences among the experimental treatments. At the same time, it can be determined how much difference there is from place to place. Finally, as explained below, this design is the only one that allows estimation of how much variation there is from place to place in the differences among (i.e. the effects of) the experimental treatments. This is measured as the *interaction* between the two factors.

In the second design (Table 10.1b), there is no attempt to measure the interactions between the two factors. The places have been assigned at random to the experimental treatments as a nested factor. The experiment is well suited to an analysis of how much difference there is among the experimental treatments over and above the intrinsic differences that occur from one place to another. This fully nested design is valid provided that the sets of places assigned to each treatment are each representative samples of the population of places that are appropriate to sample for the experimental hypothesis to be tested. This was explained in Chapter 9. Otherwise, the means of the sets of places are not equal at the start of the experiment – it is confounded. Any differences among treatments might be due to the treatments (as specified in the hypothesis) or due to intrinsic differences among the places sampled in each treatment.

Table 10.1. Orthogonal, nested and confounded arrangements of two experimental factors. Factor A is a manipulative treatment with three levels (E, experimental treatment; PC, procedural control; C, control). Factor B is a set of places included to provide some generality to the experimental results

(a) There are four places (1 . . . 4) arranged orthogonally with the treatments. Every treatment is present in each and every place. In each place, each treatment is replicated n times

Treatment		
E	PC	C
1	1	1
2	2	2
3	3	3
4	4	4

(b) Each of the treatments is set up in four different places (with n replicates in each place). Places 1–4, 5–8, 9–12 are fully nested in the treatments

Treatment		
E	PC	C
1	5	9
2	6	10
3	7	11
4	8	12

(c) A confounded design. The treatments are done in four places, but each place is used for only two of the treatments. There are n replicates of any treatment in each of the 6 places.

Treatment		
E	PC	C
1	1	3
2	2	4
3	5	5
4	6	6

The third case (Table 10.1c) is completely confounded. Different combinations of places have been assigned to each experimental treatment, but the sets of places are not independently sampled from one treatment to another. As a result, this design has none of the useful properties of the previous two. The differences among places cannot be averaged out, as in the nested design in Table 10.1b. For example, if one place happens, by chance, to have an unusually large number of species A, regardless of any effects of B, it will influence the means in two of the treatments, making them more like each other and more different from the third treatment than would be expected by chance, independent sampling as in the nested design.

In contrast, the orthogonal design is able to cancel out, by averaging, any differences among places so that they cannot influence the means of the three treatments. Any intrinsic differences among places are equally represented in each of the three treatments. If, as considered above, one of the places happens to have an above-average number of species A, this will be included in the average results of all three experimental treatments and therefore cannot affect the differences among them (although it will probably make the variation among places larger than would otherwise have been the case).

The non-orthogonal, non-nested design in Table 10.1c is inappropriate and should not be used. The differences among a set of places must either be cancelled by being orthogonally represented (Table 10.1a) or be averaged out by independent nested sampling (Table 10.1b). Some half-hearted compromise will always be a problematic and inferior attempt.

10.2 Partitioning of variation when there are two experimental factors

There are several ways to approach an understanding of partitioning a matrix of data to analyse an experiment with two factors. First, consider the entire analysis as though it were a single factorial experiment with *ab* experimental treatments, as shown in Tables 10.2 and 10.3. The data are as in Table 10.2, where the means of each relevant combination of experimental treatments are computed. The data can obviously be considered as a single factorial experiment, because there is no reason not to start with a conceptual single set of treatments. The null hypothesis would be:

$$\text{H}_0: \quad \mu_{ij} = \mu \quad \text{for all } i = 1 \ldots a, j = 1 \ldots b \qquad (10.1)$$

The data are partitioned as in Table 10.3a, which identifies the overall pattern for the analysis and also calculates how many degrees of freedom are involved in the analysis.

Table 10.2. Data from an experiment with two experimental factors. A has levels $1 \ldots i \ldots a$; B has levels $1 \ldots j \ldots b$, arranged orthogonally with levels of A. There are n replicates in each combination of A_i and B_j

		Level of factor A					
Level of factor B		1	2	…	i	…	a
1		X_{111}	X_{211}		X_{i11}		X_{a11}
		X_{112}	X_{212}		X_{i12}		X_{a12}
		⋮	⋮		⋮		⋮
		X_{11k}	X_{21k}		X_{i1k}		X_{a1k}
		⋮	⋮		⋮		⋮
		X_{11n}	X_{21n}		X_{i1n}		X_{a1n}
	Mean	$\bar{X}_{11}$	$\bar{X}_{21}$		$\bar{X}_{i1}$		$\bar{X}_{a1}$
2		X_{121}					
		X_{122}					
		⋮					
		X_{12k}					
		⋮					
		X_{12n}					
	Mean	$\bar{X}_{12}$					
⋮							
j					X_{ij1}		
					X_{ij2}		
					⋮		
					X_{ijk}		
					⋮		
					X_{ijn}		
	Mean	$\bar{X}_{1j}$			$\bar{X}_{ij}$		
⋮							
b							X_{ab1}
							X_{ab2}
							⋮
							X_{abk}
							⋮
							X_{abn}
	Mean	$\bar{X}_{1b}$			$\bar{X}_{ib}$		$\bar{X}_{ab}$

$$\bar{X}_i = \frac{\sum_{j=1}^{b}\sum_{k=1}^{n} X_{ijk}}{bn} = \frac{\sum_{j=1}^{b} \bar{X}_{ij}}{b} \qquad \bar{X}_j = \frac{\sum_{i=1}^{a}\sum_{k=1}^{n} X_{ijk}}{an} = \frac{\sum_{i=1}^{a} \bar{X}_{ij}}{a}$$

$$\bar{X} = \frac{\sum_{i=1}^{a}\sum_{j=1}^{b}\sum_{k=1}^{n} X_{ijk}}{abn} = \frac{\sum_{i=1}^{a} \bar{X}_i}{a} = \frac{\sum_{j=1}^{b} \bar{X}_j}{b} = \frac{\sum_{i=1}^{a}\sum_{j=1}^{b} \bar{X}_{ij}}{ab}$$

Table 10.3. Partitioning of a two-factor analysis of variance. Factor A has levels $1 \ldots i \ldots a$, arranged orthogonally with b levels of factor B $(1 \ldots j \ldots b)$. In each combination (the ab 'cells' of the experiment), there are n replicates

(a) Considering the analysis as a single factor with ab treatments

Source of variation	Sum of squares	Degrees of freedom
Among all cells	$\sum_{i=1}^{a}\sum_{j=1}^{b}\sum_{k=1}^{n}(\bar{X}_{ij} - \bar{X})^2$	$ab - 1$
Residual = Within cells	$\sum_{i=1}^{a}\sum_{j=1}^{b}\sum_{k=1}^{n}(X_{ijk} - \bar{X}_{ij})^2$	$ab(n - 1)$
Total	$\sum_{i=1}^{a}\sum_{j=1}^{b}\sum_{k=1}^{n}(X_{ijk} - \bar{X})^2$	$abn - 1$

(b) Among levels of factor A, ignoring factor B

Source of variation	Sum of squares	Degrees of freedom
Among levels of A	$\sum_{i=1}^{a}\sum_{j=1}^{b}\sum_{k=1}^{n}(\bar{X}_{i} - \bar{X})^2$	$a - 1$
Residual = Within samples	$\sum_{i=1}^{a}\sum_{j=1}^{b}\sum_{k=1}^{n}(X_{ijk} - \bar{X}_{i})^2$	$a(bn - 1)$
Total	$\sum_{i=1}^{a}\sum_{j=1}^{b}\sum_{k=1}^{n}(X_{ijk} - \bar{X})^2$	$abn - 1$

(c) Among levels of factor B, ignoring factor A

Source of variation	Sum of squares	Degrees of freedom
Among levels of B	$\sum_{i=1}^{a}\sum_{j=1}^{b}\sum_{k=1}^{n}(\bar{X}_{j} - \bar{X})^2$	$b - 1$
Residual = Within samples	$\sum_{i=1}^{a}\sum_{j=1}^{b}\sum_{k=1}^{n}(X_{ijk} - \bar{X}_{j})^2$	$b(an - 1)$
Total	$\sum_{i=1}^{a}\sum_{j=1}^{b}\sum_{k=1}^{n}(X_{ijk} - \bar{X})^2$	$abn - 1$

Now, start again and ignore any differences among the data that might be due to factor B. If B is ignored, the entire analysis is now equivalent to a single-factor analysis of variance of means of the levels of factor A. It has *a* treatments, each replicated *bn* times. So, the data can be used to test the null hypothesis of no differences among the means of levels of factor A, ignoring any influence of factor B, as in Table 10.3b.

Obviously, an entirely symmetrical argument can be made to analyse the data as a single factor, B, ignoring any influences or effects of A. This has *b* treatments, each with *an* replicates, as in Table 10.3c.

Clearly, the differences among means of the *ab* combinations of treatments identified in Table 10.3a have not all been accounted for by the two analyses in Tables 10.3b and c. The remaining differences can be identified purely empirically as:

Sum of squares among all treatments – Sum of squares among levels of factor A – Sum of squares among levels of factor B

and algebraically as:

$$\sum_{i=1}^{a}\sum_{j=1}^{b}\sum_{k=1}^{n}(\bar{X}_{ij}-\bar{X})^2-\sum_{i=1}^{a}\sum_{j=1}^{b}\sum_{k=1}^{n}(\bar{X}_{i}-\bar{X})^2 -\sum_{i=1}^{a}\sum_{j=1}^{b}\sum_{k=1}^{n}(\bar{X}_{j}-\bar{X})^2 \qquad (10.2)$$

This can be shown (see e.g. Winer *et al.*, 1991) to be equal to:

$$\sum_{i=1}^{a}\sum_{j=1}^{b}\sum_{k=1}^{n}(\bar{X}_{ij}-\bar{X}_{i}-\bar{X}_{j}+\bar{X})^2 \qquad (10.3)$$

and, later, will be demonstrated to be a useful measure of variability due to interactions. It is also worth noting that the partitioning of sums of squares into these sources of variation matches the degrees of freedom (Table 10.3). Noting the operational rule for determining degrees of freedom (Section 3.10), the variability among levels of factor A requires a calculation on *a* means (the $\bar{X}_i$ values) and knowledge of the overall mean $\bar{X}$. Thus, there are $(a-1)$ degrees of freedom. The calculation for *B* is self-evident.

The degrees of freedom for the interaction term are slightly more complex. They involve *ab* means for each treatment in the analysis (the $\bar{X}_{ij}$ values) and knowledge of the *a* means for levels of factor A (the $\bar{X}_i$ values) and the *b* means for levels of factor B (the $\bar{X}_j$ values). Thus, the degrees of freedom should be $(ab-a-b)$. Knowing all *a* values of $\bar{X}_i$

does, however, reduce the need to know all values of $\bar{X}_j$. Once all of the $\bar{X}_i$ values are known, their mean ($\bar{X}$) or total are known and therefore the mean and total of the $\bar{X}_j$ values is already known. Consequently, only $(b-1)$ of the $\bar{X}_j$ values need to be known; the final one is already calculable from $\bar{X}$. Thus, the number of entities in the set is ab and we need to know a and $(b-1)$ means in the set in order to calculate the sum of squares. The degrees of freedom are therefore $(ab - a - (b-1)) = (ab - a - b + 1)$. This equals $(a-1)(b-1)$ as in Table 10.4.

This final form of degrees of freedom leads to the usual terminology for describing interactions and, in more complex models, for calculating the degrees of freedom more simply. The degrees of freedom are the product of those for factor A and those for factor B. Thus, we call the interaction the A $\times$ B interaction.

Any orthogonal combination of experimental treatments (however many factors are involved) will potentially cause an interaction in the analysis. The degrees of freedom for each interaction are the products of the degrees of freedom in the factors making the interaction.

In the case of two factors, there is only one possible interaction: A $\times$ B. The degrees of freedom, calculated from the formula used to calculate the sum of squares of the interaction, are also useful to confirm that the partitioning done is appropriate and complete. However the data are to be divided into sources of variation, the degrees of freedom for the entire set (the total degrees of freedom) must always be the same and must always be totally identifiable. None should go missing.

In Table 10.3, various options were used as part of the partitioning. The total and residual degrees of freedom were identified in Table 10.3a. The other degrees of freedom were $(ab-1)$ associated with differences among all means sampled for all combinations of experimental treatments. Of these $(a-1)$ were identified as being associated with differences among levels of factor A (Table 10.3b) and $(b-1)$ were associated with factor B (Table 10.3c). This leaves:

$$(ab-1) - (a-1) - (b-1) = ab - a - b + 1 = (a-1)(b-1) \qquad (10.4)$$

which matches exactly the degrees of freedom calculated above to be associated with the sum of squares of the interaction term. Therefore, the analysis is complete – the identified sources of variation and their associated degrees of freedom sum, respectively, to the total sum of squares and the total degrees of freedom. The analysis provides an independent estimation of the variability associated with each source of variation.

Table 10.4. Analysis of variance of an experiment with two fixed experimental factors. Factor A has levels $1 \ldots i \ldots a$; factor B has levels $1 \ldots j \ldots b$. There are n replicates in each combination of levels i of factor A and j of factor B. Every set of replicates samples a population with variance σ_e^2 (i.e. all ab populations have homogeneous variances)

Source of variation	Sum of squares	Degrees of freedom	Mean square estimates
Among A	$\sum_{i=1}^{a}\sum_{j=1}^{b}\sum_{k=1}^{n}(\bar{X}_i - \bar{X})^2$	$a-1$	$\sigma_e^2 + bnk_A^2$
Among B	$\sum_{i=1}^{a}\sum_{j=1}^{b}\sum_{k=1}^{n}(\bar{X}_j - \bar{X})^2$	$b-1$	$\sigma_e^2 + ank_B^2$
$A \times B$	$\sum_{i=1}^{a}\sum_{j=1}^{b}\sum_{k=1}^{n}(\bar{X}_{ij} - \bar{X}_i - \bar{X}_j + \bar{X})^2$	$(a-1)(b-1)$	$\sigma_e^2 + nk_{AB}^2$
Residual	$\sum_{i=1}^{a}\sum_{j=1}^{b}\sum_{k=1}^{n}(X_{ijk} - \bar{X}_{ij})^2$	$ab(n-1)$	σ_e^2
Total	$\sum_{i=1}^{a}\sum_{j=1}^{b}\sum_{k=1}^{n}(X_{ijk} - \bar{X})^2$	$abn-1$	–

$$k_A^2 = \frac{\sum_{i=1}^{a}(A_i - \bar{A})^2}{(a-1)} = \frac{\sum_{i=1}^{a}A_i^2}{(a-1)}, \text{ because } \sum_{i=1}^{a}A_i = 0$$

$$k_B^2 = \frac{\sum_{j=1}^{b}(B_j - \bar{B})^2}{(b-1)} = \frac{\sum B_j^2}{(b-1}, \text{ because } \sum_{j=1}^{b}B_j = 0$$

$$k_{AB}^2 = \frac{\sum_{i=1}^{a}\sum_{j=1}^{b}(AB_{ij} - \overline{AB} - (\bar{A}_i - \bar{A}) - (\bar{B}_j - \bar{B}))^2}{(a-1)(b-1)} = \frac{\sum_{i=1}^{a}\sum_{j=1}^{b}AB_{ij}^2}{(a-1)(b-1)}, \text{ because}$$

$$\sum_{j=1}^{b}AB_{ij} = 0 \text{ for all } i \text{ and } \sum_{i=1}^{a}AB_{ij} = 0 \text{ for all } j$$

10.3 Appropriate null hypotheses for a two-factor experiment

The preceding account of how to partition the variability among the data allows a formal statement of relevant null hypotheses for the experiment. The partitioning was done on the basis that differences among levels of factor A could be examined by ignoring any effects of factor B. This requires that any influence of different levels of factor B must be equal for all levels of factor A. An alternative way of stating this is that the influence of factor B (i.e. whether levels of B differ or not) must be independent of the differences among levels of A (if there are any). So, for factor A there are a populations with means $\mu_1, \mu_2 \ldots \mu_i \ldots \mu_a$ and:

$$H_{01}: \quad \mu_1 = \mu_2 \ldots = \mu_i \ldots = \mu_a \tag{10.5}$$

assuming that there is independence between factors A and B. Simultaneously, for factor B:

$$H_{02}: \quad \mu_1 = \mu_2 \ldots = \mu_j \ldots = \mu_b \tag{10.6}$$

assuming that there is independence between factor B and factor A.

The validity of posing and testing these null hypotheses obviously depends on the assumption of independence being true. To validate this assumption requires a test of the further hypothesis that factors A and B are independent:

H_{03}: Differences among levels of factor A are independent of differences among levels of factor B

Any dependence between factors A and B can be estimated from the mean square for the A × B interaction. It is therefore essential to test the interaction first. Then, if there is no evidence for dependence between A and B, the differences among levels of A and among levels of B are logically analysable. Otherwise, they are not.

This is a matter of great importance because there is no logical validity in any test for differences among levels of either factor if there is significant interaction between them. How to deal with interactions and how to proceed to test the appropriate null hypotheses when interactions are present are discussed later (in Sections 10.5 and 10.6). Note that construction of tests and interpretation of any factor in a multifactorial analysis is totally and unambiguously dependent on there being no interaction between that source of variation and any other component of the analysis.

10.4 A linear model and estimation of components by mean squares

The model for an experiment with two orthogonal factors is:

$$X_{ijk} = \mu + A_i + B_j + AB_{ij} + e_{k(ij)} \tag{10.7}$$

where μ represents the mean of all the populations being sampled. X_{ijk} is the kth replicate in the combination of the ith level of factor A and the jth level of factor B (as in Table 10.2). A_i represents the difference between the overall mean (μ) and the mean of all (b) populations sampled in the ith level of factor A, if the null hypothesis is true and there *are* differences among the levels of factor A. This term is entirely analogous to that in the one-factor analysis, as in Chapter 7. B_j is the corresponding term for differences among levels of factor B if they exist. AB_{ij} represents an *interaction* component for the combination of the ith level of factor A and the jth level of factor B. This will be explained below.

Finally, $e_{k(ij)}$ identifies the individual residual or error term specifying the deviation of X_{ijk} from the mean of the population from which it was sampled. The brackets indicate that each replicate (k) is nested completely in the combination of treatments i and j (see Section 9.13). Thus, the experiment includes a proper, nested residual term. The experiment is independently replicated for all combinations of treatments.

The nature of the various terms and their relationships are illustrated in Figures 10.1 and 10.2. As before, various means of cells and levels of factors can be calculated from this linear equation:

$$\bar{X}_{ij} = \mu + A_i + B_j + AB_{ij} + \bar{e}_{ij} \tag{10.8}$$

$$\bar{X}_i = \mu + A_i + \bar{B} + \overline{AB}_i + \bar{e}_i$$

$$\bar{X}_j = \mu + \bar{A} + B_j + \overline{AB}_j + \bar{e}_j$$

$$\bar{X} = \mu + \bar{A} + \bar{B} + \overline{AB} + \bar{e}$$

where $\bar{A}$ and $\bar{B}$ are the means of differences among levels of factor A and factor B, respectively, if either relevant null hypothesis is false. $\overline{AB}_i$ is the average of AB_{ij} terms across the b levels of factor B in level i of factor A. $\overline{AB}_j$ is the average of the a different AB_{ij} terms in level j of factor B. $\overline{AB}$ is the mean of all AB_{ij} terms over the entire experiment (i.e. over all a levels of factor A and b levels of factor B).

$\bar{e}_{ij}$ is the average of residual terms over the n replicates in the combination of the ith level of factor A and the jth level of factor B. $\bar{e}_i$ is the average of the bn residual terms in the ith level of factor A, $\bar{e}_j$ is the average of the an residual terms in the jth level of factor B, $\bar{e}$ is the average of the abn residual terms in the entire experiment.

Provided the data are all independently sampled within and among samples and that the variances of the populations sampled in the experiment are all equal, it is quite straightforward to determine what the sums of squares and therefore the mean squares estimate. The principles

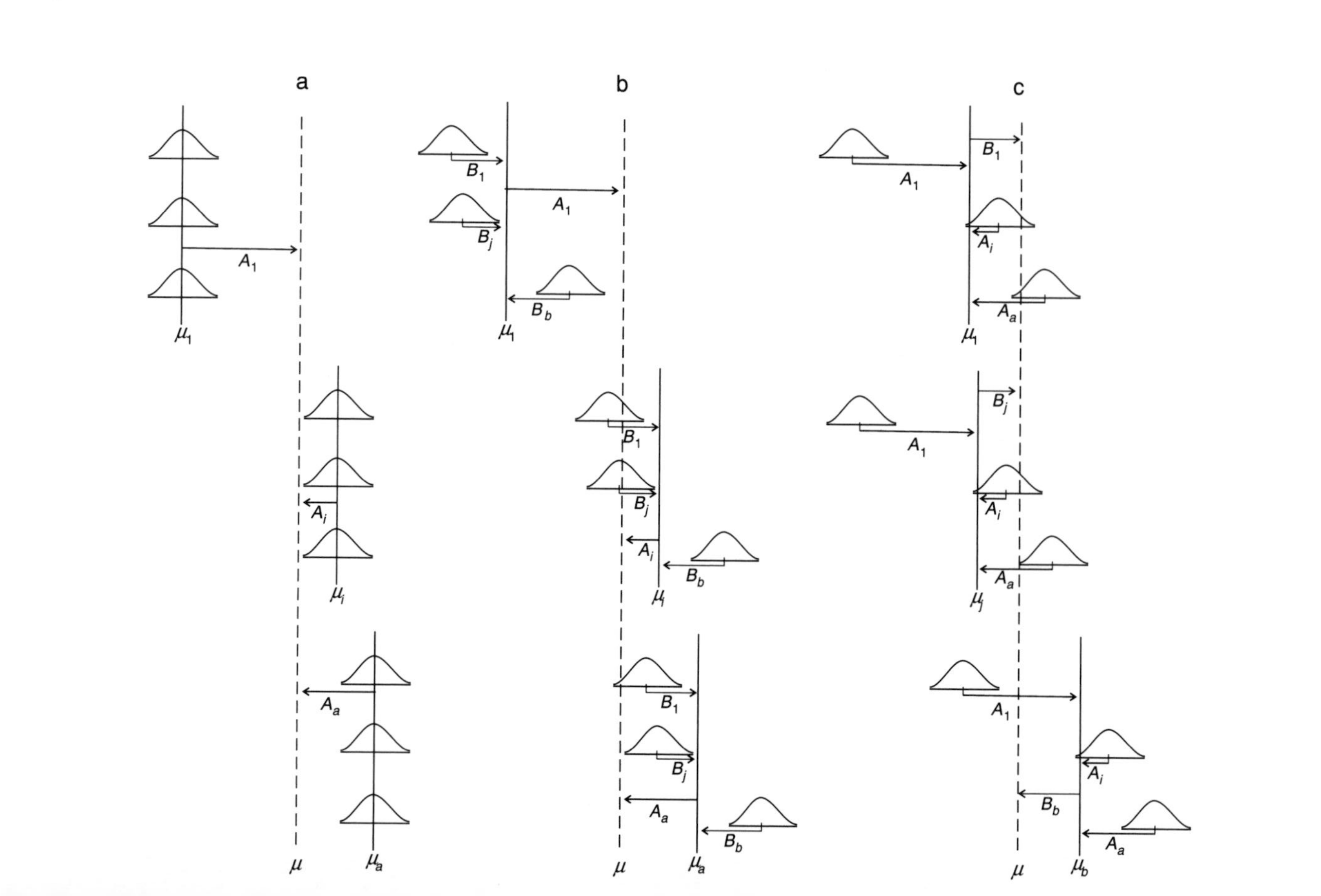

a
b
c
A_1
A_i
A_a
B_1
B_j
B_b
μ_1
μ_i
μ_j
μ_a
μ_b
μ

are as follow:

Residual sum of squares

$$= \sum_{i=1}^{a} \sum_{j=1}^{b} \sum_{k=1}^{n} (X_{ij} - \bar{X}_{ij})^2$$

$$= \sum_{i=1}^{a} \sum_{j=1}^{b} \sum_{k=1}^{n} [(\mu + A_i + B_j + AB_{ij} + e_{ijk})$$

$$- (\mu + A_i + B_j + AB_{ij} + \bar{e}_{ij})]^2 \quad (10.9)$$

and measures:

$$\sum_{i=1}^{a} \sum_{j=1}^{b} \sum_{k=1}^{n} (e_{ijk} - \bar{e}_{ij})^2 \quad (10.10)$$

Sum of squares among levels of factor A

$$= \sum_{i=1}^{a} \sum_{j=1}^{b} \sum_{k=1}^{n} (\bar{X}_i - \bar{X})^2$$

$$= \sum_{i=1}^{a} \sum_{j=1}^{b} \sum_{k=1}^{n} [(\mu + A_i + \bar{B} + \overline{AB}_i + \bar{e}_i)$$

$$- (\mu + \bar{A} + \bar{B} + \overline{AB} + \bar{e})]^2 \quad (10.11)$$

and measures:

$$\sum_{i=1}^{a} \sum_{j=1}^{b} \sum_{k=1}^{n} (A_i - \bar{A})^2 + \sum_{i=1}^{a} \sum_{j=1}^{b} \sum_{k=1}^{n} (\overline{AB}_i - \overline{AB})^2 + \sum_{i=1}^{a} \sum_{j=1}^{b} \sum_{k=1}^{n} (\bar{e}_i - \bar{e})^2 \quad (10.12)$$

Figure 10.1. Graphical illustration of differences due to two orthogonal experimental factors, A with $A_1 \ldots A_i \ldots A_a$ effects and B with $B_1 \ldots B_j \ldots B_b$ effects. (a) There are differences among levels of factor A, but not among levels of factor B. The overall mean (μ) and the means of levels of factor A (μ_i) are shown. There is no interaction between A and B. The populations being sampled in each combination of treatments are shown as normal frequency distributions. (b) There are differences among levels of A and, independently among levels of B. There is no interaction between A and B. The overall mean (μ) and the means of levels of factor A (μ_i) are shown. Note that B_j values are the same for every level of factor A. (c) Exactly as in (b), but the means of levels of factor B (μ_j) are shown. Note that A_i values are identical for every level of factor B.

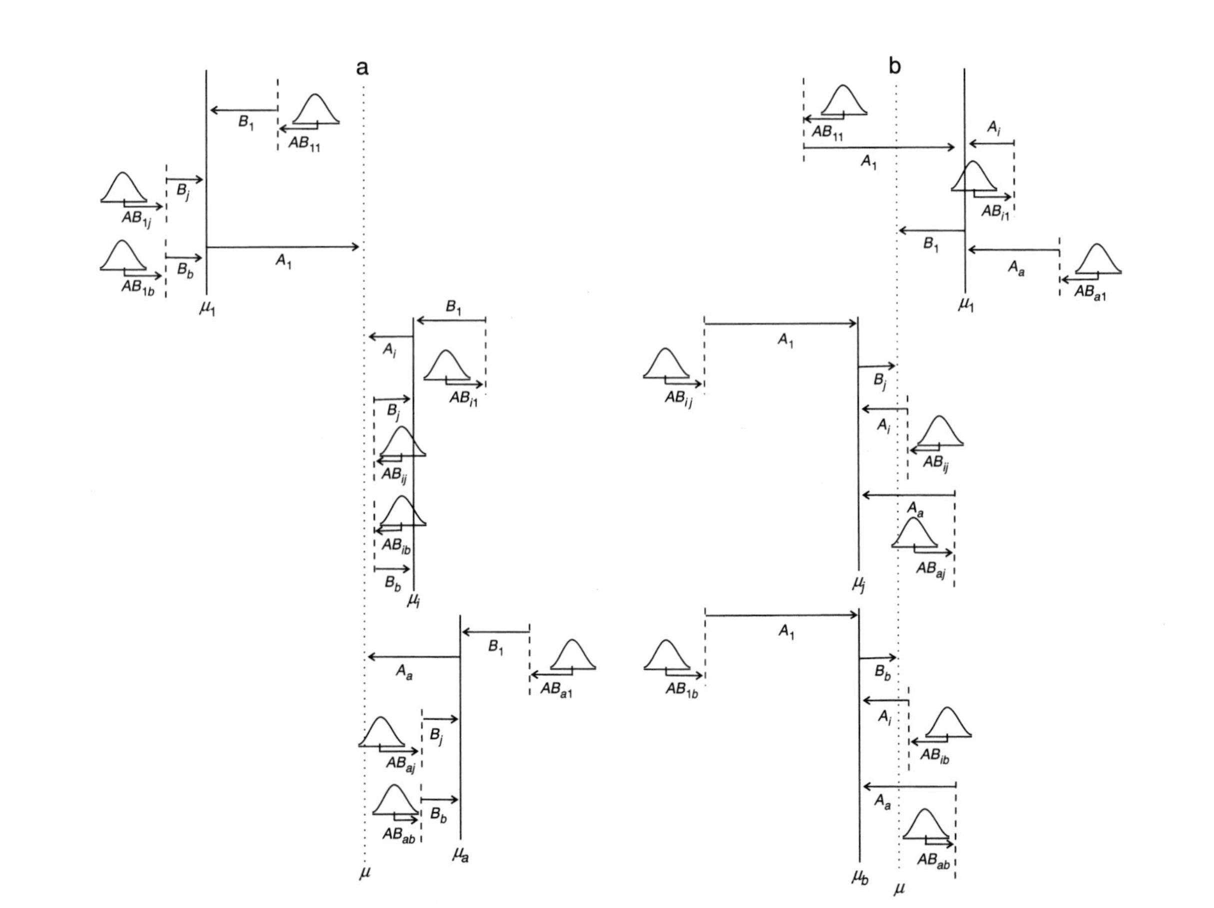

a
b

because all cross-products are zero provided that data are all independently sampled (see Section 7.5).

Sum of squares among levels of factor B

$$= \sum_{i=1}^{a} \sum_{j=1}^{b} \sum_{k=1}^{n} (\bar{X}_j - \bar{X})^2$$

$$= \sum_{i=1}^{a} \sum_{j=1}^{b} \sum_{k=1}^{n} [(\mu + \bar{A} + B_j + \overline{AB}_j + \bar{e}_j) - (\mu + \bar{A} + \bar{B} + \overline{AB} + \bar{e})]^2 \tag{10.13}$$

and measures:

$$\sum_{i=1}^{a} \sum_{j=1}^{b} \sum_{k=1}^{n} (B_j - \bar{B})^2 + \sum_{i=1}^{a} \sum_{j=1}^{b} \sum_{k=1}^{n} (\overline{AB}_j - \overline{AB})^2 + \sum_{i=1}^{a} \sum_{j=1}^{b} \sum_{k=1}^{n} (\bar{e}_j - \bar{e})^2 \tag{10.14}$$

Sum of squares for interaction

$$= \sum_{i=1}^{a} \sum_{j=1}^{b} \sum_{k=1}^{n} (\bar{X}_{ij} - \bar{X}_i - \bar{X}_j + \bar{X})^2$$

$$= \sum_{i=1}^{a} \sum_{j=1}^{b} \sum_{k=1}^{n} [(\mu + A_i + B_j + AB_{ij} + \bar{e}_{ij}) - (\mu + A_i + \bar{B} + \overline{AB}_i + \bar{e}_i) - (\mu + \bar{A} + B_j + \overline{AB}_j + \bar{e}_j) + (\mu + \bar{A} + \bar{B} + \overline{AB} + \bar{e})]^2 \tag{10.15}$$

and measures:

$$\sum_{i=1}^{a} \sum_{j=1}^{b} \sum_{k=1}^{n} (AB_{ij} - \overline{AB}_i - \overline{AB}_j + \overline{AB})^2 + \sum_{i=1}^{a} \sum_{j=1}^{b} \sum_{k=1}^{n} (\bar{e}_{ij} - \bar{e}_i - \bar{e}_j + \bar{e})^2 \tag{10.16}$$

Figure 10.2. Graphical illustration of differences due to interactions between two orthogonal experimental factors as in Figure 10.1. (a) The overall mean (μ) and the means of levels of factor A are shown. (b) The overall mean (μ) and the means of levels of factor B are shown. There are differences among levels of factor A and among levels of factor B.

The exact form of these is complicated by whether either or both of factors A and B are fixed or random (see Section 8.2). This is explained in full below. Here, consider the simplest case, where A and B are both fixed factors. As described in detail below, when A is a fixed factor:

$$\sum_{i=1}^{a} A_i = 0 \text{ and, for every level of } B_j, \sum_{i=1}^{a} AB_{ij} = 0 \tag{10.17}$$

Thus, $\bar{A}$ is zero and $\overline{AB}_j$ values are zero for all $j = 1 \ldots b$, all levels of factor B.

Conversely, when B is a fixed factor:

$$\sum_{j=1}^{b} B_j = 0 \text{ and, for every level of } A_i, \sum_{j=1}^{b} AB_{ij} = 0 \tag{10.18}$$

Thus, $\bar{B}$ is zero and $\overline{AB}_i$ values are zero for all $i = 1 \ldots a$, all levels of factor A. From this, the mean squares estimate the components shown in Table 10.4, noting the use of k_A^2 when A represents a fixed factor.

Obviously, the appropriate test for the null hypothesis concerning factor A ($\mu_1 = \mu_2 = \ldots = \mu_i = \ldots = \mu_a$) can be tested by an F-ratio:

$F =$ Mean square A/Mean square residual, with $(a-1)$ and $ab(n-1)$ degrees of freedom

Similarly, the test for differences among levels of factor B (H_0: $\mu_1 = \mu_2 = \ldots = \mu_j = \ldots = \mu_b$) is:

$F =$ Mean square B/Mean square residual, with $(b-1)$ and $ab(n-1)$ degrees of freedom

These tests should not be done until it is clear that there are no interactions between A and B, because otherwise there is no valid logical basis for testing nor interpreting the outcome of the tests, as explained above. The preliminary test for interactions is:

$F =$ Mean square A $\times$ B/Mean square residual, with $(a-1)(b-1)$ and $ab(n-1)$ degrees of freedom

The issue of how to interpret the outcome of tests is taken up below.

10.5 Why do a factorial experiment?

There are two great advantages of arranging experiments as factorial, orthogonal combinations. Broadly speaking, the first can be described as the advantages accruing because factorial experiments provide information not available from experiments with single factors. The second advantage is that of efficiency. These are considered below.

10.5.1 Information about interactions

As already discussed, factorial experiments allow (indeed, require) the calculation of information about how independent or dependent the effects of one factor are relative to the effects of another factor. The interpretation of statistical interactions is considered in detail below, but a simple example will suffice to show why they are important.

Consider the case where there are two experimental factors. Observations have revealed that the density of a prey species is smaller where their predators are abundant, but this effect is not so pronounced where another secondary prey is itself abundant. The model proposed to explain these observations is that predators reduce abundance of the primary prey species, but, where other prey are numerous, predators have less influence on the primary prey. This model needs to be tested because there are, obviously, other possible explanations for the observations. For example, the primary prey may have requirements for food similar to those of the other prey species and requirements for microhabitats (shelter) different from those of the predators. Thus, there would be a positive correlation in numbers of the two prey species and both would be negatively correlated with the abundance of the predator.

So, three hypotheses are proposed to test the model. First, the removal of predators will lead to greater densities of prey than will be found in control areas where predators are normally active. The one-tailed null hypothesis is that there will be no difference between control or experimental areas or there will be smaller abundance of prey where predators are removed.

The second hypothesis is that there will be smaller densities of prey where abundance of the second species of prey is experimentally decreased. Three densities were chosen to span the naturally observed range: (i) the density of the second species was reduced to 50%, (ii) the density was reduced to 80%, (iii) the density of the second species remained unaltered. The null hypothesis is that there will be no influence on abundance of the first species due to alteration of the second species' abundance or that decreases of the second species will result in larger numbers of the first species.

The third hypothesis is, however, the most important. It is that there will be an interaction between the two factors. If the model is correct, there should be *no* influence of reducing abundances of the second prey on numbers of the first species where predators are removed. But there should be a greater difference in abundance of the primary prey between treatments with or without predators where abundances of the second species is small. This absolutely requires there to be an interaction between the two factors. The influence of the second species will occur only where predators are present. Thus, the influence of the abundance of the second prey *cannot* be independent of the abundance of the predator. Simultaneously, the model and third hypothesis *require* that the influence of the predator is *dependent* on the density of the second species of prey. The predicted interaction cannot be detected properly unless there is a fully orthogonal combination of all levels of the two factors, each independently replicated (see Figure 10.3).

As a second case, consider a simpler experiment. Previous observations have verified the pattern that densities of two species of plant are negatively correlated in any site examined. Where B is present, A is locally smaller in number. The model proposed is that competition for resources in the soil could explain this pattern. It is therefore hypothesized that removal of species B will lead to greater final densities of A than in controls where B is left alone. This experiment could be done as a single-factor one, in one site, with appropriate replication of the two treatments. The original observations were, however, in several sites and the proposed model is supposed to apply generally. Accordingly, the experiment should be set up in several sites and examined as a factorial experiment (an alternative approach is considered in Section 13.14).

The model cannot be correct if there is no influence of removal of the putative competitor in some sites. This would lead to marked interaction between sites and treatments – because the differences between treatments must then depend on which sites are examined. In some sites, there would be no differences.

Even if the model is correct, there may be very marked differences in intensity of competition from one site to another due to differences in densities of the two species, or in amounts and quality of resources. This would lead to some interaction between the two factors and the experiment should not ignore the implications of such site dependence for any sensible interpretation.

Finally, of course, the model may be globally correct in the simplest possible way – everywhere species B is removed, species A shows the same difference in density compared with controls. Thus, there really is no site dependence; the competitive interaction is equally intense everywhere. This would lead to no statistical interaction between the

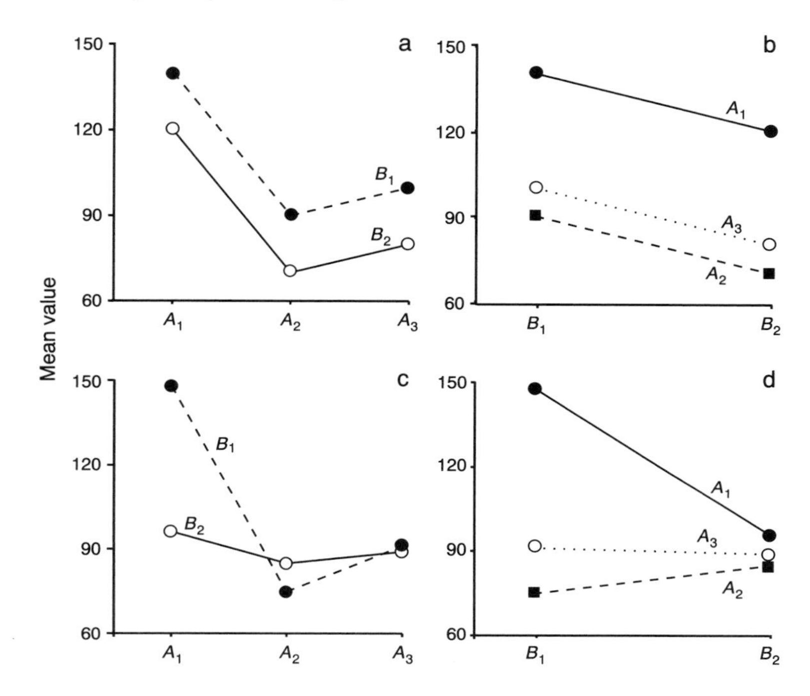

Figure 10.3. Results of an analysis of variance of two fixed orthogonal experimental factors A and B. $a = 3$ and $b = 2$. There were n replicates in each of the ab combinations of experimental treatments. In (a) and (b), there is no interaction between A and B. The differences from one level of A to another are equal for each level of B. The differences between the two levels of B are identical for all three levels of A. The graphs are parallel. (a) The trends of B against different levels of A. (b) The same data as trends of A against each level of B. In (c) and (d), there is marked interaction between A and B. There is virtually no effect of different levels of A on the mean in level 2 of B. In contrast, there is a major influence of different levels of A on level 1 of B. The differences between B_1 and B_2 are not the same at each level of A. The differences from A_1 to A_2, A_1 to A_3, A_2 to A_3 are not the same for B_1 and B_2. (c) and (d) are plotted like (a) and (b), respectively.

two factors and therefore a very strong basis for support of the generality of the model.

Orthogonal combinations of the two (or more) factors are the only really coherent way to examine the interactions between them. In some cases, as described later, an analysis of covariance could be done instead, but, as shown in Chapter 13, this requires more restrictive assumptions about the data. For example, in the first case considered above, areas with predators present or predators removed could be considered, with density of the second species of prey used as a covariate. This design would not require orthogonal arrangements of two factors, but, because an interaction is expected between removal of predator and density of the second prey, the data will not be expected to have homogeneous slopes. See Chapter 13 for further explanation. The factorial analysis of interaction is almost always better.

10.5.2 Efficiency and cost-effectiveness of factorial designs

The second reason for combining experimental treatments in factorial designs is that this makes for great efficiency and, often, for greater power to detect differences among treatments than is the case for separate experiments. Here is an example of a series of experiments of one factor done separately at each level of a second factor. The experiment is to test the hypothesis that removal of a predator will lead to greater survival of a prey species than will occur in plots with appropriate controls for the procedures of removal or in untouched controls. This hypothesis has been proposed, separately, for two different habitats (say woodland and open fields). So, it is perfectly reasonable to do two quite separate experiments, one in each of the habitats. Suppose this is done, as in Table 10.5a, where the two habitats are identified as B_1 and B_2.

Now, instead, because the same hypothesis is being examined in each habitat, the experiment could well be done as a two-factor one, with habitats as a second fixed factor. Assume that there are no interactions, so that tests of main effects are interpretable (see Section 10.3). As a result, an F-ratio to test for differences among the means of the levels of factor A (the experimental treatments) is now possible with less replication per treatment per habitat. This comes about because the means of the three experimental treatments (the levels of factor A) are now averaged over both habitats. So, the same degrees of freedom (2 and 24) are available with only $n = 5$ per habitat rather than with $n = 9$ as required for the separate experiments.

Thus, the same size of test can be achieved with five-ninths of the costs of replicates (i.e. a total of 10 rather than 18 for each treatment over both habitats). In addition, for any chosen probability of Type I error, size of intrinsic variance among replicates (σ_e^2) and effect size (k_A^2), the power of this test will be greater than for the single-factor experiments. This is explained in detail in Section 8.4. Note that the multiplier of k_A^2 in the two tests is 9 and 10, respectively, in Table 10.5a and b.

There are, of course, many circumstances where putting several treatments into the same experiment will result in there being interactions

Table 10.5. Demonstration of efficiency of an experiment with two fixed factors compared with single factor experiments at each level of the second factor. Factor A is fixed with $a = 3$ levels. In (a), it is done, separately, at each of $b = 2$ levels of the second fixed factor, B. In each experiment, there are $n = 9$ replicates of each treatment. In (b), the experiment is done as an orthogonal combination of the two factors, with a reduced number of replicates ($n = 5$ for each combination of treatments)

(a) Two single-factor experiments, one at B_1, the other at B_2

	Experiment 1 at B_1		Experiment 2 at B_2	
	Degrees of freedom	Mean square estimates	Degrees of freedom	Mean square estimates
A	2	$\sigma_e^2 + 9k_A^2/2$	2	$\sigma_e^2 + 9k_A^2/2$
Residual	24	σ_e^2	24	σ_e^2
Total	26	–	26	–

Test for A is $F = MS_A/MS_{residual}$ with 2 and 24 degrees of freedom. Cost is for 54 replicates over six treatments ($an = 3 \times 9 = 27$ in each of two experiments)

(b) A two-factor experiment at levels B_1 and B_2

Residual	Degrees of freedom	Mean square estimates
A	2	$\sigma_e^2 + 10k_A^2/2$
B	1	
A × B	2	
Residual	24	σ_e^2
Total	29	

If there is no interaction: Test for A is $F = MS_A/MS_{residual}$ with 2 and 24 degrees of freedom. Cost is for 30 replicates over six treatments($abn = 2 \times 3 \times 5 = 30$).

among them. As is explained in Section 10.8, multiple comparisons of levels of one factor in each level of a second factor will still be more efficient and will often be more powerful than the equivalent multiple comparisons in single-factor experiments.

Nevertheless, there is no justification for arbitrarily putting unrelated experimental treatments into a more complex experiment. Results would have no meaning – there are no relevant models, hypotheses or null hypotheses being tested. This sin (which Stuart Hurlbert (personal communication) is calling 'pseudofactorialism') would be a pointless way of proceeding.

Under the circumstances of the sort described above, assembling related tests of appropriately similar hypotheses into multifactorial experiments is an advantage in terms of power and cost in addition to the information gained.

10.6 Meaning and interpretation of interactions

Understanding, interpreting and dealing with interactions among experimental factors is crucial to any sensible study of quantitative ecology (as introduced in Section 10.5). The simplest start is to consider only interactions between two fixed factors, A and B which have a and b levels, respectively. As an example, consider an investigation of causes of mortality of males and females of some species when at different densities. It has been observed that, when crowded, animals tend to die faster, but there are differences from place to place that appear correlated with the relative numbers of females in the populations in the places examined. Where there is a greater proportion of females, the rate of dying is apparently larger. From this, two models are proposed. First, for either gender, there is increased mortality where densities are large as a direct result of the large densities. This might be due to increased competition for resources at larger densities, increased likelihood of contracting diseases, increased chance of being killed by a predator, etc. There are many possible causes. For example, it is also possible that greater mortality happens to occur in the places that generally support greater density. Thus, the pattern could simply be a correlation due, for example, to some feature in the habitat that allows more standing stocks of food plants (thus increasing average density) and greater chance of dying from unfavourable aspects of local climate. In this last scenario, density per se is *not* the cause of greater mortality. So, rather than spend time and resources on a study of competition or predation or diseases, it would be sensible to determine first whether increased mortality is, for whatever reason, directly related to increased density. This serves as the general model to be examined.

The hypothesis is that if patches of large density are thinned out experimentally to reduce densities to the smaller numbers seen elsewhere, rate of mortality will decrease relative to naturally dense areas. Control treatments will be needed for the process of thinning and for any procedure used to prevent emigration or immigration that would negate the experimental treatment.

At the same time, to explain why differences occur from place to place at any density, the obvious model is that the two genders differ in their rates of mortality. Because females die faster, there is greater mortality where females are in greater numbers. The appropriate hypothesis is that if populations are established at different densities, as just described, there will be greater mortality of females than males.

The experiment will therefore consist of a different densities, chosen to span the range naturally observed (and thus a fixed factor). There will also be $b = 2$ sets of each density, in one of which mortality of males is recorded. In the other set, females only are examined. This design ensures that mortality of the two genders is measured independently, which would not be the case if they were both examined in the same populations. There need to be n replicates of each combination of the densities, so $2n$ replicates of each of the a densities are established, properly representatively scattered about. In each density, n are randomly chosen to be the ones sampled for mortality of females. In each replicate, at whatever density, the animals are thinned out to create the same proportion of females in the area. In this way, there will be no confounding of density and proportion of females in any interpretation of rates of mortality.

The experiment has two fixed factors. There are now several possible patterns of outcome, depending on the relationship between the two factors. These are illustrated in Figure 10.4. In the first case (a), there are no differences in proportional mortality (measured over a specified experimental period) for either gender, nor at any density. In case (d), there is no effect of density, but females die faster than males. In (b), however, there is no difference between the genders, but animals die more quickly at larger densities.

In case (e), there is greater mortality with increased density *and* females die faster than males. In three cases ((b), (d) and (e)) where there are differences, the effects of the two factors are independent of one another. As illustrated in Figure 10.4d and e, the difference in mean mortality between males and females is the same at all three densities. In Figure 10.4b and e, the average difference in mortality from one density to another is the same for each gender. Thus, there is no interaction between the two factors.

In the other two cases illustrated (Figure 10.4c and f), there is an interaction. In both cases, with increasing density there is increasing mortality of females. In the first case (Figure 10.4c), there is no influence of density on mortality of males. In the second case, males are also affected by increased density, but to a lesser extent than are the females. Thus, the interaction is a measure of the fact that the difference between males and females is not the same at each density. Of course, because of symmetry

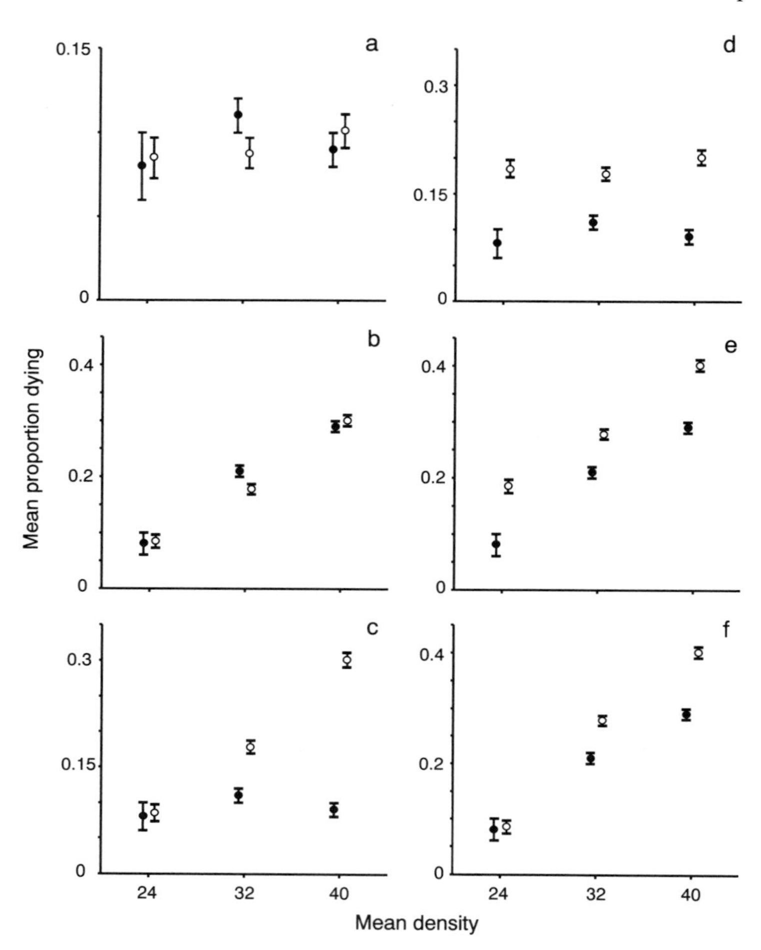

Figure 10.4. Several possible outcomes of an experimental manipulation of densities to test hypotheses about mortality of males and females (see the text). There are $a = 3$ densities (untouched, reduced to 80% or 60% of natural large density to create the range of densities originally observed). For illustration, no controls are described. There are $n = 5$ replicate set-ups examined for males and another five for females at each density. Empty circles, females; filled circles, males. (a), (c), (d) No effect of density on males; (a), (d) no effect on females. (b), (e) Differences among densities are the same for males and females. In (d) and (e), mortality of females is greater than that of males. (c), (f) Influence of increased density is greater for females than for males and there is a significant interaction.

Table 10.6. Patterns of significance in an experiment with two fixed factors as described in Section 10.6 and illustrated in Figure 10.4. Several patterns of differences are shown as (a) to (f), as in Figure 10.4. ns indicates not significant, $P > 0.05$; the asterisks denote significance, $P < 0.05$

Source of variation	Degrees of freedom	a	b	c	d	e	f
Among densities = A	$a - 1 = 2$	ns	*		ns	*	
Between genders = B	$b - 1 = 1$	ns	ns		*	*	
Interaction = A × B	$(a - 1)(b - 1) = 2$	ns	ns	*	ns	ns	*
Residual	$ab(n - 1) = 24$						
Total	$abn - 1 = 29$						

(i.e. orthogonality) of the two factors, it is simultaneously a measure of the fact that the difference in mortality between any two densities is not the same for each of the genders.

The pattern of significance in the illustrated cases is shown in summaries of the analyses in Table 10.6. Only where there is no interaction is the outcome of tests for the main effects (i.e. the differences among levels of each of the two factors) shown. This is because, once a significant interaction is known to be present, there is no longer the possibility that the two factors are independent. Thus, the null hypotheses for the two main effects are not logically valid, as explained in Sections 10.3 and 12.5.

In the case of this experiment, as is usually the case when there are two factors, all possible outcomes of tests on the three sources of variation (A, B and A × B) are simply interpretable, given the models and hypotheses proposed. There is no reason for confusion or sloppiness of thought, nor any excuse for not interpreting an interaction correctly.

Of course, in this case, if there is no interaction, but there is a difference between genders, it is immediately obvious which gender has the larger mean. In contrast, significance of differences among densities requires multiple comparisons (as in Section 8.6) to identify which densities differ. The appropriate procedures are described below (Section 10.8). Where there is a significant interaction, interpretations usually call for multiple comparisons. Sometimes, of course, as described for the example in Section 10.5.1, all that is necessary to draw proper conclusions from an experiment is that there *is* an interaction.

It is worth re-iterating a final time why interpretations of main effects are impossible or, at best, unreliable when there is an interaction. Consider the data summarized in Figure 10.5, where there are, again, three levels of factor A and two levels of B and a significant interaction between them. In

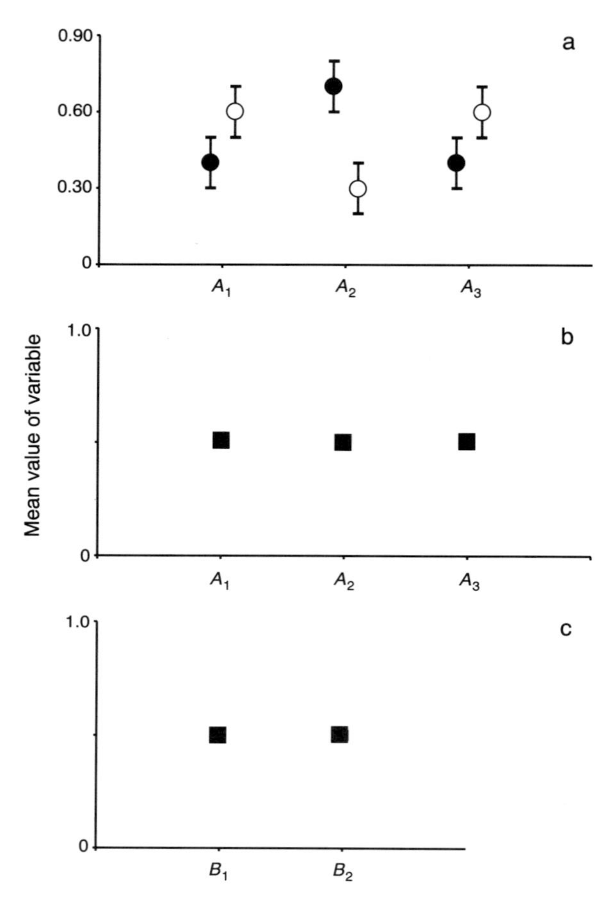

Figure 10.5. An interaction between two orthogonal factors in an experiment. Factor A has $a = 3$ levels; factor B has $b = 2$ levels; each combination is replicated n times. Data are means of the treatments. (a) For each combination of treatments; filled circles is level B_1 (20% cover), empty circles is level B_2 (60% cover) of factor B, with standard errors. (b) The means of levels of factor A averaged across levels of factor B. (c) The means of levels of factor B averaged across levels of factor A.

this case, there are opposite effects on the two levels of B caused by the middle level of A. The means of the three levels of factor A do differ for each of levels B_1 and B_2 of factor B, but this cannot appear in a test for differences among levels of A. Such a test is on the mean value of all data at each level of factor A, averaged over both levels of B. These means do not differ! Because of the interaction, the conclusion would be reached that factor A causes no change in the measured variable – despite the evidence so clearly shown in the graph (Figure 10.5a).

In this example, as shown in Figure 10.5c, there is also no difference between the two levels of factor B, when averaged over the three levels of factor A. In one sense this seems reasonable – there is no average effect of factor B. It is, however, completely wrong. There is a marked difference between the means of the two levels of factor B at all three levels of A. The means of the two levels of factor B are *always* different.

This sort of problem is *always* present when there is an interaction between two or more factors (see Section 10.10). It provides evidence that main effects are not independent. The interpretation of the interaction is done by multiple comparisons as described below.

10.7 Interactions of fixed and random factors

This discussion was of the interaction between two fixed factors. There can, of course, also be an interaction between two random factors or between a fixed and a random factor. These introduce some complications in the calculations of F-ratios, because they influence what is estimated by mean squares in an analysis.

This is an important point and must be fully understood before it is possible to complete an analysis properly. The problem is explained by considering the calculation of sums of squares for the differences among levels of an experimental factor (as in Equations 10.11 to 10.14).

The sum of squares of factor A is:

$$\sum_{i=1}^{a}\sum_{j=1}^{b}\sum_{k=1}^{n}(\bar{X}_i - \bar{X})^2$$

$$= \sum_{i=1}^{a}\sum_{j=1}^{b}\sum_{k=1}^{n}(\bar{e}_i - \bar{e})^2 + \sum_{i=1}^{a}\sum_{j=1}^{b}\sum_{k=1}^{n}(\overline{AB}_i - \overline{AB})^2$$

$$+ \sum_{i=1}^{a}\sum_{j=1}^{b}\sum_{k=1}^{n}(A_i - \bar{A})^2$$

$$= bn\sum_{i=1}^{a}(\bar{e}_i - \bar{e})^2 + bn\sum_{i=1}^{a}(\overline{AB}_i - \overline{AB})^2 + bn\sum_{i=1}^{a}(A_i - \bar{A})^2 \quad (10.19)$$

The first of these is an estimate of:

$$(a-1)bn\left(\frac{\sigma_e^2}{bn}\right) = (a-1)\sigma_e^2 \quad (10.20)$$

where σ_e^2 is the variance in all of the sampled populations. This was explained for previous models (in Sections 7.7 and 9.5).

The second, when factor B is a *fixed* factor, is defined to be zero. Because factor B is fixed, the interaction terms (AB_{ij} values) are also fixed in each level of factor A. Thus (see also Equation 10.18):

$$\sum_{j=1}^{b} AB_{ij} = 0 \quad \text{for } i = 1 \ldots a \tag{10.21}$$

This must be so, otherwise the differences between levels of factor B could not themselves be fixed and would not sum to zero. A non-zero sum of the AB_{ij} values in any level of factor A would lead to non-zero sums of the difference measuring the B_j terms.

Thus, in any level of factor A:

$$\overline{AB}_i = \frac{\sum_{j=1}^{b} AB_{ij}}{b} = 0 \tag{10.22}$$

and, over all levels of factor A:

$$\overline{AB} = \frac{\sum_{i=1}^{a} \overline{AB}_i}{a} = \frac{\sum_{i=1}^{a} \sum_{j=1}^{b} AB_{ij}}{ab} = 0 \tag{10.23}$$

Thus, the sum of squares in Equations 10.11 and 10.19 estimates:

$$(a-1)\sigma_e^2 + bn \sum_{i=1}^{a} (A_i - \bar{A})^2 \tag{10.24}$$

In contrast, when the levels of factor B are randomly sampled from a population of such levels (as described in Section 8.2), they do not sum exactly to zero in any particular experiment. Consequently, because any interaction terms (AB_{ij} values) in any level of factor A must also be randomly chosen, they also cannot sum to zero. They must be random because, if B_j values are random, the associated AB_{ij} values cannot possibly be fixed. They cannot be all possible levels that exist – there are numerous other AB_{ij} values associated with the many values of B_j not sampled in the experiment.

So, the sum of squares for factor A:

$$bn \sum_{i=1}^{a} (\bar{e}_i - \bar{e})^2 + bn \sum_{i=1}^{a} (\overline{AB}_i - \overline{AB})^2 + bn \sum_{i=1}^{a} (A_i - \bar{A})^2 \tag{10.25}$$

measures:

$$(a-1)bn\sigma_e^2 + (a-1)n\sigma_{AB}^2 + bn \sum_{i=1}^{a} (A_i - \bar{A})^2 \tag{10.26}$$

where σ^2_{AB} is the variance of the distributions of interaction terms being sampled in the experiment. There is a distribution of such terms in association with each level of factor A. Each distribution is sampled by b values in each level of factor A. The variances of these distributions are estimated as shown in Equation 10.26.

The rationale behind this is shown in Table 10.7. In (a), both factors are fixed, all interaction terms are fixed and therefore the average of interaction terms is zero for every level of both factors. When factor B is random (Table 10.7b), the interaction terms for each level of factor A are a random sample of the possible ones – dictated entirely by which levels of factor B happen to be in the experiment. So, in each level of factor A, the average of interaction terms is not zero. In each level of factor B, however, the average of interaction terms is still zero because the values are fixed (to sum to zero) by the definition of factor A as fixed. Now, the difference between any two means of factor A is not simply the difference due to the relevant levels of treatment A (the A_i values). It also involves the difference between the average interaction terms ($\overline{AB}_i$ values). In Table 10.8b, the difference between $\bar{X}_1$ and $\bar{X}_2$ (i.e. 30.5 and 20, respectively) is due to A_1, A_2, $\overline{AB}_1$ and $\overline{AB}_2$. Hence, in the sum of squares and mean square for differences among levels of factor A, there is a component due to differences among $\overline{AB}_i$ values; this is measured as σ^2_{AB} in the mean square (as in Equation 10.26 and Table 10.8). For factor B, however, the difference between two levels is only a function of B_j values and no σ^2_{AB} component is involved in the mean square.

When A *is* also random, differences between A_i and B_j values both involve AB terms (as in Table 10.7c). Consequently, the mean squares for both factors involve σ^2_{AB}. Thus, the mean square for either factor includes a mean square for the interaction between them if the *other* factor is random! Inspection of Table 10.8 indicates that the nature of the two factors *must* affect what is an appropriate procedure to test for differences among their levels.

The implications of this for a 'mixed model', i.e. when one factor is fixed and one is random, will be considered as a special case in Section 10.11. Here, note that for two fixed factors, all three sources of variation may be tested against the residual mean square. Of course, as made clear earlier (Section 10.6), there is no logically valid structure to tests of the main effects (factors A and B) if there is an interaction between them. So, the null hypothesis of no interaction is examined first and, if retained, the null hypotheses of no differences among levels of factors A and B are each tested using the residual mean square as the divisor for the F-ratio.

Table 10.7. Explanation of components of mean squares associated with fixed or random factors. In all cases, there is an experiment with $a = 3$ levels of factor A, $b = 2$ levels of factor B, arranged orthogonally with n replicates of each combination of treatments. μ (the overall mean in any experiment) is 20. A_i, B_{ij}, AB_{ij} terms (see Equation 10.7) refer to effects of levels of factor A, B and their interactions, respectively. $\hat{A}_i$, $\hat{B}_j$, $\widehat{AB}_{ij}$ refer to estimates of these terms from means ($\bar{X}_i$, $\bar{X}_j$, $\bar{X}_{ij}$, $\bar{X}$) as appropriate (see Chapter 7)

(a) Factors A and B fixed; AB_{ij} values are fixed for each level of A_i and B_j

		A_1	10	A_2	−2	A_3	−8	$\overline{AB}_j$	$\bar{X}_j$	$\hat{B}_j$
B_1	+6	AB_{11}	4	AB_{21}	3	AB_{31}	−7	0		
		$\bar{X}_{11}$	40	$\bar{X}_{21}$	27	$\bar{X}_{31}$	11		26	6
$\widehat{AB}_{i1}$			4		3		−7			
B_2	−6	AB_{12}	−4	AB_{22}	−3	AB_{32}	7	0		
		$\bar{X}_{12}$	20	$\bar{X}_{22}$	9	$\bar{X}_{32}$	13		14	−6
$\widehat{AB}_{i2}$			−4		−3		7			
$\overline{AB}_i$			0		0		0			
$\bar{X}_i$			30		18		12	$\overline{AB}$	$\bar{X}$	
$\hat{A}_i$			10		−2		−8	0	20	

(b) Factor A fixed, B random; AB_{ij} values are fixed in each level of B_j, but random in each level of A_i

		A_1	10	A_2	−2	A_3	−8	$\overline{AB}_j$	$\bar{X}_j$	$\hat{B}_j$
B_1	5	AB_{11}	3	AB_{21}	2	AB_{31}	−5	0		
		$\bar{X}_{11}$	38	$\bar{X}_{21}$	25	$\bar{X}_{31}$	12		25	4
$\widehat{AB}_{i1}$			3.5		1		−4.5			
B_2	−3	AB_{12}	−4	AB_{22}	0	AB_{32}	4	0		
		$\bar{X}_{12}$	23	$\bar{X}_{22}$	15	$\bar{X}_{32}$	13		17	−4
$\widehat{AB}_{i2}$			−3.5		−1		4.5			
$\overline{AB}_i$			−0.5		1		−0.5			
$\bar{X}_i$			30.5		20		12.5	$\overline{AB}$	$\bar{X}$	
$\hat{A}_i$			9.5		−1		−8.5	0	21	

If there is an interaction, the relevant null hypotheses are examined by the appropriate multiple comparisons (Section 10.8).

For an analysis of two random factors, the procedures are also fairly straightforward. The presence of interaction is tested by an F-ratio of the interaction versus the residual mean square (Table 10.8). If significant, the patterns of difference among means of each factor are examined by multiple comparisons at each level of the other factor, as described in Section 10.8.

Table 10.7 (*cont.*)

(c) Factors A and B random; AB_{ij} values are random for each level of A_i and B_j

		A_1	8	A_2	−6	A_3	−4	$\overline{AB}_j$	$\bar{X}_j$	$\hat{B}_j$
B_1	5	AB_{11}	6	AB_{21}	3	AB_{31}	3	3		
		$\bar{X}_{11}$	39	$\bar{X}_{21}$	19	$\bar{X}_{31}$	24		27.33	6
$\widehat{AB}_{i1}$			2.5		−3		0.5			
B_2	−3	AB_{12}	−3	AB_{22}	2	AB_{32}	−2	−1		
		$\bar{X}_{12}$	22	$\bar{X}_{22}$	13	$\bar{X}_{32}$	11		15.33	−6
$\widehat{AB}_{i2}$			−2.5		0.5		−0.5			
$\overline{AB}_i$			1.5		1.0		0.5			
$\bar{X}_i$			30.5		16.0		17.5	$\overline{AB}$	$\bar{X}$	
$\hat{A}_i$			9.17		−5.33		−3.83	1	21.33	

If there is no interaction, the main effects are examined by F-ratios of each mean square divided by the mean square for the interaction (see Table 10.8b and c). Note that there are usually few degrees of freedom for these tests. *Post-hoc* pooling (described in Section 11.7) will often be a useful procedure. If there is some reason to do so, significance of either F-ratio can be followed by the appropriate multiple comparison of the means of the levels of each factor.

There is, however, often little purpose to be gained by detailed comparison of the patterns of difference among means of random factors. This was discussed in Section 9.6. Usually, levels of random factors are examined in experiments, as representatives of a set of many more possible levels that could have been included in the experiment. The issue is whether they differ (i.e. σ_A^2 or σ_B^2 is greater than zero), rather than an examination of some interpretable pattern of differences. There is little or no point in examining the details of differences among the particular levels of a factor if these levels were arbitrarily (i.e. randomly) incorporated into the experiment.

There are relatively few convincing examples of experiments with two random factors. They do, however, have an important place in studies of temporal and spatial variability, particularly at early stages of an ecological investigation. A more structured example (about growth of snails) is examined in detail in Chapter 11.

Suppose the behaviour of some invertebrate (a beetle, a crab in a mangrove forest, anything) has been observed to vary from time to time and place to place. Perhaps the time spent feeding varies. Before examining

Table 10.8. Result of calculations of sums of squares and mean squares in two-factor experiments with fixed or random factors (see Section 10.7). For definitions of k_A^2, k_B^2 and k_{AB}^2 see Section 10.2.

(a) Both factors fixed

Source of variation	Sum of squares	Degrees of freedom	Mean square estimates	F-ratio versus
Among levels of A = A	$(a-1)\sigma_e^2 + bn\sum_{i=1}^{a}(A_i - \bar{A})^2$	$a-1$	$\sigma_e^2 + bnk_A^2$	Residual
Among levels of B = B	$(b-1)\sigma_e^2 + an\sum_{j=1}^{b}(B_j - \bar{B})^2$	$b-1$	$\sigma_e^2 + ank_B^2$	Residual
A × B	$(a-1)(b-1)\sigma_e^2 + n\sum_{i=1}^{a}\sum_{j=1}^{b}(AB_{ij} - \overline{AB}_i - \overline{AB}_j + \overline{AB})^2$	$(a-1)(b-1)$	$\sigma_e^2 + nk_{AB}^2$	Residual
Residual	$ab(n-1)\sigma_e^2$	$ab(n-1)$	σ_e^2	

(b) A fixed, B random

Source of variation	Sum of squares	Degrees of freedom	Mean square estimates	F-ratio versus
Among levels of A = A	$(a-1)\sigma_e^2 + (a-1)n\sigma_{AB}^2 + bn\sum_{i=1}^{a}(A_i - \bar{A})^2$	$(a-1)$	$\sigma_e^2 + n\sigma_{AB}^2 + bnk_A^2$	$A \times B$
Among levels of B = B	$(b-1)\sigma_e^2 + (b-1)an\sigma_B^2$	$(b-1)$	$\sigma_e^2 + an\sigma_B^2$	Residual
A × B	$(a-1)(b-1)\sigma_e^2 + (a-1)(b-1)n\sigma_{AB}^2$	$(a-1)(b-1)$	$\sigma_e^2 + n\sigma_{AB}^2$	Residual
Residual	$ab(n-1)\sigma_e^2$	$ab(n-1)$	σ_e^2	

(c) Both factors random

Source of variation	Sum of squares	Degrees of freedom	Mean square estimates	F-ratio versus
Among levels of A = A	$(a-1)\sigma_e^2 + (a-1)n\sigma_{AB}^2 + (a-1)bn\sigma_A^2$	$a-1$	$\sigma_e^2 + n\sigma_{AB}^2 + bn\sigma_A^2$	A × B
Among levels of B = B	$(b-1)\sigma_e^2 + (b-1)n\sigma_{AB}^2 + (b-1)an\sigma_A^2$	$b-1$	$\sigma_e^2 + n\sigma_{AB}^2 + an\sigma_B^2$	A × B
A × B	$(a-1)(b-1)\sigma_e^2 + (a-1)(b-1)n\sigma_{AB}^2$	$(a-1)(b-1)$	$\sigma_e^2 + n\sigma_{AB}^2$	Residual
Residual	$ab(n-1)\sigma_e^2$	$ab(n-1)$	σ_e^2	

models to explain what influences the initiation and cessation of feeding, it would be sensible to determine some general influences. So, the variability observed could be due to explicitly temporal differences – behaviour in all places is the same on a given day, but you observe it in different places each day. Or it could be site specific (i.e. temporally invariant in each place), but you see different places on different days. It could be both (temporal variations and spatial variations occur) such that behaviour in every place alters from day to day in the same way. Finally, it could vary from time to time differently from one place to another.

The hypotheses derived from these models are that if behaviour is observed in a set of places, on a set of different days, there will be differences only among days, only among places, among days and among places or an interaction between times and places, respectively. Knowing which is correct would massively simplify any study about causes of the behavioural pattern.

So, you randomly pick some sites in the study-area and examine the behaviour of n independent, randomly chosen individuals on each of a series of randomly chosen days using a different sample in each place on each day. Both factors (sites and days) are random. There are numerous other days and places that could have been studied. They are, however, orthogonal; every place is examined on every day.

The two-factor analysis is done. Whichever pattern of differences is significant (days, sites, days and sites, days × sites interaction) will support one of the above models and eliminate the others.

Whatever source of variation is significant, there is *no* valid or useful reason to pursue multiple comparisons. Discovering that the first day differs from the third, but not the others, demonstrates no useful property of the system. Apart from the lack of any hypothesis about the different days, the times examined were entirely arbitrarily chosen. All that matters is whether or not there are differences.

Sometimes, people are tempted to try to unravel *why* there are differences (it rained on the first day, was windy on the second day and sunny on the final day). This creates an unreplicated 'pseudo-experiment' to compare three patterns of weather (a fixed factor) – without replication. There is only one day with each type of weather. If you want to do this sort of thing, propose the model that specified that differences in weather are the explanation for temporal differences. Then do the experiment to test the hypothesis that behaviour differs from rainy to windy to sunny days using several randomly chosen days of each type of weather (for an example, see Underwood & Kennelly, 1990).

Similarly, for places, there is no valid way of retrospectively deciding what feature of each place caused the differences. That needs a properly stated model, hypothesis and test.

Because of the general nature of random factors, multiple comparisons are described for fixed factors. If you *want* to do them for random factors, the same logic, structure, procedures and algebra apply.

10.8 Multiple comparisons for two factors

10.8.1 When there is a significant interaction

If the interaction is significant, interest focuses on comparing the means of one factor separately at each level of the other factor and vice versa. The analyses available are then exactly the same as described for the single-factor experiment (Section 7.7). If there have been *a priori* defined patterns of differences, *a priori* procedures should be used. Note that the *a priori* defined patterns should also have predicted the significant interaction! Otherwise, *a posteriori* comparisons are appropriate. All of the comments about Type I error still apply and the use of Student–Newman–Keuls (SNK) or Ryan's test needs to be thought about carefully.

Essentially, for each level of one factor, a multiple comparison is done of the means of the other factor. The residual mean square is used to calculate the standard errors for the means, exactly as in the case when there is one factor. An example is given in Table 10.9. The outcome is clear. There is significantly greater mortality with increasing density of limpets, but this is a greater trend where cover of algae is greater. Furthermore, there is no influence of decreased cover of algae where densities are small, but a significant effect of altered cover where densities are larger. The multiple comparisons have revealed a perfectly interpretable pattern of differences that explain why there is an interaction.

10.8.2 When there is no significant interaction

When there is no interaction, multiple comparisons are done on each significant main effect. The means of each factor can, however, be averaged over all the levels of the other factor. The lack of interaction confirms that the differences among levels of one factor are independent of the other factor, so the levels of the other factor can be averaged together.

An example is shown in Table 10.10. This is an experiment similar to that described in Table 10.9, but in this case there is no interaction. There are three densities, but four levels of cover of algae. The standard

Table 10.9. Multiple comparisons of a two-factor experiment with interactions. Factor A is three densities of limpets; factor B is two average percentage covers of algae. These were chosen to test the hypothesis that mortality of limpets would vary with intraspecific density and with cover of algae. Either could cause increased or decreased mortality (greater average shelter from desiccation could be provided when limpets are crowded or surrounded by algae; greater competition may occur when limpets are crowded or more space is occupied by algae). Both factors are fixed; the levels were chosen from observed patterns of mortality

(a) Data are mean proportional mortality of limpets ($n = 5$ replicate areas, 400 cm^2, for each combination of treatments)

		Density per 400 cm^2		
		10	15	20
Percentage cover	20	0.12	0.11	0.24
	60	0.17	0.47	0.65

(b) Analysis of variance

Source of variation	Sum of squares	Degrees of freedom	Mean square	F-ratio	
Densities	0.450	2	0.225		
Percentage covers	0.560	1	0.560		
$D \times P$	0.190	2	0.095	13.57	$P < 0.01$
Residual	0.168	24	0.007		
Total	1.368	29			

Standard error for means $= \sqrt{(0.007/5)} = 0.037$

error for the means of each level of factor A is calculated in exactly the same way as before. It is the square-root of the mean square of the residual (i.e. within samples) divided by the size of sample from which each mean is calculated. The means are, however, now calculated from samples of n replicates in each of b levels of factor B – a total of bn replicates. For the levels of factor B, each mean is calculated from an replicates (n in each of a levels of factor A). Thus, the calculations of standard errors are as in Table 10.10.

There is a significant increase in mean mortality when density per 400 cm^2 is 20 compared with 15 and 10, which do not differ. Independently, there is greater mortality of limpets where percentage cover of algae is greater than 40%. Mortality increases when cover is increased from 40% to 60% and, again, to 80%.

Table 10.9 (*cont.*)

(c) Student–Newman–Keuls tests (note that there is no excessive Type I error requiring Ryan's procedures; see Section 8.6.5.4); asterisks indicate $P < 0.05$

Effect of density at each cover

	Density at 20% cover			Density at 60% cover			D
Rank	1	2	3	1	2	3	
Mean	0.11	0.12	0.24	0.17	0.47	0.65	
Difference	$^{3-1}$0.13*			$^{3-1}$0.48*			0.131
	$^{2-1}$0.01	$^{3-2}$0.12*		$^{2-1}$0.30*	$^{3-2}$0.18*		0.108

Effect of percentage cover at each density

	Percentage cover at						
	Density 10		Density 15		Density 20		D
Rank	1	2	1	2	1	2	
Mean	0.12	0.17	0.11	0.47	0.24	0.65	
Difference	$^{2-1}$0.05		$^{2-1}$0.36*		$^{2-1}$0.41*		0.108

10.8.3 Control of experiment-wise probability of Type I error

As discussed in Section 8.6.2, these multiple comparisons must be done carefully to guard against excessive probability of Type I error. The procedures discussed in Section 8.6 will protect against excessive error in each SNK test. This does, however, protect only the 'family-wise' rate of error. This is the probability of error in a given set of tests as part of the whole experiment. This is not the probability of Type I error over the whole experiment. There are five separate tests in the case illustrated in Table 10.9c and two tests in Table 10.10c.

The probability of error can be adjusted to compensate for the number of tests (Day & Quinn, 1989; Rice, 1989). Possibly, the safest course is to use the Bonferroni adjustment according to how many multiple comparisons will be done. Thus, in the example in Table 10.9, there are five SNK tests. At a nominal probability of Type I error of $\alpha = 0.05$, each should be done with $P = 0.01$ (i.e. $\alpha/5$). This will preserve the rate of error over the whole experiment (the experiment-wise error; see Section 8.6.2). In each test, the appropriate adjustments will be needed to keep $P = 0.05$ for each test when there is a risk of it being excessive (as described for Ryan's tests in Section 8.6.5.4), using $P = 0.01$ as the nominal probability for each paired comparison.

Table 10.10. Multiple comparisons in a two-factor experiment with no interactions. The experiment is as described in Table 10.9, except that there are $a = 3$ densities and $b = 4$ covers of algae; $n = 5$ replicates areas of 400 cm^2 for each treatment

(a) Mean proportional mortality of limpets

Percentage cover	Density per 400 cm^2			
	10	15	20	Mean
20	0.12	0.14	0.38	0.21
40	0.14	0.17	0.37	0.23
60	0.28	0.30	0.54	0.37
80	0.43	0.46	0.67	0.52
Mean	0.24	0.27	0.49	0.33

(b) Analysis of variance

Source of variation	Sum of squares	Degrees of freedom	Mean square	F	
Densities	0.463	2	0.231	10.51	$P < 0.01$
Percentage covers	0.903	3	0.301	13.68	$P < 0.01$
$D \times P$	0.083	6	0.014	0.63	$P > 0.05$
Residual	1.056	48	0.022		
Total	2.505	59			

Standard error for mean at each density = $\sqrt{(0.022/20)} = 0.033$
Standard error for mean at each cover = $\sqrt{(0.022/15)} = 0.039$

(c) SNK tests; asterisks indicate $P < 0.05$

					D
Density					
Rank	1	2	3		
Mean	0.24	0.27	0.49		
	$^{3-1}0.25^*$				0.11
	$^{2-1}0.03$	$^{3-2}0.22^*$			0.09
Percentage cover					
Rank	1	2	3	4	
Mean	0.21	0.23	0.37	0.52	
	$^{4-1}0.31^*$				0.14
	$^{3-1}0.16^*$	$^{4-2}0.29^*$			0.13
	$^{2-1}0.02$	$^{3-2}0.14^*$	$^{4-3}0.15^*$		0.11

As stated earlier (Section 8.6), it is more important to be sensible than to finish up with a test with no power. Miller (1981) has discussed the problem of determining when a set of tests must be considered as a 'family' and therefore must have adjustments of rates of error. There are no simple rules and slavish acceptance of some decision that Type I error is a problem will inevitably lead to excessive Type II error.

10.9 Three or more factors

There is no reason to limit experiments to two factors. The only new feature of an experiment with three factors is that there is more than one interaction. With three factors orthogonal to one another, there are four interactions. Consider an experiment like that described in Section 10.8 (and Table 10.9), but done in spring and autumn because of an hypothesis about seasonal differences. Now there are three factors – density, cover of algae and season. In spring, every combination of the first two (with three densities and two covers) is replicated five times. The entire experiment is then done again (with new, independently scattered replicates) in autumn.

The model for the data is:

$$X_{ijkl} = \mu + D_i + P_j + S_k + DP_{ij} + DS_{ik} + PS_{jk} + DPS_{ijk} + e_{l(ijk)} \quad (10.27)$$

where X_{ijkl} represents the data for the lth replicate ($l = 1 \ldots 5$) in the kth season ($k = 1, 2$), in the jth percentage cover ($j = 1, 2$) and ith density ($i = 1 \ldots 3$). D_i, P_j, S_k represent the ith level of density, jth level of percentage cover and kth level of season. DP_{ij}, DS_{ik}, PS_{jk}, DPS_{ijk} represent interactions and $e_{l(ijk)}$ represents the individual error associated with that replicate in the specified combination of treatments.

The analysis of this fixed factorial experiment is shown in Table 10.11 – without further explanation. Nevertheless, remember this is not a recipe for *any* experiment with three factors. It is only for three fixed orthogonal factors. There are many different designs with three factors: some fixed, some random, some nested. Some are described in Chapter 11. The formulae for this particular design are simple extensions of those used for two-factor experiments.

10.10 Interpretation of interactions among three factors

What matters here is the interpretation of the various interactions. Each of the first-order interactions (D × P, D × S, P × S) has exactly the same sort

Table 10.11. Analysis of three orthogonal fixed factors: A_i $(i = 1 \ldots a)$, B_j $(j = 1 \ldots b)$, C_k $(k = 1 \ldots c)$ all combinations replicated n times $(l = 1 \ldots n)$. The 'machine formulae' are algebraic equivalents to the formulae for sums of squares, but do not generate large rounding errors

Source of variation	Sum of squares	Machine formulae	Degrees of freedom	Mean square estimates
A	$\mathrm{SS_A} = \sum_{i=1}^{a}\sum_{j=1}^{b}\sum_{k=1}^{c}\sum_{l=1}^{n}(\bar{X}_i - \bar{X})^2$	$\dfrac{\sum_{i=1}^{a}\left(\sum_{j=1}^{b}\sum_{k=1}^{c}\sum_{l=1}^{n}X_{ijkl}\right)^2}{bcn} - \mathrm{Cf}$	$a - 1$	$\sigma_e^2 + bcnk_\mathrm{A}^2$
B	$\mathrm{SS_B} = \sum_{i=1}^{a}\sum_{j=1}^{b}\sum_{k=1}^{c}\sum_{l=1}^{n}(\bar{X}_j - \bar{X})^2$	$\dfrac{\sum_{j=1}^{b}\left(\sum_{i=1}^{a}\sum_{k=1}^{c}\sum_{l=1}^{n}X_{ijkl}\right)^2}{acn} - \mathrm{Cf}$	$b - 1$	$\sigma_e^2 + acnk_\mathrm{B}^2$
C	$\mathrm{SS_C} = \sum_{i=1}^{a}\sum_{j=1}^{b}\sum_{k=1}^{c}\sum_{l=1}^{n}(\bar{X}_k - \bar{X})^2$	$\dfrac{\sum_{k=1}^{c}\left(\sum_{i=1}^{a}\sum_{j=1}^{b}\sum_{l=1}^{n}X_{ijkl}\right)^2}{abn} - \mathrm{Cf}$	$c - 1$	$\sigma_e^2 + abnk_\mathrm{C}^2$
A × B	$\mathrm{SS_{AB}} = \sum_{i=1}^{a}\sum_{j=1}^{b}\sum_{k=1}^{c}\sum_{l=1}^{n}(\bar{X}_{ij} - \bar{X}_j - \bar{X}_j + \bar{X})^2$	$\dfrac{\sum_{i=1}^{a}\sum_{j=1}^{b}\left(\sum_{k=1}^{c}\sum_{l=1}^{n}X_{ijkl}\right)^2}{cn} - \mathrm{Cf} - \mathrm{SS_A} - \mathrm{ss_B}$	$(a-1)(b-1)$	$\sigma_e^2 + cnk_\mathrm{AB}^2$
A × C	$\mathrm{SS_{AC}} = \sum_{i=1}^{a}\sum_{j=1}^{b}\sum_{k=1}^{n}\sum_{l=1}^{n}(\bar{X}_{ik} - \bar{X}_i - \bar{X}_k + \bar{X})^2$	$\dfrac{\sum_{i=1}^{a}\sum_{k=1}^{c}\left(\sum_{j=1}^{b}\sum_{l=1}^{n}X_{ijkl}\right)^2}{bn} - \mathrm{Cf} - \mathrm{SS_A} - \mathrm{SS_C}$	$(a-1)(c-1)$	$\sigma_e^2 + bnk_\mathrm{AC}^2$

B × C	$SS_{BC} = \sum_{i=1}^{a}\sum_{j=1}^{b}\sum_{k=1}^{c}\sum_{l=1}^{n}(\bar{X}_{jk} - \bar{X}_j - \bar{X}_k + \bar{X})^2$	$\frac{\sum_{j=1}^{b}\sum_{k=1}^{c}\left(\sum_{i=1}^{a}\sum_{l=1}^{n}X_{ijkl}\right)^2}{an} - Cf - SS_B - SS_C$	$(b-1)(c-1)$	$\sigma_e^2 + ank_{BC}^2$
A × B × C	$SS_{ABC} = \sum_{i=1}^{a}\sum_{j=1}^{b}\sum_{k=1}^{c}\sum_{l=1}^{n}(\bar{X}_{ijk} - \bar{X}_{ij} - \bar{X}_{ik} - \bar{X}_{jk} + \bar{X}_i + \bar{X}_j + \bar{X}_k - \bar{X})^2$	$\frac{\sum_{i=1}^{a}\sum_{j=1}^{b}\sum_{k=1}^{c}\left(\sum_{l=1}^{n}X_{ijkl}\right)^2}{n} - Cf - SS_A - SS_B - SS_C - SS_{AB} - SS_{AC} - SS_{BC}$	$(a-1)(b-1) \times (c-1)$	$\sigma_e^2 + nk_{ABC}^2$
Residual	$\sum_{i=1}^{a}\sum_{j=1}^{b}\sum_{k=1}^{c}\sum_{l=1}^{n}(X_{ijkl} - \bar{X}_{ijk})^2$	$\sum_{i=1}^{a}\sum_{j=1}^{b}\sum_{k=1}^{c}\sum_{l=1}^{n}X_{ijkl}^2 - \frac{\sum_{i=1}^{a}\sum_{j=1}^{b}\sum_{k=1}^{c}\left(\sum_{l=1}^{n}X_{ijkl}\right)^2}{n}$	$abc(n-1)$	σ_e^2
Total	$\sum_{i=1}^{a}\sum_{j=1}^{b}\sum_{k=1}^{c}\sum_{l=1}^{n}(X_{ijkl} - \bar{X})^2$	$\sum_{i=1}^{a}\sum_{j=1}^{b}\sum_{k=1}^{c}\sum_{l=1}^{n}X_{ijkl}^2 - Cf$	$abcn - 1$	–
		$Cf = \frac{\left(\sum_{i=1}^{a}\sum_{j=1}^{b}\sum_{k=1}^{c}\sum_{l=1}^{n}X_{ijkl}\right)^2}{abcn}$		

of structure and interpretation as described for the interaction between two factors (Section 10.6). Thus, D × P measures the interaction between density and percentage cover, exactly as before. If the differences in mean mortality from one experimental density to another are affected by the differences in percentage cover of algae, there will be this interaction. If, however, the differences among densities or those between the two percentage covers differ from one season to another, there will be D × S or P × S interactions, respectively.

The final, second-order interaction (D × P × S) is interpretable as an interaction between D × P and S (and simultaneously D × S and P, P × S and D). Thus, it measures how consistent is the interaction between two factors from one to the next level of the third factor. As a result, like the situation with the single interaction between two factors in a two-factor experiment, there is no possible value in examining D × P, D × S, P × S (or, for that matter, D, P or S) if there is a D × P × S interaction.

In the case discussed, various outcomes are illustrated in Figure 10.6. In (a) and (d) are the data plotted for each season. There is a D × P interaction (exactly as described in Section 10.8.1), but it is constant across seasons. There is, nevertheless, a *difference* between the two seasons, but it affects all the means (and their interaction) in exactly the same way. So, there is no D × P × S interaction; the graphs of means plotted for the three densities and two percentage covers are absolutely parallel in the two seasons.

In contrast, in Figure 10.6b and e, there is a bigger difference between the means in the two percentage covers (mortality was generally greater in autumn, but only for the larger cover). This creates a P × S interaction, in addition to the D × P one already present in spring (Figure 10.6b). There is no D × S interaction (the differences in means at each density are the same in the two seasons). Nor is there a D × P × S interaction, because the D × P plots are still the same shape between seasons (therefore D × P is similar in the two seasons).

Finally, in Figure 10.6c and f, there is marked interaction of D × S (differences among densities vary with season), P × S (the difference between the two percentage covers clearly varies with season) and D × P × S. The last is caused by the fact that the two sets of graphs (plotting D × P in each season) are clearly not parallel.

An ecological interpretation for the pattern in Figure 10.6c and f would be that, in spring, mortality increases with large density, but the effect is much enhanced by a greater cover of algae. In autumn, however, there is little effect of density and a much smaller difference due to cover of algae.

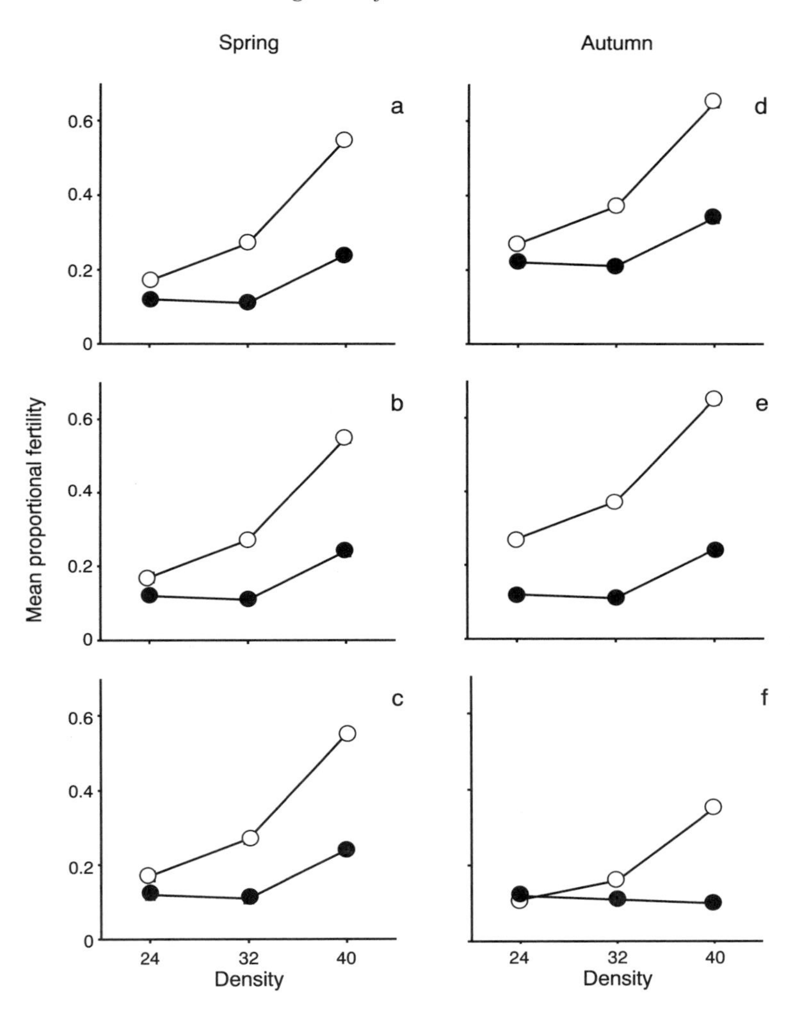

Figure 10.6. Three possible outcomes of an experiment with three factors. Densities (D) and percentage covers (P) are as in Table 10.10. The third factor is seasons (S) (with two levels – spring and autumn); the whole experiment was done twice. In all cases, there is a significant interaction between density and percentage cover (empty circles, 60%; filled circles, 20% cover of algae). In the first case ((a) and (d)), there are no interactions with seasons (D × S, P × S, D × P × S are all non-significant). In the second case ((b) and (e)), there are interactions of density and percentage cover with seasons (S) (D × S, P × S are significant). The final example, (c) and (f) is as for (b) and (e), but the interaction between density and percentage cover is different in each season (P × D × S is significant).

An appropriate model to explain these results is that food is much more abundant in autumn than in spring, so crowding does not have as much effect. Alternatively, predators may be less voracious, active or numerous in autumn than in spring.

Thus, interpretations of complex interactions are quite easy – provided that proper thought has been give to the models and hypotheses before the experiment is done. Only when adequate anticipatory thought has been given to the predictive hypothesis(es) is it straightforward to interpret results of *any* experiment. Experiments with complex sets of interactions are no exception and add no new difficulties. It must be remembered, however, that higher-order interactions frustrate any tests of hypotheses (i.e. any statements of hypotheses to test) about lower-order interactions or main effects.

10.11 Power and the detection of interactions

When using factorial experiments, it will often be desirable to estimate, in advance, the power of the experiment to detect interactions anticipated as part of an hypothesis. For example, consider the situation where numbers of a plant are sometimes much smaller near shade than out in the open, but in some areas are very similar in the two habitats. A grazing insect is often present, feeding on the plants in shaded areas, but not in the sunny areas. A sensible model to explain the variation in pattern from place to place is that the insect is entirely responsible for the difference in densities of plants between sunny and shaded areas, but the insect is not present in some places, so there is no difference in densities of plants between sunny and shaded spots in such areas. From this, the hypothesis can be proposed that, if insects are prevented from foraging on the plants in some sunny and shaded areas, there will be no difference in densities of plants. If the insects are allowed to continue to forage in other areas, there will eventually be a reduction of numbers of plants in the shaded as opposed to the sunny areas.

An experiment to test this hypothesis could be established with several replicate plots of the following treatments: in sunny areas, with or without cages to prevent insects attacking the plants; in shaded areas, the same treatments. A better experiment would also have treatments to control for the effects of cages to exclude the insects, but let us keep it simple here for the purpose of considering power. The experiment has two fixed, orthogonal factors (sunny versus shaded, with two levels; with versus without insects, with two levels). The hypothesis predicts an interaction between these two factors. At the end of the experiment, in plots where insects have been excluded, there will be a greater difference between the densities of plants in the sunny as opposed to the shaded areas than is the case where insects have been allowed to feed. There must be more plants in shaded areas where the insects are excluded and the density of plants in such areas will match that in sunny areas with or without insects. Where insects are allowed to feed, there will be fewer plants in the shaded plots. Without such an interaction appearing in the analysis, the model cannot be correct.

Calculation of the power of the experiment to detect such an interaction is quite straightforward. Note the estimates provided by the mean squares in Table 10.4. The appropriate F-ratio to test the null hypothesis of no interaction is the mean square for the interaction divided by the mean square of the residual. The determination of power for such a test is identical with that described for a fixed factor in an experiment with only one factor, as in Section 8.3. The only difference is that the construction of the alternative hypothesis is slightly more complicated because it involves determination of how large the interaction will be between the levels of the two factors. In the particular example described here, this is easily accomplished given the magnitudes of differences between densities in shaded and sunny areas available from the original observations.

In some cases, it may be more difficult to construct a quantified assessment of the alternative hypothesis. Note also for this case that if there is no interaction in the analysis of the experimental results, the model must be wrong, regardless of the statistical significance or otherwise of the two separate factors. There would be no point in becoming confused by tests of the main effects in the absence of interaction. The results already demonstrate, by the lack of an interaction, that the model is wrong. You must now go back to the beginning and think of a new explanation for the observations – your original ones and the ones just gained from the experiment. Significance or not of the main effects will not help this process.

Similarly, the determination of power for the interaction between random factors is also straightforward. There is no new feature compared with the description of a random effects model in Section 8.4. The interaction between two random factors (or, as described later in Section 10.14, the interaction between a fixed and a random factor) is itself a random effect. Therefore, the calculation of power to detect the interaction between two random factors in a two-factor experiment is identical with that described in Section 8.4. The F-ratio to test for the presence of an interaction would be the mean square for the interaction divided by the mean square for the residual. Each has the same components of variance as was described for the case of a single factor. Again, as described for the calculation of power for random factors, it will usually be somewhat more difficult to construct the alternative hypothesis about the variance of some random experimental factor than is the case for a fixed factor. Nevertheless, there is no new feature to the determination of power for an interaction between two factors.

10.12 Spatial replication of ecological experiments

There are some serious statistical problems with mixed models for analysis of variance, i.e. those with a mixture of fixed and random components. There are two general classes of these for designs with two factors. In one, a fixed factor (see Section 8.2) is to be investigated, but the replication is a random factor. In the second case, the replication is a random factor, but the different levels of the fixed factor are, literally, all applied to the same randomly chosen levels of the random factor. The latter case is an unreplicated repeated measures design and is discussed in Chapter 12. The former is discussed here because of its importance in ecological studies.

Consider an experiment with one fixed factor, for example density of plants. Their well-being (growth, reproduction, survival, whatever) has been observed to vary from place to place (at a distance of, say, a kilometre), apparently because density varies. This model (differences in density cause the observed differences in, say, growth) leads to the hypothesis that reductions in density will cause increased growth. There will be two levels of density (10 and 20 plants per m^2), created by thinning out randomly chosen clumps of plants. This is factor A, with $a = 2$ levels and is fixed. The treatments are replicated in $n = 4$ scattered plots over a site some hundred metres on a side.

Now, however, thought must be given to the purpose of the experiment. The original observations were of differences in places about a kilometre apart. The experimental site is, effectively, one such place. We need replicated sites spaced about a kilometre apart. These should be randomly chosen out of the places where the plant occurs. Thus, we have a random factor B, with, for example, $b = 3$ places. Each of the treatments is thus randomly and independently scattered in each of the places. You must have the replicate plots of each treatment in each place. Otherwise, if there were only one plot of each treatment in a place, the growth of plants may differ from place to place because of small-scale variation, nothing to do with the treatment. This type of variation can be estimated from the replicates of each treatment in each plot. The model for the analysis is:

$$X_{ijk} = \mu + A_i + B_j + AB_{ij} + e_{k(ij)} \qquad (10.28)$$

where X_{ijk} represents the average growth of plants in the kth replicate plot of treatment i in place j. A_i, B_j and AB_{ij} represent the effect of the ith treatment (if density does influence growth), the difference from average conditions prevailing in the jth place (if places differ) and their interaction. The A_i terms are fixed and therefore assumed to sum to zero. The B_j terms are assumed to be from a normal distribution with mean zero and variance σ^2_B (as in Figure 9.4). The interaction terms are also randomly sampled in each level of factor A – the AB_{ij} terms that arrive in your experiment are a feature of the plots that you happened to include. They are, for each experimental treatment, a random sample of possible interaction terms. Thus, the interaction of a fixed and a random factor is, itself, a random component of the analysis (Searle *et al.*, 1992).

The experimental design just described should not be confused with some others that are often described as mixed models. It is not an unreplicated randomized blocks experiment (as discussed in Section 12.1), nor an unreplicated repeated measures design (as in Section 12.5). Nor is it the same as the replicated repeated measures design used by Scheffé (1959) to illustrate the properties of mixed models. In his design, there were a machines (a fixed factor A) and b workers (randomly chosen as factor B). Each worker was assigned to each machine for n replicate days. The analysis therefore involved repeated sampling of the same worker over the n days on each machine (with yet more potential non-independence among samples, as also discussed in Section 7.14). Nor is it the similar example illustrated by Winer *et al.* (1991) as a mixed model with repeated measures.

Here, there are independently sampled levels of B and independently scattered replicates of the levels of A in every level of B. The analysis proceeds exactly as illustrated in Table 10.8b. If there is no interaction between factors A and B, main effects should be tested. If the estimated interaction between A and B is sufficiently small, its mean square can be pooled with the residual mean square (see Sections 9.7 and 11.7). Note the caution expressed earlier. The interaction must be sufficiently small that it fails to cause rejection of the null hypothesis (that there is no interaction) at probability of Type I error $P = 0.25$ This will provide a powerful test of the differences among levels of the experimental treatments averaged over the set of places examined. Remember that, for complex designs, the decision about which sources of variation to pool should be made in advance of the experiments to avoid the temptation to keep adjusting the analysis until some main effect becomes statistically significant.

If there is significant interaction between the random places and the fixed treatments, further analysis is done as comparisons of treatments in each place (as in Section 10.8.1). In this case, no test is required of the overall differences among treatments because these cannot be averaged over the various places. The outcome of the experiment is that the treatments have different influences from place to place. All further tests will be multiple comparisons on means in each place.

The remaining situation is problematic, as discussed in the next section.

10.13 What to do with a mixed model

If the interaction is not significant, but cannot be pooled, there is a difficulty with these analyses. There is a test available for each of the main effects, places and treatments, as in Table 10.12. The test for differences among treatments (A in Table 10.12) has relatively few degrees of freedom. It is, in general, unlikely to be very powerful. So, unless differences among levels of the fixed factor are large, there is no great likelihood that they will be detected. There is a further problem. The F-ratio test for factor A over the A × B interaction is only approximately an F-ratio, as discussed by Wilk & Kempthorne (1955), Scheffé (1959), Winer *et al.* (1991) and, more recently, by Searle *et al.* (1992).

For this test (of A over A × B) to be underpinned by the central distribution of F, there must be particular properties of symmetry in the matrix of variances and covariances of the data in the different levels of factor A. Symmetry requires that the degree to which there are correlations among the data, i.e. among the replicates in each sample and among the randomly chosen places, is the same for each level of the experimentally applied treatment. Thus, even though data may not be strictly independently sampled, there is no problem with testing null hypotheses about the treatments provided the degree of non-independence is similar for all treatments.

This is largely tested by the test for interaction. If the differences among treatments are correlated in different ways with the differences among

Table 10.12. Analysis of a mixed model with two factors. Factor A is fixed with $1 \ldots i \ldots a$ levels; factor B is random with $1 \ldots j \ldots b$ levels. There are n replicates in each combination

(a) Analysis of variance

Source of variation	Degrees of freedom	Mean square estimates
A	$(a-1)$	$\sigma_e^2 + n\sigma_{AB}^2 + bnk_A^2$
B	$(b-1)$	$\sigma_e^2 + an\sigma_B^2$
A × B	$(a-1)(b-1)$	$\sigma_e^2 + n\sigma_{AB}^2$
Residual	$ab(n-1)$	σ_e^2

(b) F-ratios when A × B interaction can be pooled with the residual

A	$F = MS_A/MS_{pooled}$, with $(a-1)$ and $ab(n-1)+(a-1)(b-1)$ degrees of freedom
B	$F = MS_B/MS_{pooled}$, with $(a-1)$ and $ab(n-1)+(a-1)(b-1)$ degrees of freedom

(c) F-ratios when A × B interaction is not significant, but cannot be pooled

A	$F = MS_A/MS_{AB}$, with $(a-1)$ and $(a-1)(b-1)$ degrees of freedom
B	$F = MS_B/MS_{residual}$, with $(b-1)$ and $ab(n-1)$ degrees of freedom

places or the variations among replicates (i.e. the different A_i terms are correlated differently with the B_j or the $e_{k(ij)}$ terms), there will usually be significant interaction. When there is interaction, you should not be testing the main effects of A and B anyway, so there is no problem with the test for A because you do not do it. A lack of interaction implies there is independence (no correlation) of effects of treatments from one place to another. Thus, if the interaction term is not significant and can be pooled, there would also be no problem with a test for A. It is now an ordinary F-ratio. The problem comes when you are in no-man's land, caught between a non-significant test for interaction and not being able to pool the interaction term.

There is relatively little problem for balanced data (i.e. designs where there is the same number of replicates in every combination of treatments), but serious and sometimes insuperable dilemmas where data are unbalanced (Searle *et al.*, 1992). Burdick & Graybill (1992, p. 170) made this very plain: 'unlike the situation encountered with the balanced mixed model, tests of hypotheses and confidence intervals on estimable functions of fixed effects parameters are not simple to develop'. Yet again, there are convincing reasons to attempt to keep the data balanced in all experiments.

So, it is important to be able to test for interactions and to be able to pool the interaction term if this seems reasonable, given the issues discussed earlier (Sections 9.7 and 11.7). This is the reason unreplicated randomized blocks and unreplicated repeated measures analyses are problems in biological experiments (as discussed in Chapter 12). Otherwise, it is generally preferable to analyse the outcome of the experiment by multiple comparisons in each level of the random factor (B or, here, places). You need replicates of each treatment in each block to be able to test for interactions.

10.14 Problems with power in a mixed analysis

Calculation of the power of an experiment is the other issue that is problematic in a mixed analysis with one (or more) fixed and one (or more) random factors in orthogonal combinations. The F-ratio using the mean square of the fixed factor (A) divided by the mean square for interaction (A $\times$ B) comes from the central F-distribution when the null hypothesis is true. The central F-distribution is, however, only appropriate under a set of strict assumptions about the nature of variances and covariances (Scheffé, 1959; Searle *et al.*, 1992).

Various simulations of data when the assumptions are violated (see e.g. Box, 1954b) suggest that the use of an F-ratio distribution will not be too bad an approximation to whatever is the real distribution of the test statistic. Under some conditions of departure from the assumptions for use of F, Box (1954b) provided a conservative test, which may be preferable (for its use, see Winer *et al.*, 1991). Box's (1954b) test does, however, reduce the power of the test for differences among levels of the fixed factor.

There is, nevertheless, a major problem with any attempt to calculate power. According to Winer *et al.* (1991), unless the assumption of symmetry of covariances of the data in the different levels of the fixed factor is met, the distribution of F under some alternative to the null hypothesis (that there are no differences among levels of factor A) is not explicitly calculable (see also Burdick & Graybill, 1992). It is not a non-central F-distribution as described in Section 8.3. There is correlation between sampled estimates of σ^2_{AB} and k^2_A as components of the mean square for factor A. Nor can the distribution of F under an alternative to the null hypothesis be calculated as though it were a random factor (as in Section 8.4). So, no alternative distribution of F is available and power cannot be calculated explicitly before the experiment. Retrospectively, if it were worth doing, power could be estimated by simulations using the values of mean squares and estimates of components of variation obtained in the experiment.

All of this suggests three things about the use of mixed models. First, they *must* be used in ecology and other areas of biology. That is the only way to get tests of hypotheses about processes operating over spatial scales requiring experimental treatments to be placed in randomly chosen parts of the area.

Second, replicate everything! Then, tests on interactions can be done properly. This will allow some aspects of the restrictive assumptions to be evaluated. It will also dictate whether tests on the main fixed factor are possible (the interaction can be pooled) or unwise (it cannot be) or are irrelevant (the interaction is significant).

Finally, if you need to calculate the power of the test for the fixed factor to detect some specified alternative, you will need to do it in two ways. Use the non-central distributions of F to calculate power assuming that pooling of the interaction will be possible. This is done exactly as described in Section 8.3. Also use the central F-ratio to calculate power of the test of a random component in the analysis (if the interaction can be pooled) or the interaction itself (if it cannot be). Use the procedures described in Section 8.5. The latter will require much more thought about the specification of an alternative to the null hypothesis of no interaction (see the discussion in Underwood, 1993).

Above all, keep calm. Maintain your biological view of the worth of the hypothesis and the importance of the experiment. Then, get some professional statistical advice!

10.15 Magnitudes of effects of treatments

It is quite often desirable to estimate the sizes of the variability attributable to different treatments in an analysis. Unfortunately, this has become a popular activity in some areas of biology. For example, Welden & Slauson (1986) proposed that the importance of a particular process acting on an organism can be measured as the proportion of total variability in some variable that is associated with that process. Thus, in an ecological experiment analysing the reduction of growth due to increased intraspecific density, there would be a series of densities, each replicated several times. The measure of magnitude of effect of density as proposed by Welden & Slauson (1986) would then be the proportion of total variability in growth attributable to differences among the different densities. This is the sum of squares for differences among densities divided by the total sum of squares. This is not a useful nor a comparative measure of effects of density (see Underwood & Petraitis, 1993). The flaws and difficulties are summarized below (Section 10.15.2).

It is, however, worth considering some methods that will help to achieve a measure of the relative importance of different components. They will only be introduced briefly here and you are directed to the detailed accounts by Burdick & Graybill (1992) and Searle *et al.* (1992) for recent syntheses.

10.15.1 Magnitudes of effects of fixed treatments

Consider, for example, a two-factor experiment on the influences of crowding (at several chosen densities) and starting size (several chosen sizes) on growth of animals. It may be of interest to determine the 'importance' or relative importance of the two different factors on the mean growth. For example, even if both factors cause significant differences among mean rates of growth of the animals, it may be of value to know that starting size is less of an influence than is density.

The effect of density can be estimated as in Table 10.13. So can the effect of starting size and the interaction between the two. In theory, this allows a measure of the relative importance of one or other factor. In fact, several measures are available (see e.g. Fleiss, 1969). As Winer *et al.* (1991) pointed out, *relative* importance of one or other factor can be considered only relative to some reference or base. It has been proposed that this can be done by calculating:

$$\omega^2_{\mathrm{A}} = \frac{\hat{\theta}^2_{\mathrm{A}}}{\hat{\theta}^2_{\mathrm{A}} + \hat{\theta}^2_{\mathrm{B}} + \hat{\theta}^2_{\mathrm{AB}} + \sigma^2_e} \qquad (10.29)$$

$$\omega^2_{\mathrm{B}} = \frac{\hat{\theta}^2_{\mathrm{B}}}{\hat{\theta}^2_{\mathrm{A}} + \hat{\theta}^2_{\mathrm{B}} + \hat{\theta}^2_{\mathrm{AB}} + \sigma^2_e}$$

$$\omega^2_{\mathrm{AB}} = \frac{\hat{\theta}^2_{\mathrm{AB}}}{\hat{\theta}^2_{\mathrm{A}} + \hat{\theta}^2_{\mathrm{B}} + \hat{\theta}^2_{\mathrm{AB}} + \sigma^2_e}$$

10.15.2 Some problems with such measures

There are, however, several problems with this (see also the detailed critique in Underwood & Petraitis, 1993). First, is the problem that

Table 10.13. Estimating the magnitudes of effects of fixed factors in a two-factor experiment. Factor A is fixed with a levels; factor B is fixed with b levels; there are n replicates in each combination of treatments

Effects of factor A are defined as:
$$\theta_A^2 = \frac{\sum_{i=1}^{a}(A_i - \bar{A})^2}{a} = \frac{(a-1)k_A^2}{a}$$

Effects of factor B are defined as:
$$\theta_B^2 = \frac{\sum_{j=1}^{b}(B_j - \bar{B})^2}{b} = \frac{(b-1)k_B^2}{b}$$

Magnitude of interaction A ×B is defined as:
$$\theta_{AB}^2 = \frac{\sum_{i=1}^{a}\sum_{j=1}^{b}(AB_{ij} - \bar{A}_i - \bar{B}_j + \overline{AB})^2}{ab} = \frac{(a-1)(b-1)k_{AB}^2}{ab}$$

where k_A^2, k_B^2 and k_{AB}^2 are as in Table 10.4

Source of variation	Mean square estimates
A	$\sigma_e^2 + abn\theta_A^2/(a-1)$
B	$\sigma_e^2 + abn\theta_B^2/(b-1)$
A × B	$\sigma_e^2 + abn\theta_{AB}^2/(a-1)(b-1)$
Residual	σ_e^2

θ_A^2 is estimated as $\hat{\theta}_A^2 = \frac{(\text{Mean square A} - \text{Mean square residual})(a-1)}{abn}$

θ_B^2 is estimated as $\hat{\theta}_B^2 = \frac{(\text{Mean square B} - \text{Mean square residual})(b-1)}{abn}$

θ_{AB}^2 is estimated as

$$\hat{\theta}_{AB}^2 = \frac{(\text{Mean square AB} - \text{Mean square residual})(a-1)(b-1)}{abn}$$

σ_e^2 is, of course, estimated by the mean square residual

such comparisons involve an incommensurable component. The three estimates of fixed effects are not at all the same thing as the estimate of variance within populations obtained from the residual mean square. The latter is a sample variance; the former are measures of squared differences among constants. At best, they are 'variance-like' formulae for measuring differences among treatments. They are not variances –

they represent measures of the defined fixed effects. To measure them relative to some base or reference that includes an estimate of variance among replicates does not lead to any particularly valid or justifiable measure. It is akin to assessing the relative importance of apples or oranges by a reference to the combined worth of apples, oranges and bananas (but bananas are not evaluated with the same precision).

The second problem is that these measures are, of course, relative to each other. One enormous advantage of an analysis of variance is to construct independent, orthogonal measures of the magnitude of differences among mean values of a variable subject to different factors. The end result is then to put the measures together for the sole purpose of creating non-independent, relative measures. This makes no sense at all – despite its increasing popularity in recent ecological literature. Such relative measures cannot be used for very much – if anything.

For example, there is no way to compare the magnitude of the importance of density in the experiment considered above in two such experiments in different places. These will only be comparable if the magnitude of effects of starting size, interaction and variance within populations are all equal between the two places. Otherwise, any comparison of relative importance of density will be inextricably mixed up with differences in importance of any of the other factors and components of measure. The estimate of the individual measure ($\hat{\theta}^2_A$) could, however, be compared in its own right. For some reason, this is rarely done.

It seems that the only value in such comparative measures is *within* an experiment. It still seems quirky to want to know the magnitude of effect of density or of starting size in some measure that is affected by the other factor. If, however, you want to do this, go ahead using some formula like that above. It may, nevertheless, be a more logically justifiable procedure to measure the relative magnitude of effects of the factors and the interactions by calculating them only over the sum of effects of factors, without the variance within populations. Thus:

$$\omega'^2_A = \frac{\hat{\theta}^2_A}{\hat{\theta}^2_A + \hat{\theta}^2_B + \hat{\theta}^2_{AB}} \tag{10.30}$$

and so on for the other measures.

There are, however, further, insurmountable difficulties which are considered below (Section 10.16).

Table 10.14. Estimating the magnitudes of components of variation due to two random factors in an experiment. Factor A is random with a levels; Factor B is random with b levels; there are n replicates in each combination of treatments

(a) Analysis of variance

Source of variation	Degrees of freedom	Mean square estimates
A	$(a-1)$	$\sigma_e^2 + n\sigma_{AB}^2 + bn\sigma_A^2$
B	$(b-1)$	$\sigma_e^2 + n\sigma_{AB}^2 + an\sigma_B^2$
A × B	$(a-1)(b-1)$	$\sigma_e^2 + n\sigma_{AB}^2$
Residual	$ab(n-1)$	σ_e^2

(b) Estimates of components of variation

$$\sigma_A^2 \text{ is estimated as } \hat{\sigma}_A^2 = \frac{(\text{Mean square A} - \text{Mean square AB})}{bn}$$

$$\sigma_B^2 \text{ is estimated as } \hat{\sigma}_B^2 = \frac{(\text{Mean square B} - \text{Mean square AB})}{an}$$

$$\sigma_{AB}^2 \text{ is estimated as } \hat{\sigma}_{AB}^2 = \frac{(\text{Mean square AB} - \text{Mean square residual})}{n}$$

$$\sigma_e^2 \text{ is estimated as } \hat{\sigma}_e^2 = \text{Mean square residual}$$

10.15.3 Magnitudes of components of variance of random treatments

Estimating components of variation for random factors follows the same general principle used for the fixed case (Section 10.15.1). Now, however, it is also possible to calculate the confidence interval for each estimate. It is also the case that in a completely random effects experiment, all components including variation within samples are of the same form. They are all estimates or are sums of estimates of variances. This, at least, makes a relative measure of each component a rational one because all components are measured comparably. There is no incommensurability as in the case of fixed factors.

Again, the components in a two-factor experiment are illustrated (Table 10.14). Factor A is times of sampling, randomly chosen during spring. Factor B is plots of flowering plants scattered at random over a study-site. In each plot, there are n replicate sticky traps to collect bugs. The original observations were of differences from place to place and time to time in the abundances of bugs on flowering plants. The initial model to explain this is that there are consistent differences from place to place because local patches of habitat support consistent populations of bugs. Differences from time to time are due to variations in weather and therefore affect all patches in the same way. Alternative explanations all involve inconsistent patterns of temporal change from one patch to another. Thus, the hypothesis is that there will be differences from one time of sampling to another and from one plot to another, but no interaction between them. The plots are randomly chosen to represent different parts of the habitat. The times are randomly chosen because this avoids any cyclic or other structured aspects of the dynamics of the bugs' populations that would cause temporal correlation between samples. Obviously, the same plots must be sampled, otherwise the consistency of time-courses cannot be examined at all!

The components of variance attributable to each source of variation are shown in Table 10.14. Each is, as is obvious from their calculation, dependent on the precision with which other components are estimated. As with fixed factors in Section 10.15.1, these could be calculated relative to each other, but this would reduce their value. When the magnitudes of various components are assessed relative to each other, they are no longer independent and become entangled. If $\hat{\sigma}^2_{\mathrm{B}}$ is large, $\hat{\sigma}^2_{\mathrm{A}}$ and $\hat{\sigma}^2_{\mathrm{AB}}$ will be relatively small. This will occur regardless of the actual sizes of effects of A and A $\times$ B.

For random factors (i.e. variance component models), there are two alternative approaches to expressing the magnitudes of effects so that they can be compared to each other. First, it is possible to construct variances (and therefore confidence intervals) for the estimates. The general principles and available formulae are well described by Burdick & Graybill (1992) and Searle *et al.* (1992). By constructing confidence intervals for the variance components, it is possible to compare the magnitudes of components directly, without calculating them relative to each other.

The other approach is the use of the ratio of each estimated magnitude of variance over the estimate of residual variance. This is θ, the term used in calculations of power for detecting differences among levels of a random factor (Section 8.4). Once θ has been calculated, its confidence intervals can be calculated for each component (Burdick & Graybill, 1992; Searle *et al.*, 1992). This provides a measure of size of each component relative to the same reference – the residual variance. Each is comparable using confidence intervals.

Consider the case of an experiment with one random factor, A, with *a* levels, each replicated *n* times, leading to the analysis in Table 10.15. In

Table 10.15. Estimation of components of variance and their variances in a single-factor experiment

(a) Analysis of variance

Source of variation	Degrees of freedom	Mean square estimates
A = Among levels of factor A	$a-1$	$\sigma_e^2 + n\sigma_A^2$
Residual	$a(n-1)$	σ_e^2

(b) Estimates of components of variance

$$\hat{\sigma}_A^2 = \frac{(\text{Mean square A} - \text{Mean square residual})}{n}$$

$$\text{with variance } (\sigma_A^2) = \frac{2}{n^2}\left[\frac{(\text{Mean square A})^2}{(a-1)} + \frac{(\text{Mean square residual})^2}{a(n-1)}\right]$$

$$\hat{\sigma}_e^2 = \text{Mean square residual}$$

$$\text{with variance } (\hat{\sigma}_e^2) = \frac{2\ (\text{Mean square residual})^2}{a(n-1)}$$

turn, the variances associated with A and the residual are estimated as $\hat{\sigma}_A^2$ and $\hat{\sigma}_e^2$, also in Table 10.15. Define their ratio, θ, as in Section 8.4:

$$\theta = \frac{\hat{\sigma}_A^2}{\hat{\sigma}_e^2} \tag{10.31}$$

and choose the probability for a confidence interval (e.g. $\alpha = 0.05$ for a 95% confidence interval; see Section 5.3). From the distribution of F and knowing that random effects lead to central F-distributions (Section 8.4), it is possible to construct the following confidence interval for θ:

$$\text{Prob. } (L \leq \theta \leq U) = 1 - \alpha \tag{10.32}$$

where L is the lower and U is the upper limit.

$$L = \frac{1}{n}\left(\frac{F_{\text{obs}}}{F_{1-\alpha/2}} - 1\right) \tag{10.33}$$

$$U = \frac{1}{n}\left(\frac{F_{\text{obs}}}{F_{\alpha/2}} - 1\right)$$

where F_{obs} is the F-ratio calculated in the analysis of variance (i.e. mean square A/mean square residual). n is the sample size. $F_{1-\alpha}$ and F_{α} are the tabulated values of F with $(a - 1)$ and $a(n - 1)$ degrees of freedom for probabilities $(1 - \alpha)$ and α, respectively (0.95 and 0.05 here). These formulae come from Winer *et al.* (1991).

This may be a useful procedure to include in many ecological studies, particularly those where variability in space or time is assessed at several scales to determine which scales are associated with the greatest variability.

In any experiment with several random sources of variation, it will be possible for there to be negative estimates of magnitude of some sources. For example, in Table 10.14, it is possible for the mean square for A × B to be larger than that for A, leading to a negative estimate of $\hat{\sigma}^2_A$. This causes difficulties for determining the degrees of freedom that are associated with estimates of magnitudes of variance components (Satterthwaite, 1946; Gaylor & Hopper, 1969). Some authors consider that negative estimators could be declared to be zero, but, if you really want to pursue descriptions and estimations of your data, this is biased and not a reliable procedure (Scheffé, 1959; Winer *et al.*, 1991). A negative estimate is one that underestimates a true value that might be positive, not zero. For the purposes of constructing confidence intervals, this would be important. There has been considerable discussion of the procedures available to deal with analyses in which some of the components produce negative estimates of variances.

One course of analysis to remove this problem is to use an alternative analytical structure to estimate the components of variation so that they must become positive or minimally zero. This would avoid having negative estimates of parameters that are defined to be positive. Two methods of estimation worth some thought are maximal likelihood (ML; see McLean *et al.*, 1991) and restricted maximal likelihood (REML). It is well known that, given balanced data, REML estimators produce estimates identical with those provided by analysis of variance (e.g. Searle *et al.*, 1992). Furthermore, for balanced data, analysis of variance produces better quadratic, unbiased estimators than ML estimators (Graybill & Hultquist, 1961) and REML estimators are generally superior to those provided by ML (reviewed in Robinson, 1987). For some models, ML estimators do not converge on a unique solution.

In fact, for balanced data, REML produces the same results as those obtained in an analysis of variance having eliminated from the model those terms that produce negative estimates of variance (D. Fletcher, personal communication and unpublished simulations of various models;

see also the similar outcome via a different method in Thompson and Moore, 1963). Thus, for balanced data, REML offers no advantage over setting a source of variation to zero in an analysis of variance.

For unbalanced analyses, negative estimates of variance can be handled by REML, but still have all the problems of interpretation associated with unbalanced data. They are not considered further here (see the earlier discussion of the desirability of getting balanced data); you are referred to the review by Robinson (1987) and the books by Searle *et al.* (1992) and Burdick & Graybill (1992).

10.16 Problems with estimates of effects

However the data are handled, there are several severe problems for the *logical* use of estimates of sizes of effects of treatments in ecological experiments. No doubt many of these also apply to other areas of biology. Some of them apply generally. Despite the views of many statisticians that testing hypotheses is a secondary pursuit and that 'estimation' is the aim of quantitative science, these problems have not been given much practical consideration.

10.16.1 Summation and interactions

As explained in Section 7.20, it must be assumed that the average effect of the levels of a fixed factor is zero. This has no effect on tests of the null hypothesis of no differences among levels, or on an estimate of magnitude of differences among levels if the null hypothesis is rejected (as in Section 10.15.1). It does, however, affect estimation of the magnitude of individual differences among levels (A_i terms for a factor, A). This was discussed in Section 7.20.

A greater dilemma is imposed when there are at least two orthogonal factors, generating interactions. Suppose an interaction A $\times$ B, between two orthogonal factors A and B, is significant. It demonstrates that differences between levels of factor A (i.e. effects of A) are dependent on which level of B is examined and vice versa. Now it is *impossible* to specify the magnitudes of effects of either A or B. They do not have an influence that is separate from each other. Nor is it possible to determine them in any meaningful way as a relative measure (such as ω^2_{A} in Section 10.15.1). Nor, of course, is it possible to measure them from mean squares or sums of squares assessing either their average differences or the interaction between them.

Interactions will cause this problem whether they are between fixed, random or a mixture of sources of variation. There is no point in measuring the purported effects of A and B unless it can be shown that their interaction is, effectively, zero. All attempts to do so are problematic, to say the least.

It may be desirable to simulate or model interaction and main effects to determine some joint influence they have on a variable, but that is not the topic here. Presumably such modelling would lead to testable hypotheses about the magnitudes of the combined influences of main effects and their interactions. Such hypotheses would then require an appropriate experimental design and analysis.

10.16.2 Comparisons among experiments or areas

Another problem with estimates of magnitudes of effects is in attempts to use them to compare experiments or habitats. It is a common desire amongst ecologists to wish to compare the relative importance of such processes as predation or competition from one study or habitat or assemblage to another. In theory, such comparisons can be done using quantitative estimates of the importance of these factors from experiments in each area.

This is, however, fraught with great hazards, as reviewed by Underwood & Petraitis (1993). As a summary here, the magnitude of differences among levels of a factor is a function of the nature of the levels. If two experiments on a fixed factor (such as density in an experiment about competition) have a different range of densities, their estimates of effects will be very different. In any attempt to use estimates of relative magnitude (as in Section 10.15.1), a comparison is realistic *only* if the magnitudes of effects of the other factors and the size of the residual are equal between the two experiments.

This all leads directly to the notion that attempts to compare relative importance of various ecological processes among areas or experiments must be designed explicitly with such comparisons in mind. It is next to impossible to use data gathered for other purposes because the comparisons are always confused by differences in the ways the experiments were designed.

So, before comparing processes in different habitats or systems, specific hypotheses about the influences of those processes should be proposed and tested by appropriate experiments in each habitat. To allow the comparison between habitats, the experiments need to be designed so that they can be compared.

10.16.3 Conclusions on magnitudes of effects

The foregoing provides a cautionary brake on the rush to estimate components of variance or relative magnitudes of effects of fixed factors. Although sympathetic to the need for better descriptions of sampling and experiments (see e.g. Andrew & Mapstone, 1987), I urge caution in the use of inappropriate descriptions.

The best way to be sure of the relative importance of experimental factors is to contrast them in appropriately designed experiments. The best way to have any certainty about variances or confidence intervals of estimates of magnitude of components of variance is to get estimates from independent experimental investigations of those components (see e.g. Underwood & Chapman, 1996).

11

Construction of any analysis from general principles

11.1 General procedures

Having waded through the principles of nested and orthogonal, fixed and random factors, we are now in a position to consider the general principles of experimental design in relation to tests of hypotheses (Table 11.1). It is assumed that the logical basis for a test is well defined and the purpose of any experiment is very clear before any attempt is made to do it (see Chapters 2 and 4). If this is so, the appropriate controls and replicates will have been thought through properly. All of the logic having been sorted out, the steps in designing any experiment involving analysis of numerous means are the same, however complex it must be.

First, the relevant linear equation or linear model for the design must be formulated. Second, the appropriate degrees of freedom must be identified and the estimates to be obtained from mean squares defined. Finally, the formulation of sums of squares must be determined.

These steps are illustrated below for two examples. The first is an experiment about growth of plants under two different experimental factors. It has been observed that plants of a particular species grow taller in some areas than in others. It has been suggested as a model that this is due to limitations of nitrogen in the soil of some areas. Alternatively, it has been suggested (as a second model) that attacks by insects vary in intensity from place to place, causing variations in the growth of the plants. A third explanation is that both processes occur. These are based on knowledge of the variability in numbers of insects and in availability of nitrogen from place to place.

The relevant hypotheses are, first, that addition of nitrogen to patches with relatively short plants will cause them to grow taller, compared with controls and, second, that exclusion of insects from plots with relatively short plants will lead to increases in size. The various treatments are set

Table 11.1. Steps in designing a complex experiment

1. Define models, hypotheses, null hypotheses.
   Define populations and variables to sample.
2. Determine relevant factors (from hypotheses).
   Define whether fixed or random.
3. Define necessary replication:
   (a) to create generality;
   (b) to ensure no confounding.
4. Organize all necessary controls (to avoid confounding and other logical errors).
5. Construct linear model.
6. Calculate degrees of freedom.
7. Calculate the mean squares estimates and what are the appropriate F-ratios.
8. Where appropriate, do pilot experiments.
9. Determine power of tests and where problems of lack of degrees of freedom might be resolved.
10. If necessary, redesign experiment from steps 3 to 8.
11. Determine how to calculate the sums of squares correctly.

up orthogonally to each other. Thus, factor A is 'Nitrogen', with 3 levels: added nitrogen (in solution), added water without nitrogen, untouched control. Factor B is 'Exclusion' with three levels: cages to exclude insects, half-open cages to control for the effects of mesh, and untouched controls. Each of the nine treatments is set up in four replicate plots, i.e. 36 plots scattered independently over the study-area. In each plot, five plants are measured.

So, there are two orthogonal fixed factors (nitrogen and exclusion), with a third factor (plots) nested in the combinations of these two. Plots represent a random factor.

The second case concerns observed patterns of different rates of growth of snails on large boulders on different rocky shores. Two primary models can explain why snails grow faster in some places. Either the snails on different shores are different (or become different once they start to grow). Alternatively, the physical conditions prevailing or the amount of food available on the different shores influence the rate of growth. Of course, a combination of the two forms a third model.

From these, the following hypotheses should be examined:

1. If snails from several shores (where rates of growth are likely to differ) are put on a sample of boulders on each of the shores, they will grow at different rates on each shore if model 1 is correct. If the model is incorrect, they will grow at similar rates because their origin does not matter, but their destination does.

2. If model 2 is correct, when snails from any shore are put on a sample of boulders on each of several shores, they will grow at different rates. If model 2 is not correct, they will grow at the same rate because there will be no differences among shores.
3. If origin and destination matter, both factors will be significant in the analyses. Alternatively, if origin and destination both matter, snails from several shores put onto several shores including the one from which they were taken will show an interaction in their rates of growth, such that snails of different origin grow at different rates on each shore, but do not grow at the same rate from one shore to another. There will be an interaction between origin and destination of the snails.

These hypotheses and null hypotheses can be tested by transplanting snails from and to different shores. They are not very sophisticated hypotheses. Presumably, if the rates of growth of snails on a number of shores are known, as they should be from the original observations, specific predictions about the differences from place to place can be made and the shores used would then be considered a fixed factor (because the hypotheses would relate to these particular shores). Nevertheless, this example will illustrate the procedures for analysing complex experiments. There will need to be various controls for translocation and disturbance (see Chapman, 1986). To explain the principles, such controls will be ignored in what follows – all snails are removed from their original habitat and moved onto new boulders on their original or on other shores.

To avoid confounding differences among shores with differences due to individual boulders, several boulders are used on each shore. On each boulder, fences are used to confine the snails, so that those from different origins are kept separately and independently of each other (see Section 7.14). This requires replication of cages with snails of each origin on each boulder, limiting the number of 'origins' that can be examined. Snails from two shores where rates of growth are known to differ are transplanted to three boulders on each of four shores, including the one from which they came. Two cages of each of the two types of snails, with five animals in each, are put on every boulder, i.e. four cages per boulder. Because growth is to be measured, the snails will presumably be marked and measured at the start. Thus, there are the following factors:

1. Origin, a random factor (A) with two levels ($a = 2$). It is random because many other shores could have been chosen.

2. Destination, a random factor (B) with four levels ($b = 4$). Again, many shores could have been chosen. These two are orthogonal to each other.
3. Boulders within destination shore, a random factor (C) nested in B, with $c = 3$ levels on each shore. This is nested in destination, but orthogonal to origin, because snails from each origin are present on every boulder.
4. Cages, nested in the combination of origin and boulder. This random factor has two levels for every combination. In every cage are $n = 5$ replicate snails.

11.2 Constructing the linear model

Having sorted out the hypotheses and the relevant factors, the relevant linear model should be constructed, using a subscript for each factor in the experiment. There must be a term for every source of variation in the experiment. It is important to note nested factors by the use of brackets as described in Chapter 9. Every combination of orthogonal factors must potentially generate an interaction, so the interactions must also be included in the linear model.

For the first example, the model describing the data is:

$$X_{ijkl} = \mu + N_i + E_j + NE_{ij} + P(NE)_{k(ij)} + e_{l(ijk)} \tag{11.1}$$

where N_i and E_j denote level i of the nitrogen treatment and level j of the exclusion treatment ($i = 1 \ldots a$; $j = 1 \ldots b$). These are orthogonal to each other, so there is potentially an interaction between them. P indicates the replicated plots which are in sets; each set has a particular combination (i and j) of the two treatments. Hence, the nesting described in the model ($k = 1 \ldots c$). $e_{l(ijk)}$ indicates the difference between an individual plant ($l; l = 1 \ldots n$) in plot k of combination i and j of the two experimental treatments and is the mean of the population of plants that receive that combination of treatments.

Note that the model now includes three types of sources of variation – those necessary to test the stated hypothesis (N, E), those made necessary by the requirements of replication and to avoid confounding of interpretations of differences (P, e) and any interactions between orthogonal combinations of these (N $\times$ E).

The second example follows the same principles, but has more sources of variation:

$$X_{ijklr} = \mu + O_i + S_j + OS_{ij} + B(S)_{k(j)} + OB(S)_{ik(j)} + C(OBS)_{l(ijk)} + e_{r(ijkl)} \tag{11.2}$$

where O_i and S_j refer to the ith shore of origin ($i = 1 \ldots a$) and jth destination shore ($j = 1 \ldots b$). Snails from each shore of origin will be put on each shore, so these two factors are orthogonal and potentially generate an interaction. The boulders on each shore are obviously nested in a shore and are shown as $B(S)_{k(j)}$ ($k = 1 \ldots c$ on each shore). Snails from each shore of origin are present on every boulder on each shore. Thus, the shores of origin (O) and boulders (B) are fully orthogonal to each other and may interact (OB(S)) in the model. This interaction would indicate small-scale (boulder-to-boulder) variability in the differences between average rates of growth of snails originally from different shores.

Again, there are the required experimental treatments (O and S), the factors required to create appropriate replication (B, C, *e*) and interactions among any orthogonal combinations of these (OS, OB(S)).

11.3 Calculating the degrees of freedom

The next step (see Table 11.1) is to calculate the degrees of freedom for each source of variation. It is important to do this early in the process. It is most useful for ensuring that the linear model contains all the relevant terms. Partitioning the degrees of freedom is quite simple and obeys the requirements of the operational definition (Section 3.10), but is simpler using the following guidelines.

1. For each orthogonal factor, the number of degrees of freedom is the number of levels of that factor minus 1.
2. For each nested factor, the number of degrees of freedom is the product of:
   (a) the number of levels of the nested factor in every set (i.e. in every combination) of treatments in which they are nested minus 1; and
   (b) the number of combinations (or sets) of treatments in which this factor is nested.
3. For every interaction, the number of degrees of freedom is the product of all degrees of freedom for every source of variation in the interaction.
4. The total degrees of freedom is the product of the number of levels of every factor in the experiment (including replicates) minus 1.
5. The degrees of freedom for all combinations of treatments (i.e. all the samples) is the product of the number of levels of every factor in the experiment (except replicates) minus 1.

Table 11.2. Determination of degrees of freedom for experimental manipulation of nitrogen and attacks by insects. See Section 11.1 for details

Source of variation	Degrees of freedom	
[Among all combinations of treatments]	$[abc - 1]$	[35]
Nitrogen treatments = N	$a - 1$	2
Exclusion treatments = E	$b - 1$	2
N × E	$(a - 1)(b - 1)$	4
Plots(N × E)	$ab(c - 1)$	27
Residual	$abc(n - 1)$	144
Total	$abcn - 1$	179

Table 11.3. Determination of degrees of freedom for experiment on growth of snails from two shores transplanted among several shores. See Section 11.1 for details

Source of variation	Degrees of freedom	
[Among all combinations of treatments]	$[abcd - 1]$	[47]
Original shore = O	$a - 1$	1
Shore = S	$b - 1$	3
O × S	$(a - 1)(b - 1)$	3
Boulder(S) = B(S)	$b(c - 1)$	8
O × B(S)	$(a - 1)b(c - 1)$	8
Cages(O × B(S))	$abc(d - 1)$	24
Residual	$abcd(n - 1)$	192
Total	$abcdn - 1$	239

6. The residual degrees of freedom are calculated from sets of replicates nested in the combination of all sources of variation in the model. For each set of replicates, there are $(n - 1)$ degrees of freedom. So, the residual degrees of freedom are the product of $(n - 1)$ and the numbers of levels of every factor in the experiment.

It is worth calculating the total degrees of freedom and those for the combination of all treatments first. The degrees of freedom for all combinations of treatments added to the residual degrees of freedom (i.e. among replicates or within samples) should add to the total. If not, something is wrong in the calculation and you need to start again.

For the first experiment considered here, there are to be three levels of manipulation of nitrogen and three levels of treatments to exclude insects. These are set up as nine combinations of experimental treatments, each replicated in four plots. So, there are 36 combinations of treatments or 36 total plots, with 35 degrees of freedom (Table 11.2). Because five plants are measured in each plot, there are 36×4 degrees of freedom for the residual (i.e. four degrees of freedom in each of the 36 plots) and $(36 \times 5) - 1$ total degrees of freedom. In Table 11.2, the total degrees of freedom is the sum of those for the residual and the combinations of all treatments.

The calculations for the second experiment are in Table 11.3.

11.4 Mean square estimates and *F*-ratios

The next and probably the most important process is to determine what it is that the analysis of variance will measure. What each mean square estimates can be determined by a set of procedures originally described by Cornfield and Tukey (1956) but also described extensively by Scheffé (1959) and Winer *et al.* (1991). The version described here is the modification described by Underwood (1981), which does not allow for the theoretical use of fixed nested factors. If such things were incorporated into experiments, the factors in which they were nested would be confounded and no test interpretable. If, for some reason, you want to do such experiments, read the texts by Scheffé (1959) and Winer *et al.* (1991) for the appropriate procedures.

Remember that we need a procedure to help with the construction of estimates because of the complications imposed by nested and orthogonal factors and by the algebra associated with fixed and random factors. Some texts (e.g. Zar, 1974) consider that it is best to list possible experimental designs and their mean squares. You are then supposed to match your experiment with one of these designs. This is not a particularly inspiring basis for designing experiments; there are numerous different designs of experiments (in excess of 100 unique alternatives with four factors).

It is much better strategy to become familiar with the operations necessary to construct a chosen experiment from first principles. This involves construction of mean squares, using a 'table of multipliers'. The procedures are quite simple, but describing them is not.

It is important to be able to construct the appropriate model for some experiment you wish to do. It will enable optimization of intensity of sampling for various nested components and will allow prior evaluation of the relevant *F*-ratios (e.g. to ensure they are valid). Where appropriate, it will also allow calculations of power and determination of what components might be pooled to simplify analyses and to increase power of tests. Although there are numerous statistical packages for computing the analyses, you will still need to be able to enter the appropriate model and you will be wise to know in advance how to check and verify the output.

The first step is to construct a table with a row for each source of variation in the linear model, except the mean. The table also has a column for each subscript used in the model (i.e. each term in the model that is not an interaction). Examples are shown in Tables 11.4 and 11.5. The table is then completed using the following rules, which guarantee that the correct combinations of terms will ultimately finish up in every mean square.

For each column in turn, enter the appropriate value, for each row in turn, using the following rules. Note the subscript in the column and whether it denotes a fixed or random factor. Then:

1. If the term in the row being considered has the column's subscript and:
   (a) the column's subscript is not in brackets (i.e. does not include nested components):
      (i) enter 0 if the column's subscript represents a fixed factor; or
      (ii) enter 1 if the column's subscript represents a random factor;
   (b) the column's subscript is in brackets (i.e. does include nested components): enter 1.
2. If the term in the row being considered does not have the column's subscript: enter the number of levels of the factor represented by the column's subscript.

This all sounds like instructions for catching fog – but is quite easy to do. In the first example (Table 11.4), the first column represents nitrogen treatments (i), a fixed factor with a levels. So, the first and third rows contain i, not in brackets. Their entry is 0 (rule 1(a)(i) above). The fourth and fifth rows contain i in brackets. Their entry is 1 (rule 1(b) above). The second row does not contain i. Its entry is a, the number of levels of the factor represented by column i (rule 2 above). For column k, which represents the random factor of plots, with c levels, the entries are explained in Table 11.4. k is not in the first three rows, so c (the number of levels of the factor represented by column k) is entered in the table. In the fourth row, k is present, but not in brackets, so 1 is entered because k represents a random factor. In the fifth row, k is present in brackets so, again, 1 is entered.

All entries in Table 11.4 are explained. Those in Table 11.5 can be easily understood by reference to the above rules.

The next part of the proceedings is to determine what the mean square estimates for each source of variation. This is done by determining, on a row-by-row basis, first which components belong in each mean square

Table 11.4. Table of multipliers for example 1, an experiment on effects of nitrogen and insects on plants (the linear model is Equation 11.1)

	Column headings are subscripts in the model							
Source of variation	i	Rule	j	Rule	k	Rule	l	Rule
N_i	0	1(a)(i)	b	2	c	2	n	2
E_j	a	2	0	1(a)(i)	c	2	n	2
$N \times E_{ij}$	0	1(a)(i)	0	1(a)(i)	c	2	n	2
$P(N \times E)_{k(ij)}$	1	1(b)	1	1(b)	1	1(a)(ii)	n	2
$e_{l(ijk)}$	1	1(b)	1	1(b)	1	1(b)	1	1(a)(ii)

For Rules, see p. 365.

Table 11.5. Table of multipliers for example 2, an experiment on growth of snails (the linear model is Equation 11.2)

Source of variation	i	j	k	l	r
O_i	1	b	c	d	n
S_j	a	1	c	d	n
$O \times S_{ij}$	1	1	c	d	n
$B(S)_{k(j)}$	a	1	1	d	n
$O \times B(S)_{ik(j)}$	1	1	1	d	n
$C(O \times B(S))_{l(ijk)}$	1	1	1	1	n
$e_{r(ijkl)}$	1	1	1	1	1

and then what is their magnitude.

Again, this is easily accomplished by the following rules. For each row in turn (called the 'target row') identify all rows that contain its subscripts. The mean square for the target row will (notionally) contain a component of variation representing every row that contains the subscript(s) found in the target row.

1. Identify in the target row all components of variation that belong in its mean square. There will be terms for every row which contains the subscript(s) found in the target row.
2. Each component is multiplied by the product of all entries in the row of the table which is represented by that component, omitting all columns with subscripts in the *target row*.
3. In the row representing the residual (i.e. e terms), the multiplier is 1.

These rules are difficult to write, but easy to do. Don't panic! Look at Table 11.6. For target row N_i, subscript i is found in rows N_i, NE_{ij}, $P(NE)_{k(ij)}$, $e_{l(ijk)}$. So, all of these have components in the mean square

Table 11.6. Analysis of example 1, an experiment on growth of plants with or without extra nitrogen (factor N) and where insects are present or excluded (factor E)

(a) Mean square estimates

Source of variation	i	j	k	l	Mean square estimates
N_i	0	b	c	n	$\sigma_e^2 + n\sigma_{\text{P(NE)}}^2 + (0 \times k_{\text{NE}}^2) + bcnk_{\text{N}}^2$
E_j	a	0	c	n	$\sigma_e^2 + n\sigma_{\text{P(NE)}}^2 + (0 \times k_{\text{NE}}^2) + acnk_{\text{E}}^2$
$N \times E_{ij}$	0	0	c	n	$\sigma_e^2 + n\sigma_{\text{P(NE)}}^2 + cnk_{\text{NE}}^2$
$P(N \times E)_{k(ij)}$	1	1	1	n	$\sigma_e^2 + n\sigma_{\text{P(NE)}}^2$
$e_{l(ijk)}$	1	1	1	1	σ_e^2

(b) Analysis of variance

Source of variation	Degrees of freedom	Mean square estimates	F-ratio versus	Degrees of freedom for F in this example
N_i	$a-1$	$\sigma_e^2 + n\sigma_{\text{P(NE)}}^2 + bcnk_{\text{N}}^2$	P(NE)	2,27
E_j	$b-1$	$\sigma_e^2 + n\sigma_{\text{P(NE)}}^2 + acnk_{\text{E}}^2$	P(NE)	2,27
$N \times E_{ij}$	$(a-1)(b-1)$	$\sigma_e^2 + n\sigma_{\text{P(NE)}}^2 + cnk_{\text{NE}}^2$	P(NE)	4,27
$P(N \times E)_{k(ij)}$	$ab(c-1)$	$\sigma_e^2 + n\sigma_{\text{P(NE)}}^2$	Residual	27,144
Residual = $e_{l(ijk)}$	$abc(n-1)$	σ_e^2	–	–

for N_i. Each component is a variance representing a random factor (a σ^2 term) or a sum of squared differences (a k^2 term) representing a fixed factor. They are shown for row N_i as k_{N}^2, k_{NE}^2, $\sigma_{\text{P(NE)}}^2$ and σ_e^2, respectively. σ_e^2 is multiplied by 1 – the product of entries in row $e_{l(ijk)}$ omitting column i. Column i is omitted because i is a subscript in the target row (i.e. N_i) for which the mean square is being constructed. Similarly, therefore $\sigma_{\text{P(NE)}}^2$ is multiplied by n, k_{NE}^2 is multiplied by 0 (the product of entries in row NE_{ij}, omitting column i) and vanishes. Finally, k_{N}^2 is multiplied by bcn – the product of entries in row N_i, omitting column i.

This works for all rows. Target row NE_{ij} has components from rows NE_{ij}, $P(NE)_{k(ij)}$ and $e_{l(ijk)}$ because these rows all have i and j in them. The multipliers are 1 for σ_e^2, n for $\sigma_{\text{P(NE)}}^2$ and cn for k_{NE}^2. In all cases, these are the products of the entries in the relevant rows (e, P(NE), NE, respectively), omitting columns i and j because these subscripts are present in the target row, NE.

Putting these results together with the degrees of freedom for each source of variation provides the analysis in Table 11.6b. Relevant *F*-ratios are easily identified. Under the null hypothesis that there are no differences among levels of factor X, the term k_X^2 (if X is fixed) or σ_X^2 (if X is random) will be zero. So, to test for significant non-zero values, the mean square for X must be divided by a mean square that contains all the other components except k_X^2 (or σ_X^2). In the cases of N, E and N $\times$ E in Table 11.6, such a term is P(NE). If the relevant null hypothesis is true for N or E or N $\times$ E, its mean square will be equal to that of P(NE) and dividing by the latter will give a ratio of unity. Hence, the *F*-ratios described in Table 11.6.

Note that some of these tests will be unwise or, at best, approximate, even though they can formally be constructed in a logically correct way. Any test of one component that is already involved in significant interactions is illogical, as discussed extensively elsewhere in this book (Sections 10.3, 10.10 and 10.13). Any test of a fixed component by dividing its mean square by a mean square with more than one variance component (σ^2 term) is usually, at best, only an approximate test (see Section 10.13).

Using these procedures for the second example is no more difficult. The results are in Table 11.7. For example, the mean square for the interaction O $\times$ S must contain components for *e*, C(OBS), OB(S) and OS because these rows all contain *i* and *j*. Each component is then multiplied by the products of its own row, omitting columns *i* and *j* (i.e. *l*, *n*, *dn*, *cdn*, respectively) see Table 11.5.

Again, the mean squares enable construction of appropriate *F*-ratios, as shown in Table 11.7. In this case, however, there is no test for differences among the shores on which snails are placed. As explained below (Section 11.7), the solution to this depends on the *post hoc* discovery that some appropriate source of variation is non-significant or can be pooled. Leave worrying about this until you get to that section! Of course, if you have an experimental design that leads to no available tests for important hypotheses, you may want to go back through the reasoning and design a different experiment or construct different hypotheses from the underlying model.

To ensure that the procedures for constructing mean squares and *F*-ratios are fully understood, the table of multipliers and estimates obtained in mean squares are described in the next section for two classes of design. First are the two- and three-factor models seen before. Second is a series of designs widely recommended for use in biology (in Sections 12.1 to 12.5). Working your way through these should help to increase your expertise.

Table 11.7. Mean square estimates and appropriate F-ratios for example 2, an experiment on growth of snails on different shores (see Table 11.5 and degrees of freedom in Table 11.3)

Source of variation	Degrees of freedom	Mean square estimates	F-ratio versus	Degrees of freedom for F in this example
O_i	$a-1$	$\sigma_e^2 + n\sigma^2_{C(OBS)} + dn\sigma^2_{OB(S)} + cdn\sigma^2_{OS} + bcdn\sigma^2_{O}$	OS	1, 3
S_j	$b-1$	$\sigma_e^2 + n\sigma^2_{C(OBS)} + dn\sigma^2_{OB(S)} + adn\sigma^2_{B(S)} + cdn\sigma^2_{OS} + acdn\sigma^2_{S}$	No test	–
OS_{ij}	$(a-1)(b-1)$	$\sigma_e^2 + n\sigma^2_{C(OBS)} + dn\sigma^2_{OB(S)} + cdn\sigma^2_{OS}$	OB(S)	3, 8
$B(S)_{k(j)}$	$b(c-1)$	$\sigma_e^2 + n\sigma^2_{C(OBS)} + dn\sigma^2_{OB(S)} + adn\sigma^2_{B(S)}$	OB(S)	8, 8
$OB(S)_{ik(j)}$	$(a-1)b(c-1)$	$\sigma_e^2 + n\sigma^2_{C(OBS)} + dn\sigma^2_{OB(S)}$	C(OBS)	8, 24
$C(OBS)_{l(ijk)}$	$abc(d-1)$	$\sigma_e^2 + n\sigma^2_{C(OBS)}$	Residual	24, 192
Residual $= e_{r(ijkl)}$	$abcd(n-1)$	σ_e^2	–	–

11.5 Designs seen before

11.5.1 Designs with two factors

Various experimental designs with two factors have been encountered so far. These include the nested analysis described in detail in Sections 9.3 to 9.6 and the designs with fixed and random orthogonal factors described in detail in Chapter 10. Here, their origin is summarized by construction of the tables of multipliers, which leads to mean square estimates that can be checked back to the earlier descriptions (see Table 11.8).

11.5.2 Designs with three factors

En route to here, only a few designs with three factors have wafted into view. There are, of course, many possibilities. One is with three fixed factors, orthogonally arranged, as described in Sections 10.9 and 10.10.

Table 11.8. Designs with two factors

(a) Nested analysis (see Table 9.3). Factor A is fixed, with levels $i = 1 \ldots a$. Factor B is random with levels $j = 1 \ldots b$ nested in each level of A. Each combination is replicated n times ($k = 1 \ldots n$)

Source of variation	Degrees of freedom	Table of multipliers i	j	k	Mean square estimates	F-ratio versus
A_i	$a-1$	0	b	n	$\sigma_e^2 + n\sigma_{\mathrm{B(A)}}^2 + bnk_{\mathrm{A}}^2$	B(A)
$B(A)_{j(i)}$	$a(b-1)$	1	1	n	$\sigma_e^2 + n\sigma_{\mathrm{B(A)}}^2$	Residual
Residual $= e_{k(ij)}$	$ab(n-1)$	1	1	1	σ_e^2	–

If factor A is random, k_{A}^2 is replaced by σ_{A}^2, but otherwise there is no change.

(b) Two fixed factors (see Table 10.8a). Factors A and B are fixed and orthogonal to each other. Factor A has $i = 1 \ldots a$ and Factor B has $j = 1 \ldots b$ levels. Each combination of treatments is replicated n times ($k = 1 \ldots n$)

Source of variation	Degrees of freedom	Table of multipliers i	j	k	Mean square estimates	F-ratio versus
A_i	$a-1$	0	b	n	$\sigma_e^2 + bnk_{\mathrm{A}}^2$	Residual
B_j	$b-1$	a	0	n	$\sigma_e^2 + ank_{\mathrm{B}}^2$	Residual
$A \times B_{ij}$	$(a-1)(b-1)$	0	b	n	$\sigma_e^2 + nk_{\mathrm{AB}}^2$	Residual
Residual $= e_{k(ij)}$	$ab(n-1)$	1	1	1	σ_e^2	–

Table 11.8 (*cont.*)

(c) One fixed and one random factor (Table 10.8b). Factor A is as in (b). Factor B is random with b levels. All other details are as in (b)

Source of variation	Degrees of freedom	Table of multipliers i	j	k	Mean square estimates	F-ratio versus
A_i	$a-1$	0	b	n	$\sigma_e^2 + n\sigma_{AB}^2 + bnk_A^2$	AB
B_j	$b-1$	a	1	n	$\sigma_e^2 + na\sigma_B^2$	Residual
$A \times B_{ij}$	$(a-1)(b-1)$	0	1	n	$\sigma_e^2 + n\sigma_{AB}^2$	Residual
Residual $= e_{k(ij)}$	$ab(n-1)$	1	1	1	σ_e^2	–

(d) Two random factors (Table 10.8c). Factors A and B are both random. All other details are as in (b)

Source of variation	Degrees of freedom	Table of multipliers i	j	k	Mean square estimates	F-ratio versus
A_i	$a-1$	1	b	n	$\sigma_e^2 + n\sigma_{AB}^2 + bn\sigma_A^2$	AB
B_j	$b-1$	a	1	n	$\sigma_e^2 + na\sigma_B^2$	AB
$A \times B_{ij}$	$(a-1)(b-1)$	1	1	n	$\sigma_e^2 + n\sigma_{AB}^2$	Residual
Residual $= e_{k(ij)}$	$ab(n-1)$	1	1	1	σ_e^2	–

Another is the nested analysis with more than one level of hierarchy (as described for more complicated scenarios in Sections 9.9 and 9.10).

Finally, there are other combinations of nesting (as in the example in Tables 11.4 and 11.5) and of fixed and random orthogonal factors. All the possibilities can be derived from the use of tables of multipliers. Some examples are shown in Tables 11.9 to 11.11.

As can be seen from these examples, several analytical models can share the same linear model, but differ according to whether factors are fixed or random. The only way to be sure that a given, properly structured hypothesis or series of hypotheses will be properly examined by a chosen experiment is to construct the basis of the analysis from first principles. This, of course, requires that the hypotheses have been fully defined before the process can start.

Finally, several models produce structures in which there is no test for differences in some source(s) of variation. These are considered below (Section 11.7).

Table 11.9. Examples of experimental designs with three factors. In all cases, factor A has $i = 1 \dots a$ levels; factor B has $j = 1 \dots b$ levels; factor C has $k = 1 \dots c$ levels. All combinations of factors are replicated n times ($l = 1 \dots n$). There are three orthogonal factors and the fully fixed case is in Table 10.11. (a) A, B fixed, C random; (b) A fixed, B, C random

$$X_{ijkl} = \mu + A_i + B_j + C_k + AB_{ij} + AC_{ik} + BC_{jk} + ABC_{ijk} + e_{l(ijk)}$$

	(a)						(b)					
	Table of multipliers						Table of multipliers					
Source of variation	i	j	k	l	Mean square estimates	F-ratio versus	i	j	k	l	Mean square estimates	F-ratio versus
A_i	0	b	c	n	$\sigma_e^2 + bn\sigma_{\text{AC}}^2 + bcnk_{\text{A}}^2$	AC	0	b	c	n	$\sigma_e^2 + n\sigma_{\text{ABC}}^2 + bn\sigma_{\text{AC}}^2 + cn\sigma_{\text{AB}}^2 + bcnk_{\text{A}}^2$	No test
B_j	a	1	c	n	$\sigma_e^2 + an\sigma_{\text{BC}}^2 + acnk_{\text{B}}^2$	BC	a	1	c	n	$\sigma_e^2 + an\sigma_{\text{BC}}^2 + acn\sigma_{\text{B}}^2$	BC
C_k	a	b	1	n	$\sigma_e^2 + abn\sigma_{\text{C}}^2$	Residual	a	b	1	n	$\sigma_e^2 + an\sigma_{\text{BC}}^2 + abn\sigma_{\text{C}}^2$	BC
AB_{ij}	0	1	c	n	$\sigma_e^2 + n\sigma_{\text{ABC}}^2 + cnk_{\text{AB}}^2$	ABC	0	1	c	n	$\sigma_e^2 + n\sigma_{\text{ABC}}^2 + cn\sigma_{\text{AB}}^2$	ABC
AC_{ik}	0	b	1	n	$\sigma_e^2 + bn\sigma_{\text{AC}}^2$	Residual	0	b	1	n	$\sigma_e^2 + n\sigma_{\text{ABC}}^2 + bn\sigma_{\text{AC}}^2$	ABC
BC_{jk}	a	1	1	n	$\sigma_e^2 + an\sigma_{\text{BC}}^2$	Residual	a	1	1	n	$\sigma_e^2 + an\sigma_{\text{BC}}^2$	Residual
ABC_{ijk}	0	1	1	n	$\sigma_e^2 + n\sigma_{\text{ABC}}^2$	Residual	0	1	1	n	$\sigma_e^2 + n\sigma_{\text{ABC}}^2$	Residual
Residual = $e_{l(ijk)}$	1	1	1	1	σ_e^2	–	1	1	1	1	σ_e^2	–

Table 11.10. Examples of experimental designs with three factors. In all cases, factor A has $i = 1 \ldots a$ levels; factor B has $j = 1 \ldots b$ levels; factor C has $k = 1 \ldots c$ levels. All combinations of factors are replicated n times ($l = 1 \ldots n$). There are two orthogonal factors (A, B), with C nested in A. (a) A, B fixed; (b) A fixed, B random

$$X_{ijkl} = \mu + A_i + B_j + AB_{ij} + C(A)_{k(i)} + BC(A)_{jk(i)} + e_{l(ijk)}$$

	(a)						(b)					
	Table of multipliers						Table of multipliers					
Source of variation	i	j	k	l	Mean square estimates	F-ratio versus	i	j	k	l	Mean square estimates	F-ratio versus
A_i	0	b	c	n	$\sigma_e^2 + bn\sigma_{\mathrm{C(A)}}^2 + bcnk_{\mathrm{A}}^2$	C(A)	0	b	c	n	$\sigma_e^2 + n\sigma_{\mathrm{BC(A)}}^2 + bn\sigma_{\mathrm{C(A)}}^2 + cn\sigma_{\mathrm{AB}}^2$ $+bcnk_{\mathrm{A}}^2$	No test
B_j	a	0	c	n	$\sigma_e^2 + n\sigma_{\mathrm{BC(A)}}^2 + acnk_{\mathrm{B}}^2$	BC(A)	a	1	c	n	$\sigma_e^2 + n\sigma_{\mathrm{BC(A)}}^2 + acn\sigma_{\mathrm{B}}^2$	BC(A)
AB_{ij}	0	0	c	n	$\sigma_e^2 + n\sigma_{\mathrm{BC(A)}}^2 + cnk_{\mathrm{AB}}^2$	BC(A)	0	1	c	n	$\sigma_e^2 + n\sigma_{\mathrm{BC(A)}}^2 + cn\sigma_{\mathrm{AB}}^2$	BC(A)
$C(A)_{k(i)}$	1	b	1	n	$\sigma_e^2 + bn\sigma_{\mathrm{C(A)}}^2$	Residual	1	b	1	n	$\sigma_e^2 + n\sigma_{\mathrm{BC(A)}}^2 + bn\sigma_{\mathrm{C(A)}}^2$	BC(A)
$BC(A)_{jk(i)}$	1	0	1	n	$\sigma_e^2 + n\sigma_{\mathrm{BC(A)}}^2$	Residual	1	1	1	n	$\sigma_e^2 + n\sigma_{\mathrm{BC(A)}}^2$	Residual
Residual $= e_{l(ijk)}$	1	1	1	1	σ_e^2	–	1	1	1	1	σ_e^2	–

Table 11.11. Examples of experimental designs with three factors. In all cases, factor A has $i = 1 \ldots a$ levels; factor B has $j = 1 \ldots b$ levels; factor C has $k = 1 \ldots c$ levels. All combinations of factors are replicated n times ($l = 1 \ldots n$). There is one factor (C) nested in combinations of the other two. (*a*) A, B fixed; (*b*) A fixed, B random

$X_{ijkl} = \mu + A_i + B_j + C(AB)_{k(ij)} + AB_{ij} + e_{l(ijk)}$

	(a)						(b)					
	Table of multipliers						Table of multipliers					
Source of variation	i	j	k	l	Mean square estimates	F-ratio versus	i	j	k	l	Mean square estimates	F-ratio versus
A_i	0	b	c	n	$\sigma_e^2 + n\sigma_{C(AB)}^2 + bcnk_A^2$	C(AB)	0	b	c	n	$\sigma_e^2 + n\sigma_{C(AB)}^2 + cn\sigma_{AB}^2 + bcnk_A^2$	AB
B_j	a	0	c	n	$\sigma_e^2 + n\sigma_{C(AB)}^2 + acnk_B^2$	C(AB)	a	1	c	n	$\sigma_e^2 + n\sigma_{C(AB)}^2 + acn\sigma_B^2$	C(AB)
AB_{ij}	0	0	c	n	$\sigma_e^2 + n\sigma_{C(AB)}^2 + cnk_{AB}^2$	C(AB)	0	1	c	n	$\sigma_e^2 + n\sigma_{C(AB)}^2 + cn\sigma_{AB}^2$	C(AB)
$C(AB)_{k(ij)}$	1	1	1	n	$\sigma_e^2 + n\sigma_{C(AB)}^2$	Residual	1	1	1	n	$\sigma_e^2 + n\sigma_{C(AB)}^2$	Residual
Residual $= e_{l(ijk)}$	1	1	1	1	σ_e^2	–	1	1	1	1	σ_e^2	

11.6 Construction of sums of squares using orthogonal designs

Having arrived at a satisfactory and correct structure for the experiment, there remains the problem of how to estimate the relevant mean squares. This requires construction of the appropriate sums of squares. There are several ways of doing this. First, you can use whatever statistical package happens to be on whoever's computer you are using. This may be satisfactory, but does require certainty that you have specified the model in the correct manner for that package. Some of them are Byzantine in their instructions or, at the very least, obscure. There are many ways of making mistakes.

Alternatively, you can get the 'rules' for constructing sums of squares from various texts (e.g. Scheffé, 1959; Crowder & Hand, 1990). These are fail-safe, but even more complex than the ones described in Section 11.4 to determine what is in each mean square. I doubt whether most of us would welcome that! There are simpler algorithms if you are programming the computations.

As a third option, or as an aid to check the output of your favourite (or most available) package, you could use the procedure described by Underwood (1981) to use a fully orthogonal model. These require little sophistication in their specification for computers and can easily be programmed for any number of factors. The sums of squares for your chosen model can then be recombined. No further details are needed here. Assuming that the model for the analysis has been constructed properly and that the sums of squares have been calculated properly, the rest of the analysis can proceed.

11.7 *Post hoc* pooling

This topic was introduced in Section 9.7. It is a procedure to remove terms from a model if they can be shown to be sufficiently close to zero. It includes protection from excessive Type I error in subsequent tests. It is a useful technique, particularly where tests are not powerful or there is no test for a give source of variation. The procedures have already been described in sufficient detail for you to be able to use them (see Section 9.7). It is important to consider, in advance, what may be appropriate terms to eliminate or pool if they then turn out to be non-significant (under the pooling procedure described in Section 9.7). It is not wise to keep eliminating terms in order to increase the capacity of tests to reject null hypotheses. This is tantamount to keeping probing the data until something comes out as significant. It is much more appropriate

Table 11.12. *Post hoc* pooling to provide a test for differences among shores (i.e. S) in the experiment described in Table 11.7, where mean squares are described

(a) O × S is not significant at $P = 0.25$ and can be eliminated. Eliminate O × S from the mean square for S.

$$\text{Test for } S \text{ is } F = \frac{\mathrm{MS_S}}{\mathrm{MS_{B(S)}}}$$

with $(b-1)$ and $b(c-1)$ degrees of freedom.

(b) B(S) is not significant at $P = 0.25$ and can be eliminated from the mean square for S.

$$\text{Test for } S \text{ is } F = \frac{\mathrm{MS_S}}{\mathrm{MS_{O\times S}}}$$

with $(b-1)$ and $(a-1)(b-1)$ degrees of freedom.

(c) Both O × S and B(S) are not significant at $P = 0.25$. Pool O × S and B(S) with O × B(S) and eliminate them from the mean square for S.

$$\text{Test for } S \text{ is } F = \frac{\mathrm{MS_S}}{\text{Pooled MS}[\mathrm{O \times S + B(S) + O \times B(S)}]}$$

with $(b-1)$ and $[(a-1)(b-1) + b(c-1) + (a-1)b(c-1)]$ degrees of freedom $= (b-1)$ and $(abc - a - b + 1)$ degrees of freedom.

to specify in advance which term(s) would be pooled if necessary and possible.

An example for a complex design will help. Remind yourself of the second experiment given in Table 11.7. There is no test for differences among shores (i.e. the destinations of experimental snails). Examination of the mean square estimates reveals why and also what might be done about it. First, there is no point in testing for differences among shores if there is an interaction between origin and destination in the mean growth of snails (i.e. if O × S is significant). A significant interaction would identify that the third model (a combination of influences of shore of origin and environmental influences of shore of destination) was the explanation for the original observations. If, however, this were non-significant and could be eliminated (i.e. non-significant at $P = 0.25$, see Section 9.7), there would be a test for differences among shores, as in Table 11.12.

Alternatively, if there were no significant difference (at $P = 0.25$) among boulders within each shore, B(S) could be eliminated from the model and its sums of squares pooled with OB(S), providing a test for shores (Table 11.12).

Of course, if both B(S) and O × S were pooled, a test with more degrees of freedom could be done – as is the case if OB(S) could be pooled. Construction of the mean squares allows the nature of the analysis to be explored to capitalize on ways of increasing the capacity to test things and the power of such tests. No more discussion is needed. You could examine the other cases in this book where there are no tests (see e.g. Table 11.9).

11.8 Quasi *F*-ratios

In some texts, where no test is available for some source of variation a test has been constructed by contrived combinations of mean squares. These lead to quasi *F*-ratios – approximations to valid tests. An example is from the three-factor analysis with two random factors given in Table 11.9b. There is no test for factor A. A quasi test can be constructed (see e.g. Zar, 1974; Winer *et al.*, 1991). Using the mean square estimates for this analysis, note that:

$$\mathrm{MS}_1 = \mathrm{MS}_\mathrm{A} + \mathrm{MS}_\mathrm{ABC} = 2\sigma_e^2 + 2n\sigma_\mathrm{ABC}^2 + bn\sigma_\mathrm{AC}^2 + cn\sigma_\mathrm{AB}^2 + bcnk_\mathrm{A}^2 \quad (11.3)$$

$$\mathrm{MS}_2 = \mathrm{MS}_\mathrm{AB} + \mathrm{MS}_\mathrm{AC} = 2\sigma_e^2 + 2n\sigma_\mathrm{ABC}^2 + bn\sigma_\mathrm{AC}^2 + cn\sigma_\mathrm{AB}^2 \quad (11.4)$$

Therefore, a test for $k_\mathrm{A}^2 = 0$ is $F = \mathrm{MS}_1/\mathrm{MS}_2$. This has all the problems of a mixed model (Chapter 10) and the degrees of freedom are not obvious. They can be approximated by the procedure invented by Satterthwaite (1946) and described by Winer *et al.* (1991). If MS_A has degrees of freedom f_A $(= a - 1)$, MS_{AB} has f_AB $(= (a-1)(b-1))$, etc., then:

$$\text{Degrees of freedom for } \mathrm{MS}_1 = \frac{(\mathrm{MS}_\mathrm{A} + \mathrm{MS}_\mathrm{ABC})^2}{\left[\left(\frac{\mathrm{MS}_\mathrm{A}^2}{f_\mathrm{A}}\right) + \left(\frac{\mathrm{MS}_\mathrm{ABC}^2}{f_\mathrm{ABC}}\right)\right]} \quad (11.5)$$

$$\text{Degrees of freedom for } \mathrm{MS}_2 = \frac{(\mathrm{MS}_\mathrm{AB} + \mathrm{MS}_\mathrm{AC})^2}{\left[\left(\frac{\mathrm{MS}_\mathrm{AB}^2}{f_\mathrm{AB}}\right) + \left(\frac{\mathrm{MS}_\mathrm{AC}^2}{f_\mathrm{AC}}\right)\right]} \quad (11.6)$$

These are, however, *approximate* and often result in very approximate tests. The approximations to an *F*-distribution and therefore the probabilities associated with such tests are often not very good. Don't use these tests! Use *post hoc* pooling, as described in the previous section or, when pooling of interactions cannot be done, proceed as if interaction terms were significant.

11.9 Multiple comparisons

It does not matter how complicated the experiment is, there are no new theories or complications for doing multiple comparisons to determine which means differ when F-ratios are significant. The procedures were all discussed in Chapter 8, Section 9.6 and Section 10.8 for various classes of experiment.

The only trick is to know how to calculate the correct standard error (and the associated degrees of freedom) for any set of means to be compared. This is simple. For procedures such as the Student–Newman–Keuls test (and its modifications as in Section 8.6.5), a common standard error is used for each mean to be compared. Remember that the analysis should not be done unless the variances are homogeneous, so the variances can be pooled to form a common standard error. This is calculated from the variance within those means and the total number of data-points in each mean (see also Section 10.8). For any chosen source of variation in an analysis, the standard error is calculated from the mean square used as the divisor for the F-ratio used to test for differences among those means. Thus, if differences among means in source of variation X are examined by an F-ratio of mean square X/mean square Y, the standard error for means in source of variation X is:

$$\sqrt{\left(\frac{\text{Mean square Y}}{\text{Total number of readings in each mean}}\right)} \tag{11.7}$$

Some examples will illustrate the point. In the first experiment (Table 11.6), if there is an interaction between nitrogen treatments and exclusion of grazers, you would want to examine differences among the three levels of nitrogen ($a = 3$) at each level of exclusion and among the three levels of exclusion ($b = 3$) at each level of nitrogen. The divisor for the F-ratio you used to test the interaction was the mean square for P(NE), with 27 degrees of freedom (see Table 11.6). Thus, the standard error for means of each nitrogen treatment in each level of exclusion is averaged across $n = 5$ replicates in each of $c = 4$ plots and is:

$$\text{Standard error N} \times \text{E} = \sqrt{\left[\frac{\text{Mean square P(NE)}}{cn}\right]}$$

$$= \sqrt{\left[\frac{\text{Mean square P(NE)}}{20}\right]} \tag{11.8}$$

with $ab(c-1) = 27$ degrees of freedom.

If the interaction were not significant, you would need to compare the means of the nitrogen treatments averaged over replicates, plots and exclusion treatments:

$$\text{Standard error } N = \sqrt{\left[\frac{\text{Mean square P(NE)}}{bcn}\right]}$$

$$= \sqrt{\left[\frac{\text{Mean square P(NE)}}{60}\right]} \qquad (11.9)$$

In the second experiment, suppose you have discovered there was no interaction between shore of origin (O) and destination (S) and you could pool O × S. You can now test for differences among shores (S) using the mean square for boulders (i.e. B(S)) as in Table 11.12. Further, suppose that tests for O and for S each revealed significant differences. Now, you might want to examine the mean growth on each shore of origin and destination. You probably would not because these are random factors (see Section 10.7) and their specific patterns of differences have no particular meaning. Discovery that either or both factors is significant causes rejection of one or both null hypotheses in favour of the first (shores of origin differ) or the second (shores of destination differ) or both models. We can, however, use these multiple comparisons as an example.

The mean of each shore of origin is averaged over $n = 5$ replicates in $d = 2$ cages on $c = 3$ boulders on each of $b = 4$ shores. The interaction, O × S, was eliminated, making it possible to pool O × S with O × B(S) because they now estimate the same components in their mean squares. This pooled mean square is then used to test for differences among shores of origin (F = Mean square O/Mean square pooled [O × S + O × B(S)]) with 1 and 11 degrees of freedom (see Tables 11.3 and 11.7). So, the standard error for the mean growth of snails from shores of different origins is:

$$\text{Standard error O} = \sqrt{\left\{\frac{\text{Mean square pooled [OS + OB(S)]}}{bcn}\right\}}$$

$$= \sqrt{\left\{\frac{\text{Mean square pooled [OS + OB(S)]}}{120}\right\}} \qquad (11.10).$$

with 11 degrees of freedom (i.e. the pooled degrees of freedom from O × S and O × B(S)).

For shores of destination, S, the test for differences is F = Mean square S/mean square B(S) as in Table 11.12, with $(b - 1) = 3$ and $b(c - 1) = 8$ degrees of freedom. The mean growth of snails on each shore is averaged over $n = 5$ replicates in each of $d = 2$ cages on $c = 3$ boulders for each of $a = 2$ shores of origin.

$$\text{Standard error S} = \sqrt{\left[\frac{\text{Mean square B(S)}}{acdn}\right]}$$

$$= \sqrt{\left[\frac{\text{Mean square B(S)}}{60}\right]} \qquad (11.11)$$

with 8 degrees of freedom.

By keeping alert and breathing deeply three times before you start, you should be able to work out the standard errors for any set of means in an analysis.

This ends the general considerations of how to organize an experimental design.

11.10 Missing data and other practicalities

One problem that seems to recur in all sorts of experiments is that odd bits go missing, creating a variety of unbalanced designs. This happens sometimes because of demonic intrusion – an elephant steps on one of the cages, a flash flood eliminates one of the recorders. In one of my experiments, a storm removed a rocky shore by breaking it into lumps and taking it away.

Alternatively, it may happen because the treatments are potentially fatal to the experimental animals or plants, so, in some levels of a treatment, replicates die and, inconveniently, cannot then have their heart rates measured. Such situations can often be headed off by putting extra replicates in those treatments to start with and then using a randomly chosen number of the survivors to match the numbers in other treatments.

If data are not balanced, there will be immense difficulties (see e.g. Nelder, 1977; Burdick & Graybill, 1992; Searle *et al.*, 1992) for the analysis because:

1. The multipliers of components of mean squares are not the same in different sources of variation in the analysis. This makes it impossible to derive valid F-ratio tests for some sources of variation.
2. The degrees of freedom for some sources of variation are only approximate, making it very difficult to use tests based on the probabilities associated with an F-ratio.

3. Violations of assumptions about normality and homogeneity of variances are much more likely to affect probabilities of Type I error in attempts to do unbalanced analyses.

Increasing sophistication of computing allows more and more calculation of statistical tests. This does not mean that logic and careful thought about trying to get balanced samples should be abandoned simply because algebraic formulae are available to provide approximate tests.

There are, as mentioned in Chapter 10, more general frameworks for attempting tests (e.g. ML and REML procedures, see Section 10.15.3). These do not produce exact tests and, for many designs, exact tests are not possible (see e.g. Burdick & Graybill, 1992). Problems also arise because there is more than one way of partitioning data in calculation of sums of squares (Maxwell & Delaney, 1990; Burdick & Graybill, 1992) and these perform differently in tests. Shaw & Mitchell-Olds (1993) recently recommended one particular procedure as being superior. Another approach was described by Milliken & Johnson (1984). In all cases, the issues are technically complex, the analyses are not straightforward and the interpretations fraught with hazards. This requires sensible help at an early stage from experienced statistical consultants. There are also problems for interpretation of outcomes where estimates from some populations are more precise than are others – unless there is some very good reason for such differences in precision.

In a well-planned, orderly research programme, the experiments should be designed from the outset to be balanced. Then most of the problems of analysis and many of the difficulties with underlying assumptions will not cause headaches for interpretation of the results.

Some problems will require unbalanced analyses because resources are so scarce, or logistical constraints so great that costs of the experimental units are immense. For such studies, extreme caution will be necessary and virtually all of the simplicity provided by balanced, orthogonal structures will be lost. It seems probable that such large-scale or expensive or otherwise difficult experiments are done despite their difficulties because the outcomes are important. This being so, the onus is on the experimenter to provide a full and complete warning about the potential problems for interpretation.

In other experiments, the problems are caused by accidents or mishaps. These conspire to remove most of the logic in any conclusions from an analysis. Therefore, it is worth doing everything you can to ensure balanced samples. Use extra replicates; don't lose any data; practise lots of quality control to ensure that all samples or readings are valid.

11.10.1 Loss of individual replicates

Occasionally, one replicate goes missing out of one, or each of a few, of the treatments. If there are many replicates, one can be taken at random out of every other treatment to even up the data. This removes a bit of information, but is more than compensated by the ease of doing and interpreting a balanced design.

If, however, there are relatively few replicates in each combination of treatments you could replace a missing one with the average of the others in that combination. Using the average value means that there is no influence of the dummy replicate on either the estimated average nor the estimated variance of that treatment.

You must, however, reduce the degrees of freedom in the analysis to compensate for the missing (and dummied) data. Suppose one plant is missing in the experiment on nitrogen and insects (Table 11.2), so that only four plants were measured in one of the combinations of treatment. If the average of that combination is used to make a dummy, fifth replicate, a balanced analysis could be done. The degrees of freedom for the residual must be reduced by 1, making 143 (Table 11.2) – a trivial change. In this case, the analysis may well use Plots (N × E) for tests, so no great error could come from the use of one pseudo-reading. If variation among plots were found to be non-significant and could be eliminated (i.e. it is not significant at $P = 0.25$) and pooled with the residual mean square, the residual mean square would be wrong by only 1/144, which can scarcely matter. Note in this case, that Plots (N × E) could be eliminated, but no great benefit would be gained from pooling its sum of squares with the residual because the latter already had such a large number of degrees of freedom.

In contrast, in an analysis with $a = 2$, $b = 3$ and $n = 3$, losing one replicate from one treatment and doing as above causes a change of 1 in 12 (about 8%) in the residual. This is excessive. Here, however, running the analysis with two randomly chosen replicates for all the other treatments would produce a small analysis.

In this situation, I recommend running the balanced analysis with $n = 2$ and the analysis with $n = 3$, using a dummy replicate where one is missing. If you get essentially the same results, do multiple comparisons from the second analysis. If you get different results, go and do better, bigger experiments and make allowance for the likely loss of replicates.

11.10.2 Missing sets of replicates

Occasionally, whole sets of data disappear, so that no replicates remain. For example, one of the plots in the first experiment may be destroyed by a rhinoceros using it to build a nest. This is obviously more serious than before. The missing data could be made up as dummy values using the mean of the other plots in that treatment. Now there are $n = 5$ identical values of the average of that treatment. These, of course, have no variance and contribute nothing to the residual sum of squares. Four degrees of freedom must be removed to compensate, making the residual degrees of freedom equal 140 (see Table 11.2).

The variation among plots in the affected treatment is also unchanged by inventing the data. It is estimated from the remaining plots in that treatment and the estimate is unaltered by using their average. The degrees of freedom for plots must, however, be reduced by 1, making a total of 26 instead of 27.

So, sums of squares are calculated using all the dummy data for the missing plot and then degrees of freedom are reduced as necessary. The mean squares are calculated and the analysis proceeds.

There are many possible ways that data can disappear from an experiment so that a whole treatment is missing. In a two-factor experiment with $a = 3$, $b = 4$ and $n = 5$, something may cause the loss of all five replicates from one of the 12 treatments. Under these circumstances, there are numerous methods for inventing or guessing the missing data (see e.g. Federer, 1955; Mead & Curnow, 1983).

This is all very well provided there are no interactions between the two factors in the experiment. If there are no interactions, it is easy to estimate what mean to put in the missing cell, from those of the levels of A and B adjacent to the level where it belongs.

If, however, there are interactions, this cannot possibly be done. There is no way to guess what data to use without knowing the precise nature of the interaction and how it would have affected the missing cell. This cannot be estimated without data from the missing combination of treatments (see also Milliken & Johnston, 1984).

I recommend that you *do not* use any of the available procedures, unless you can demonstrate that there are no interactions between factors in the rest of the experiment. If you are then prepared to assume that the missing bit is not the bit that would have caused interactions, you could use a guess based on the available means of combinations of treatments.

In an experiment with $a = 3$ and $b = 4$, if, say, combination A_2B_3 goes missing, you could do an analysis using A_1, A_2, A_3 and B_1, B_2, B_4 ($a = 3$, $b = 3$), i.e. omitting treatment B_3 altogether. You could do a second analysis using A_1, A_3 and B_1 to B_4 ($a = 2$, $b = 4$), omitting A_2. From these, you would have a good idea of the presence of interactions. If there were none, you could proceed with any chosen method that can handle a missing cell or combination of treatments.

Of course, by the time you have done these two analyses, you have probably adequately tested your hypothesis! You will probably not need to consider what might have happened if your missing data were like the ones you guess. All of this suggests that common sense, rather than sophisticated calculations to provide guesses, might be the best strategy.

12

Some common and some particular experimental designs

Certain sorts of experimental designs recur repeatedly because, despite the specific models and hypotheses, there are common needs for spatial replication, controls, etc. Some of these designs rejoice in specific names – the taxonomy of which can become quite cumbersome. Several of them derive from their early agricultural roots (Fisher, 1935).

Rather than attempt to understand these designs in the ways they are usually presented in textbooks, it is undoubtedly a much better strategy to learn how to construct an experimental design from first principles, as in Chapter 11. In that way, you are less likely to pick an incorrect design for your experiment. You are also far less likely to be tempted to try to identify an appropriate framework for analysing data *after* you have done the experiment.

Nevertheless, some designs warrant specific mention in detail – if only to demonstrate some problems in their use. Some of the most commonly used experimental designs cause assumptions to be made about the nature of or relationships among data over and above assumptions about their distributions. These assumptions are often unlikely to be true but, sadly, are often ignored in the way the designs are explained or ignored, to the detriment of interpretations, when data are analysed. Some of the most common designs are considered below.

In addition, some consideration is given to asymmetrical experimental designs that have particular merit in some aspects of experimental ecology and environmental sampling. These are also introduced here.

12.1 Unreplicated randomized blocks design

One design that is commonly used in agricultural experiments and widely advocated for use in ecology is the randomized blocks design.

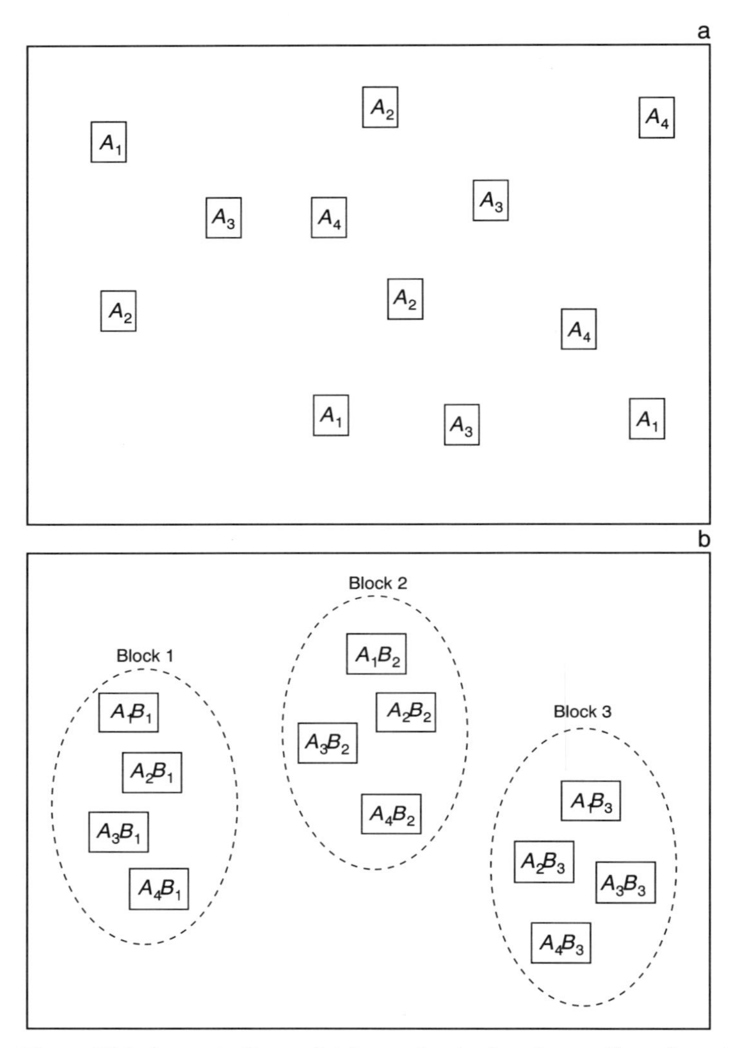

Figure 12.1. Layout of completely randomized and unreplicated random blocks experiments. There are $a = 4$ treatments (i.e. levels of factor A), each replicated $b = 3$ times. In (a), the experimental units are scattered at random. In (b), one unit of each treatment is placed in three arbitrarily chosen blocks.

The principle is simple. Suppose there is an experiment with a treatments, each replicated n times. The simplest design would scatter the replicates of each treatment independently over the study area (as in Figure 12.1a). The analysis is then a one-factor analysis of variance.

Instead, however, we might first divide the area into b arbitrarily chosen 'blocks' – i.e. we stratify the whole area into blocks. Then, one replicate of each treatment is put in a random position in each block (as in Figure 12.1b). Now we have a two-factor experiment with a treatments each represented once in b blocks. The treatments and blocks are orthogonal to one another (every treatment is present in every block).

The model for the data is:

$$X_{ijk} = \mu + A_i + B_j + AB_{ij} + e_{k(ij)} \qquad (12.1)$$

where X_{ijk} represents the data from the ith experimental treatment ($i = 1 \ldots a$) in the jth block ($j = 1 \ldots b$). μ is the overall mean; A_i and B_j represent the effects, if they exist, of the ith treatment and the jth block. AB_{ij} is the interaction between these factors. $e_{k(ij)}$ represents the variability of the particular spot at which experimental unit i is placed in block j, which is not likely to be exactly identical to everywhere else in that block. k is, essentially, a dummy variable, because variability among units of the same treatment cannot be measured in any block. There are no replicates of any treatment in each block.

The analysis is as in Table 12.1. The blocks are randomly chosen and so are simply representative pieces of the habitat: blocks is a random factor. The treatments are, in the case considered here, fixed, but could in some experiments be random.

The object of 'blocking' is to be able to remove noise from the data, where the background variation can be attributed to differences from one place (i.e. block) to another. If, for example, there is a good chance

Table 12.1. Table of multipliers and mean square estimates for an unreplicated random blocks experiment as in Equation 12.1. There are a treatments (A_i is a fixed factor; $i = 1 \ldots a$), each represented once in each of b randomly-chosen blocks ($j = 1 \ldots b$). Thus, treatments and blocks are orthogonal to one another

	Table of multipliers				
Source of variation	i	j	k	Degrees of freedom	Mean square estimates
Among treatments = A	0	b	1	$a-1$	$\sigma_e^2 + \sigma_{AB}^2 + bk_A^2$
Among blocks = B	a	1	1	$b-1$	$\sigma_e^2 + a\sigma_B^2$
'Error'= A × B	0	1	1	$(a-1)(b-1)$	$\sigma_e^2 + \sigma_{AB}^2$
Total				$ab-1$	
Note: Residual cannot be estimated= $e_{k(ij)}$	1	1	1	–	σ_e^2

that some general gradient across the study-area affects, say, the growth of plants, blocking would be an efficient way to estimate the effects of the gradient and to prevent them from masking any effects of experimental treatments.

All the treatments are represented in each block and therefore any differences from one block to another cancel out in the measurement of differences among treatments. In contrast to the completely randomized design, the randomized block design has a smaller 'error' for testing for differences among treatments. Some of the difference among randomly scattered replicates in the first design was due to differences across the study-area. These are now estimated in the blocks term, leaving a smaller 'error' representing small-scale variation among experimental units in each block.

If you wanted to know how much difference blocking makes to such an experiment, you could use the formula described by Fisher (1935), which calculates the ratio of the mean squares for error from a completely randomized and a randomized block experiment, correcting for their different degrees of freedom.

Where blocks are to be considered a random factor (as, indeed, they should because blocks are only in the experiment to provide randomized replication of treatments in space), the mean square estimates demonstrate the problem with the analysis. Most textbooks suggest that the null hypothesis of no difference among treatments can be tested using the so-called 'error' term in the analysis. An F-ratio of treatments mean square/'error' mean square is used. Note, however, that this assumes that there are no interactions in the experiment. It assumes that differences among treatments, if there are any, are not at all affected by differences among the blocks. In other words, it is assumed that there are no differences from place to place that cause different outcomes of the various experimental treatments.

This assumption is extremely unlikely in many situations (see also Milliken & Johnson, 1984). It cannot, however, be tested in such unreplicated experiments. There is no proper residual term to evaluate the interaction, because there are no independent random replicates.

Mead (1988) has also pointed out the problem if there is interaction between blocks and treatments ('if block-treatment interactions occur, then the usual linear model is inadequate'; *ibid*., p. 608). He also advocated the use of replicates of each treatment in each block, but considered that this was rarely done because it reduces the precision of comparisons of treatments within each block. Given the likelihood that there will be interactions from block to block with the differences among treatments in many, if not all, ecological experimental settings and that the interaction of differences among treatments from place to place is often the point of an ecological study, it would always be better to use an appropriate design with replication in each block.

In some descriptions of this design (e.g. Snedecor & Cochran, 1989; Winer *et al.*, 1991), it is assumed that blocks are not random, but are a fixed factor. This is done without regard to the nature of the blocks, nor to the logic of the relevant null hypothesis. It is permissible simply because it makes no differences to the analysis or interpretation if it is assumed that there is no interaction between treatments and blocks. The interpretation is unaffected because the differences among the blocks are not usually going to be interpreted.

The expected values of mean squares (as per the formulae in Section 11.4) are shown in Table 12.1. The interaction term in the so-called 'error' mean square is explicitly present and must be assumed to be zero for the analysis to have any meaning. As explained in Sections 10.3 and 10.6, if interactions are present they can cause real differences among treatments to be undetectable or they can create apparent differences where none is present. It is worth giving very careful thought to the difficulties before embarking on an unreplicated randomized blocks experiment. It is usually going to be more informative to have appropriate replication of every treatment in each block. In many ecological studies, the interaction between treatments (i.e. manipulations designed to test hypotheses about ecological processes) and blocks (different portions of a habitat) is a crucial component of the study. So, designing the experiment to be able to demonstrate how much the processes vary from place to place is a necessity, without which interpretations would mean little.

12.2 Tukey's test for non-additivity

For some particular forms of potential interaction, Tukey's (1949b) test for non-additivity provides a test for the presence of interactions, even though there is no replication. The test can detect one subset of possible interactions as a multiplicative effect of differences due to treatments and those due to blocks (or any other second factor in an experiment).

Consequently, under the assumption of no interactions between treatments (A_i; $i = 1 \ldots a$) and blocks (B_j; $j = 1 \ldots b$), the model for the

Table 12.2. Tukey's test for non-additivity in an unreplicated random blocks experiment. There are $a = 4$ treatments ($i = 1 \ldots 4$), each represented once in $b = 3$ blocks ($j = 1 \ldots 3$). Data are X_{ij}. There is significant non-additivity ($P < 0.05$), precluding further analysis of the data

	Treatment (i)					
Block (j)	1	2	3	4	$\bar{X}_j$	$\bar{X}_j - \bar{X}$
1	75	61	25	44	51.3	−4.6
2	92	77	3	104	69.0	13.1
3	16	31	87	56	47.5	−8.4
$\bar{X}_i$	61	56.3	38.3	68	$\bar{X} = 55.9$	
$\bar{X}_i - \bar{X}$	5.1	0.4	−17.6	12.1		

Source of variation	Sum of squares		Degrees of freedom		Mean square	F-ratio
Among treatments = A	$SS_A = 1443.6$		3		481.2	
Among blocks = B	$SS_B = 1055.2$		2		527.6	
A × B	9028.2		6		1504.7	
Non-additivity		$SS_{NA} = 5469.5$		1	5469.5	7.69 $P < 0.05$
Remainder		3558.7		5	711.7	
Total	11527		11		–	

$$SS_{NA} = \frac{ab\left[\sum_{i=1}^{a}\sum_{j=1}^{b}(\bar{X}_i - \bar{X})(\bar{X}_j - \bar{X})X_{ij}\right]^2}{SS_A \times SS_B}$$

unreplicated randomized blocks experiment is:

$$X_{ij} = \mu + A_i + B_j + e_{ij} \tag{12.2}$$

where X_{ij} is the datum from the ith treatment in the jth block. If there is 'multiplicative' interaction, the relevant model is:

$$X_{ij} = \mu + A_i + B_j + \lambda(A_i \times B_j) + f_{ij} \tag{12.3}$$

where $A_i \times B_j$ is the product of the effect of treatment i as applied to block j. The error terms in the two models are e_{ij} and f_{ij}, respectively.

If this particular form of interaction is present in the experiment, fitting Equation 12.3 should generate smaller errors than does fitting Equation 12.2. Therefore, the f_{ij} values should be smaller than the e_{ij} values. The difference between these two estimates of error can be used to test the null hypothesis of no interactions between the two factors in the experiment. This is constructed as a test of the null hypothesis that $\lambda = 0$. The mechanics of the test are shown in Table 12.2. It is important to note that, even where Tukey's test is non-significant, there may *still* be interactions, but of a different form from those in Equation 12.3. Thus, even the protection of Tukey's procedure is not a particularly safe method.

It is clear that unreplicated 'randomized blocks' experiments should not be the choice in ecological experiments unless there is compellingly good evidence that interactions of treatments and blocks are unlikely. Where this evidence would come from is not clear. Presumably if enough evidence already exists that some sets of treatments have entirely consistent effects in different arbitrarily chosen places, it is already known what these effects are. Consequently, there would be little point in further testing of this particular null hypothesis, so future analyses without replication would not be necessary!

Several other designs have the same general problem that they require restrictive assumptions about lack of certain interactions – yet these assumptions are ignored. Some of the these designs are considered below.

12.3 Split-plot designs

Related to the unreplicated randomized blocks design is the unreplicated split-plot experiment. It is most usually used in an experiment with two factors, applied in orthogonal combinations in random blocks or plots.

There are two completely different designs known as split-plot experiments. Here, the first considered is as described by Winer *et al.* (1991). As an example, consider an experiment to investigate hypotheses about

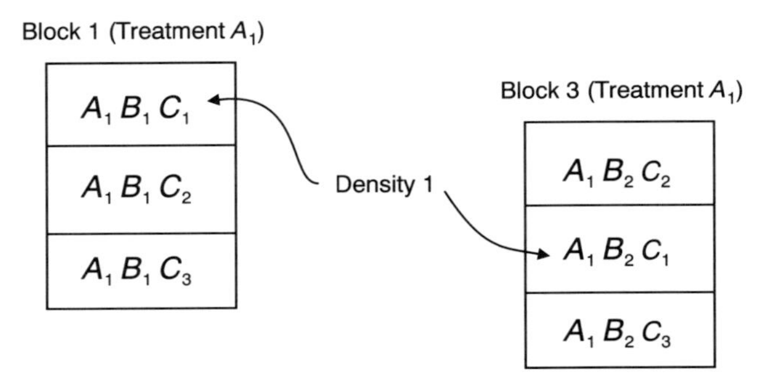

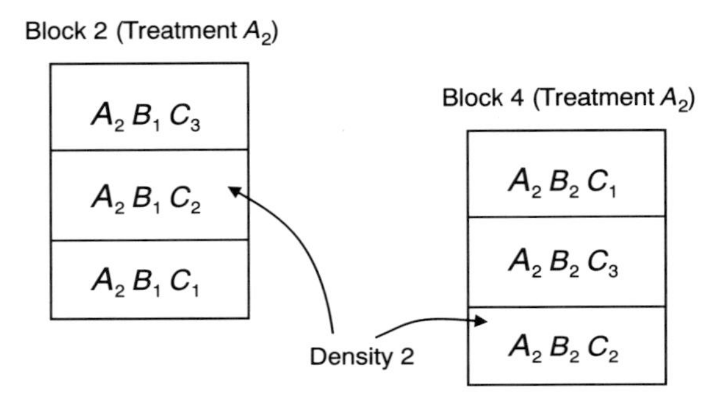

Figure 12.2. Layout of a split-plot experiment with two orthogonal fixed factors. A_i is watering ($i = 1, 2$) and C_k is density ($k = 1 \ldots 3$). The six treatments are established in four blocks or plots. Two blocks are subject to normal watering and two get reduced watering. Blocks are therefore nested in watering treatments. $B(A)_{j(i)}$ represents blocks within watering treatments; $j = 1, 2$ in each level of A_i, so blocks 1 and 2 are B_1 in A_1 and A_2, respectively; Blocks 3 and 4 are B_2 in A_1 and A_2, respectively.

frequency of attacks by insects under different conditions of stress caused by lack of water and crowding of plants. There are two fixed factors: density of plants (three specific densities are chosen to represent small, medium and large densities of the species) and watering regime (reduced and normal). There are six orthogonal combinations of treatments. Four plots, or arbitrarily chosen blocks of ground, are available for the experiment.

In Winer *et al.*'s (1991) version of the split-plot experiment, the layout would be as in Figure 12.2. There are two fully orthogonal treatments (A and C; every level of A is present with every level of C). The blocks represent a randomly sampled factor, nested in A (there are two blocks in each level of A). Blocks in each level of A are, however, orthogonal to the remaining treatment C (every level of C is present in every Block in each level of A).

The model for the data from such an experiment is:

$$X_{ijkl} = \mu + A_i + B(A)_{j(i)} + C_k + AC_{ik} + CB(A)_{kj(i)} + e_{l(ijk)} \quad (12.4)$$

where X_{ijkl} represents the proportion of plants attacked (or whatever is the relevant variable) at density k in block j of watering condition i. A_i, C_k and $B_{j(i)}$ represent, if they exist, the effects of treatments A (watering), C (density) and plot B in treatment A, respectively. AB_{ij} and $CB(A)_{kj(i)}$ represent interactions among these factors. $e_{l(ijk)}$ represents the error due to the smaller-scale differences between samples because of spatial variability within blocks. Thus, in Figure 12.2, the data from treatment combinations $A_1B_1C_1$ and $A_1B_1C_2$ (i.e. in block 1 of watering level 1) may differ because of small-scale variation between the two groups of plants, which are obviously at slightly different positions in the block.

The analysis of such data is in Table 12.3. There is a perfectly valid test for differences among levels of factor A (i.e. the treatment applied to whole plots – the watering treatment here). This is an F-ratio of the mean square for A divided by the mean square for B(A), with $(a - 1)$ and $a(b - 1)$ degrees of freedom (see Table 12.3). This test is exactly the same as a test in a nested analysis of variance (see Chapter 9). Here, plots are nested in watering treatments and the residual (i.e. variations associated with differences from place to place within plots) is fully nested in both. So, although this last component has not been estimated, because there are no replicates, the test for A is still valid.

The interpretation of such a test is, however, completely dependent on whether or not there is an interaction between A and C (i.e. between watering and density of plants). In this experiment, such an interaction is likely. In theory, the A × C interaction and differences among levels of factor C (densities) could be examined by F-ratios of their mean squares over the mean square for the interaction C × B(A). Inspection of Table 12.3 will demonstrate why this is so. If, for example, there are no differences due to density, $k_C^2 = 0$ and mean square C = mean square C × B(A). In many texts, therefore, the mean square C × B(A) is called the within-plot or sub-plot error and is used for these tests.

The problem is that interpretation of such tests depends upon whether there are interactions of factor C with differences among plots. If there are such interactions, the effects of treatment C (i.e. differences due to different densities of plants in the present example) vary from plot to

Table 12.3. Table of multipliers and mean square estimates for an unreplicated split-plot experiment. There are a treatments (A_i is a fixed factor; $i = 1 \ldots a$), applied to the whole of b plots or blocks ($B(A)_{j(i)}$ is a random factor; $j = 1 \ldots b$, nested in each A_i). A second factor consists of c treatments (C_k is also a fixed factor; $k = 1 \ldots c$), orthogonal to A_i and $B(A)_{j(i)}$ by being represented in every plot. See Figure 12.2 for the layout. The example discussed in the text is indicated in brackets

	Table of multipliers					
Source of variation	i	j	k	l	Degrees of freedom	Mean square estimates
Among levels of factor A = A	0	b	c	1	$a-1$	$\sigma_e^2 + c\sigma_{B(A)}^2 + bck_A^2$
[watering treatments]					1	
Among levels of factor B = B(A)	1	1	c	1	$a(b-1)$	$\sigma_e^2 + c\sigma_{B(A)}^2$
[plots (watering treatments)]					2	
Among levels of factor C = C	a	b	0	1	$(c-1)$	$\sigma_e^2 + \sigma_{CB(A)}^2 + abk_C^2$
[among densities]					2	
A × C	0	b	0	1	$(a-1)(c-1)$	$\sigma_e^2 + \sigma_{CB(A)}^2 + bk_{AC}^2$
[watering × density]					2	
C × B(A)	1	1	0	1	$a(b-1)(c-1)$	$\sigma_e^2 + \sigma_{CB(A)}^2$
[density × plots(watering)]					4	
Total					$abc-1$	
[Total]					11	
Note: Residual ($e_{l(ijk)}$) cannot be estimated	1	1	1	1	–	σ_e^2

plot. Under these circumstances, there is no sensible way to interpret the outcome of a global test for differences among levels of C. The effect of this treatment must be examined on a plot by plot basis. Similarly, any interaction between A and C can only be interpreted logically if there is no interaction between C and plots (i.e. B(A)). In this experiment, there can be no test for interactions C × B(A), because the true residual variance cannot be estimated.

This point has not been emphasized in the description of such tests and designs in many books. It is, however, the case that *all* descriptions of split-plot designs include the explicit assumption that interactions between treatments and plots (C × B(A)) are zero (see the detailed accounts in such texts as Snedecor & Cochran, 1989; Winer *et al.*, 1991). The above comments about interpretations of other factors (A, C, A × C) is why the interaction C × B(A) must be zero. Note, particularly, that even though tests for A do not involve assumptions about interactions, their interpretation does depend on A × C. The latter cannot be sensibly examined without assuming no C × B(A), which cannot be tested.

There are several possible solutions to this problem, all of which avoid the use of a so-called split-plot design. First, if enough plots are available for the experiment described, the combinations of treatments can be re-arranged so that there is proper replication within plots. For example, if there were 12 plots available, two can be used for each combination of treatments A_i and C_k (i.e. all six combinations of watering and density). In each plot, each treatment could be replicated three times (or twice to save money and time). Then there would be two nested plots in each of the six orthogonal combinations of A and C and three (or two) replicates of the combination of treatments in each plot (as in Figure 12.3). The model for the analysis would be:

$$X_{ijkl} = \mu + A_i + C_k + AC_{ik} + B(AC)_{j(ik)} + e_{l(ijk)} \qquad (12.5)$$

where A_i and C_k are the levels of factors A and C, respectively; AC_{ik} represents their interaction and $B(AC)_{j(ik)}$ represents the plots in each combination of treatments and $e_{l(ijk)}$ represents differences of replicate l of the combination of treatments (ik) in plot j. The analysis is straightforward (Table 12.4a)

The design without replication in each plot is not appropriate if the model and hypothesis being examined *require* knowledge of any interactions among plots in the effects of treatments. For example, if the original observations being explained include variability from place to

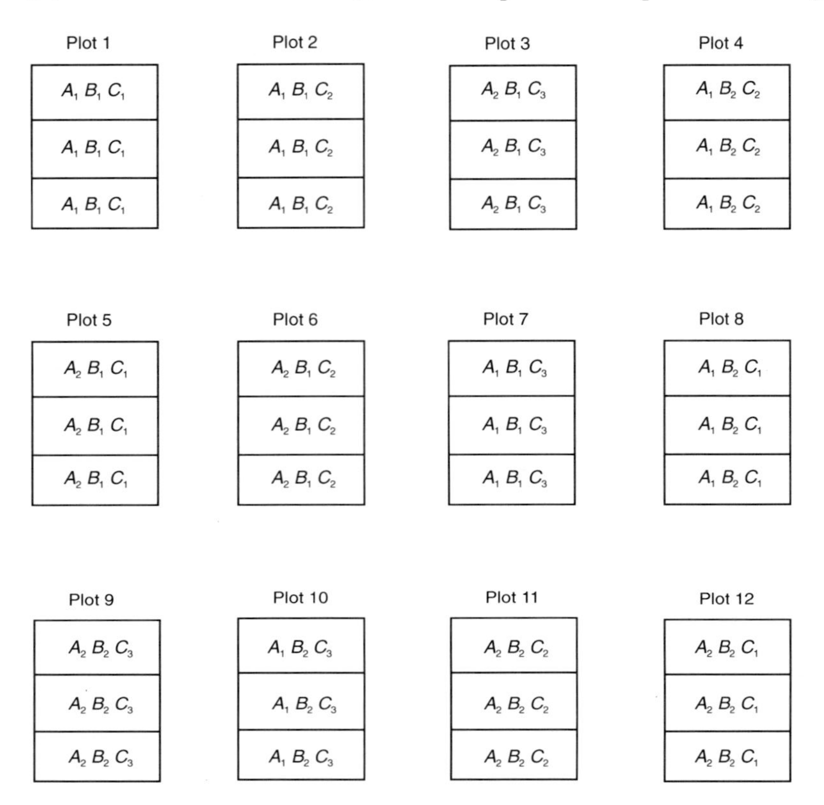

Figure 12.3. Layout of a modified experiment with two orthogonal fixed factors, as in Figure 12.1. The six treatments are replicated in 12 blocks or plots, so that there are $b = 2$ plots in each of the six combinations of the two factors. For example, plots 2 and 4 replicate combinations of treatments A_1, C_2.

place in the effects of attacks by insects, the model to explain them may include the notion that there are effects of density, that these vary according to the amount of water around (i.e. there are interactions) *and* that these processes vary according to the nutrients in the soil or the numbers of insects present or any other process that differs from place to place. Under such circumstances, interactions of treatments with plots *must* be examined in experimental tests of hypotheses derived from the model.

For this to be possible, some combinations of treatments must be replicated in plots. The plots will, however, have to be larger, or the area covered by each experimental set of plants smaller, so that at least four treatment units can be put into each plot. In this way, all three levels of factor C can still be present in every plot in each level of factor

Table 12.4. Two alternatives to an unreplicated split-plot design for the experiment analysed in Table 12.3 (and see Equation 12.5)

(a) There are 12 plots. The two levels of factor A are each applied to six whole plots. In each plot, three replicates of one level of factor C are then established. There are therefore two randomly chosen plots on each combination of the two factors (see Figure 12.3). In each plot, there are $n = 3$ replicates of a given treatment

Source of variation	Table of multipliers i	j	k	l	Degrees of freedom	Mean square estimates	F-ratio versus
A_i	0	b	c	n	$a - 1 = 1$	$\sigma_e^2 + n\sigma_{B(AC)}^2 + bcnk_A^2$	B(AC)
C_k	a	b	0	n	$c - 1 = 2$	$\sigma_e^2 + n\sigma_{B(AC)}^2 + abnk_C^2$	B(AC)
$A \times C_{ik}$	0	b	0	n	$(a-1)(c-1) = 2$	$\sigma_e^2 + n\sigma_{B(AC)}^2 + bnk_{AC}^2$	B(AC)
$B(A \times C)_{j(ik)}$	1	1	1	n	$ac(b-1) = 6$	$\sigma_e^2 + n\sigma_{B(AC)}^2$	Residual
Residual	1	1	1	n	$abc(n-1) = 24$	σ_e^2	–
Total	–	–	–	–	$abcn - 1 = 35$	–	–

(b) There are four plots with an extra replicate of one treatment in each plot, as in Figure 12.4. The residual is estimated from the replicated treatments in the various plots (i.e. from differences between the two units of $A_1B_1C_1$, $A_2B_1C_3$, $A_1B_2C_3$, $A_2B_2C_2$ in Blocks 1 to 4, respectively). All other components are as in Table 12.3; one replicate of the replicated treatment is ignored, at random, to make the design as in Table 12.3

Source of variation	Degrees of freedom	Mean square estimates	F-ratio versus
A_i	$a - 1 = 1$	$\sigma_e^2 + c\sigma_{B(A)}^2 + bck_A^2$	B(A)
$B(A)_{j(i)}$	$a(b-1) = 2$	$\sigma_e^2 + c\sigma_{B(A)}^2$	Residual
C_k	$c - 1 = 2$	$\sigma_e^2 + \sigma_{CB(A)}^2 + abk_C^2$	CB(A)
$A \times C_{ik}$	$(a-1)(c-1) = 2$	$\sigma_e^2 + \sigma_{CB(A)}^2 + bk_{AC}^2$	CB(A)
$C \times B(A)_{kj(i)}$	$a(b-1)(c-1) = 4$	$\sigma_e^2 + \sigma_{CB(A)}^2$	Residual
Residual	4	σ_e^2	–

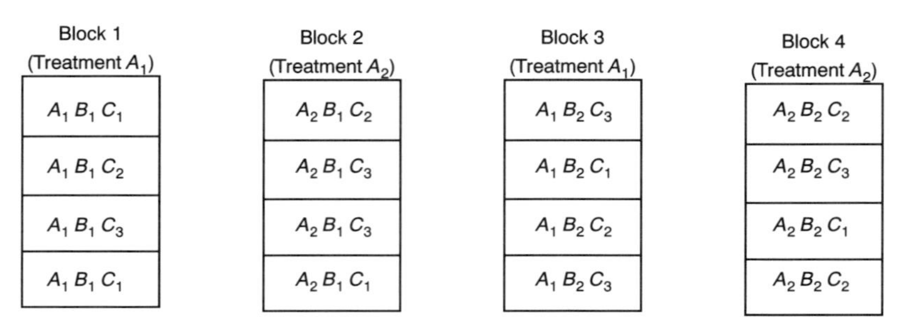

Figure 12.4. Layout of a modified split-plot experiment with two orthogonal fixed factors, as in Figure 12.1. The six treatments are established in four blocks or plots, with an extra replicate of one treatment in each plot. All other details are as in Figure 12.1.

A, maintaining the orthogonality of A with C and C with B(A). The extra unit can then replicate some treatment in each plot, to estimate variability within plots. One possible design is shown in Figure 12.4 and its analysis in Table 12.4b. There are other ways of choosing how to replicate the units, for example the same treatment (e.g. A_1C_1) could be replicated in both plots where it occurs, but this allows variation to be estimated for only a few of the combinations of treatments. If there are four plots, four of the total of six combinations could be used, as in Figure 12.4.

Of course, if plots can be any size, they could be large enough to include at least two replicate patches or units of experimental plants for every combination in the plot. If the original plots in Figure 12.2 were twice as large, all three combinations of A_iC_k in each plot could have been replicated, providing a direct estimate of residual variation for the analysis in Table 12.4.

The second version of a split-plot experiment is that described by Snedecor & Cochran (1989). Keeping the same example of watering and density of plants, the layout is shown in Figure 12.5 and the analysis is in Table 12.5. This design is, in fact, an unreplicated three-factor experiment with two fixed and one random factor. The so-called split-plot version in Table 12.5 is possible only if interactions with plots are all assumed to be zero. The analysis in Table 12.5 shows why this is necessary. The model for the data is:

$$X_{ijkl} = \mu + A_i + B_j + C_k + AB_{ij} + AC_{ik} + BC_{jk} + ABC_{ijk} + e_{l(ijk)} \quad (12.6)$$

where A_i, B_j, C_k represent the effects of watering treatment i, plot j, density k, respectively. Terms AB_{ij} to ABC_{ijk} represent the interactions.

Table 12.5. Another so-called split-plot design, as in Figure 12.5 and Equation 12.6. The 'main plot' and 'sub-plot' errors are as identified by Snedecor & Cochran (1989). Their use in F-ratios to test hypotheses about factors A and C obviously depends on the assumption that interactions with B are zero

Source of variation		Table of multipliers				Degrees of freedom	Mean square estimates	F-ratio versus
		i	j	k	l			
Watering treatment	= A	0	b	c	1	$a-1=1$	$\sigma_e^2 + c\sigma_{AB}^2 + bck_A^2$	'Main-plot' error
Plots	= B	a	1	c	1	$b-1=3$	$\sigma_e^2 + ac\sigma_B^2$	'Main-plot' error
'Main-plot' error	= A × B	0	1	c	1	$(a-1)(b-1)=3$	$\sigma_e^2 + c\sigma_{AB}^2$	
Densities	= C	a	b	0	1	$c-1=2$	$\sigma_e^2 + a\sigma_{AC}^2 + abk_C^2$	'Sub-plot' error
Watering × Densities	= A × C	0	b	0	1	$(a-1)(c-1)=2$	$\sigma_e^2 + \sigma_{ABC}^2 + bk_{AC}^2$	'Sub-plot' error
'Sub-plot' error						$a(b-1)(c-1)=12$	See below	
Total						$abc-1=23$		
But 'sub-plot' error is sum of:								
Densities × plots	= B × C	a	1	0	1	$(b-1)(c-1)=6$	$\sigma_e^2 + a\sigma_{BC}^2$	
Watering × densities × plots	= A × B × C	0	1	0	1	$(a-1)(b-1)(c-1)=6$	$\sigma_e^2 + \sigma_{ABC}^2$	

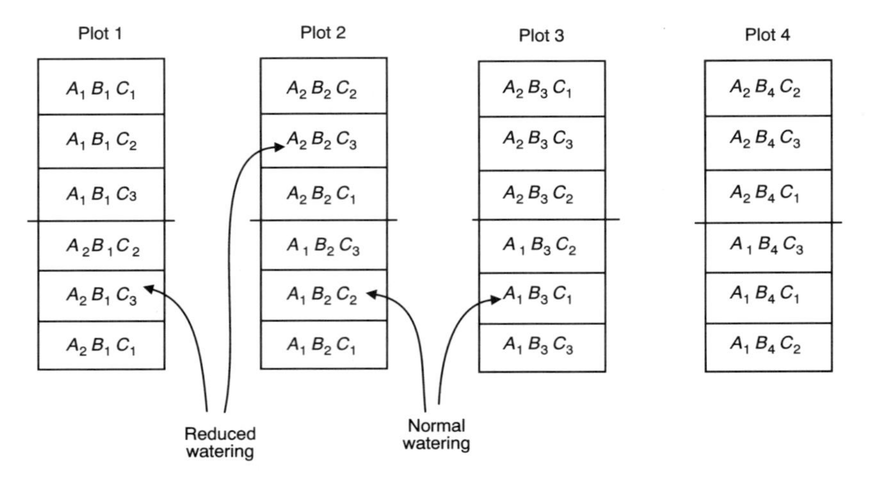

Figure 12.5. A different type of split-plot design for an experiment with two fixed, orthogonal factors (as described in Figure 12.1). A_i is watering ($i = 1, 2$), C_k is density ($k = 1 \ldots 3$). The six treatments are arranged in four plots (or blocks), B_j, fully orthogonal to the treatments ($j = 1 \ldots 4$).

$e_{l(ijk)}$ is the error associated with the individual unit ($A_iB_jC_k$ in Figure 12.4) being different from the mean response of the plot to which it belongs (B_j). This is not measurable separately from the ABC interaction because there are no replicates.

The tests all require interactions A × B, B × C, A × B × C to be zero to make them valid (those that use 'sub-plot' errors) or to make them interpretable (all tests), as explained above.

The solutions to the problems in this design are, again, to create proper replication of all (or, if not possible, some) treatments and to make these properly independent of one another.

Whichever strategy or design is used, there always remains the problem that most layouts of split-plot designs may cause spatial non-independence among treatments or replicates because the different units are so close. It is common (as shown in Figures 12.2 to 12.4) for the treatments to be applied to whole sub-plots or rows of plants or whatever units are being examined. They are then touching or very near each other, making them potentially capable of influencing nearby units. Suppose, for example, that attacks by insects are greater where density of plants is large and there is reduced watering. The insects will probably accumulate in such areas. In the arrangements in Figures 12.2 and 12.4, large accumulations of insects in combination A_2C_3 (reduced watering/ large density) may spill over into adjacent units in blocks 2 and 4. In

Figure 12.3, in contrast, large numbers of insects may spread out to the other replicates in plots 3 and 9, creating non-independence among replicates within treatments. The problems of different types of non-independence were discussed in Chapter 7.

All in all, it would probably be better *not* to create plots or blocks unless the problems of lack of replication and the potential problems of non-independence can be solved.

12.4 Latin squares

Another very common type of design that supposedly enhances the capacity to detect differences among experimental treatments is a so-called Latin square. There is, yet again, a baroque growth of these to take into account various types of complexity of design and degree of orthogonality of factors (Fisher & Yates, 1953; Cochran & Cox, 1957).

The principle behind the designs is simple, but failure to conform to the restrictive assumptions makes them extremely unsuitable for most ecological experiments. As an example, consider an experiment with $a = 4$ levels of experimental factor A (which here is considered fixed). If the experiment is to be replicated $n = 4$ times, the 16 units can be scattered around the study-site at random. The residual variation (i.e. among replicates within each treatment) might then be large because of the potentially large spatial variation across the study-area.

Suppose, therefore, that the experiment is confined to a smaller area in order to reduce the variability among replicates. This may be completely inappropriate if the original observations were at a larger spatial scale. Otherwise, the experiment might be done in a relatively small area and the experimental units will, essentially, be arranged in a grid.

Alternatively, the experiment may be in aquaria or a glass-house or ponds or pens, or other units, that are already in a grid. Either way, a Latin square would consist of arranging the four replicates of each treatment so that every treatment is equally represented in the two directions of the grid that may have gradients of variability running across them. Such an arrangement is shown in Figure 12.6.

The model used for the analysis should be a three-factor design, with all factors fixed. The treatments were chosen to be fixed. The rows and columns in the grid are fixed (there are only four of each and all are in the experiment; see Chapter 8). Thus, the model should be:

$$X_{ijk} = \mu + A_i + R_j + C_k + AR_{ij} + AC_{ik} + RC_{jk} + ARC_{ijk} + e_{ijk} \qquad (12.7)$$

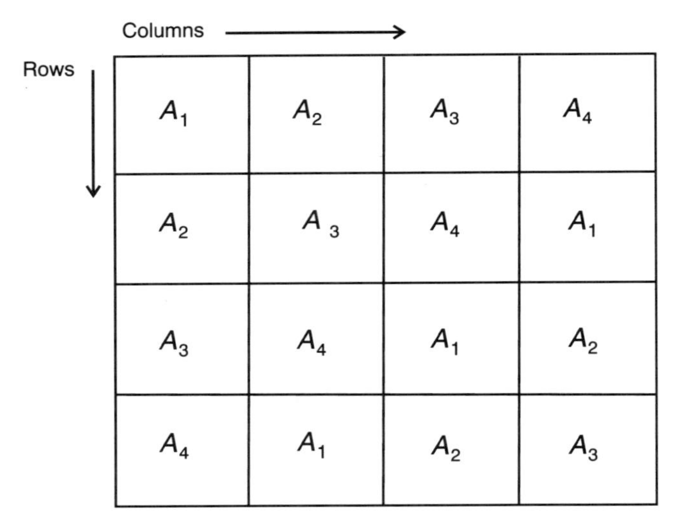

Figure 12.6. A Latin square layout for an experiment with $a = 4$ treatments (A_i; $i = 1 \ldots 4$), each replicated $n = 4$ times. The treatments are arranged in a grid, such that every treatment is represented once in each row and once in each column of the grid.

where A_i represents the effect of treatment i ($i = 1 \ldots a$) and R_j, C_k are the effects of being in row j and column k ($j = 1 \ldots a$; $k = 1 \ldots a$), if any of these exist. AR_{ij}, AC_{ik}, RC_{jk} and ARC_{ijk} represent interactions. e_{ijk} represents intrinsic variation unique to an experimental unit that makes it potentially different from any other unit that might have been placed at that particular row and column and subject to that treatment.

The often recommended analysis is that in Table 12.6a, where it must be assumed that all interactions are zero, so that the error term really only measures residual variation (i.e. due to e values in Equation 12.7). This analysis has the following model for the data:

$$X_{ijk} = \mu + A_i + R_j + C_k + e_{ijk} \tag{12.8}$$

with notation as for Equation 12.7.

The assumption of no interaction is inherently unlikely for most experiments. Where there are gradients running across a study-site and influencing some variable of interest, it is usually unlikely that they will be two simple gradients running independently of one another, at right angles and lined up with the layout of the experiment.

Where the assumption is, unknown to the experimenter, false, the error mean square in Table 12.6a is increased by components representing

Table 12.6. Analysis of a Latin square experiment with $a = 4$ levels of factor A ($i = 1 \ldots a$), each replicated $n = 4$ times

(a) Analysis as a Latin square. According to this, differences among treatments can be examined as an F-ratio of mean square A/mean square error, with three and six degrees of freedom

Source of variation	Degrees of freedom	Mean square estimates
Among treatments = A	3	$\sigma^2_{\text{error}} + nk^2_{\text{A}}$
Among rows	3	$\sigma^2_{\text{error}} + n\sigma^2_{\text{rows}}$
Among columns	3	$\sigma^2_{\text{error}} + n\sigma^2_{\text{columns}}$
Error	6	σ^2_{error}
Total	15	–

(b) Actual analysis. Obviously, the analysis is unworkable unless it is assumed that all interactions are zero

Source of variation		Degrees of freedom
Among treatments	= A	3
Among rows	= R	3
Among columns	= C	3
A × R A × C R × C A × R × C Residual	= Error	6

interactions (as in Table 12.6b), thus making σ^2_{error} larger in the error than in the among treatments mean square. Thus, there will be increased Type II error in the analysis because the ratio of the two σ^2_{error} terms will be <1 and larger differences among treatments will be necessary to cause significantly large values of F.

Solutions to the problem of not having independent estimates of intrinsic variation, i.e. the difficulty of determining the existence or magnitude of interactions are as before. Design the experiment to have proper, independent replicates.

12.5 Unreplicated repeated measures

In many experiments, repeated measurements are taken of the same sampled units. This is particularly prevalent where the establishment of experimental treatments is relatively time-consuming or expensive compared with the cost of taking data from the units in which treatments have been established.

Thus, in many environmental projects, the same areas or plots are repeatedly visited and data taken from them. In fact, there are programmes where this is the recommended procedure – despite the problems it causes. As explained earlier (Section 7.14), such repeated sampling of the same plots introduces a considerable risk of temporal non-independence. This is non-independence among times of sampling, leading to increased or decreased probability of Type I error in assessment of differences among times, depending on whether there is negative or positive serial correlation.

Other areas of ecological or biological experimentation in which repeated measures analyses are widespread are behavioural and some physiological studies. In these cases, repeated serial observations on the same individuals are easy to make or are inevitable because of the machinery (video-films, telemetry) used to take measurements. Sophisticated monitoring equipment leads to continuous streams of data being available. This leads to the notion that it 'would be a pity not to use all the data'.

The problem is probably best identified by an example concerning the effects of fire on plants. The observations leading to this experiment are that many ground-covering plants and small shrubs are killed when a fire occurs. Some plants do, nevertheless, recover to revegetate an area. The plants are also known to be killed by smothering with dust, which cuts off the light. Again, many may recover. There are, however, differences in rates of recovery from place to place. Two models proposed to account for the differences are (a) that fire alone causes all the mortality, or (b) that fire itself spares many plants but the subsequent cover of ash is fatal. Hence, under the first model, the hypothesis is that there will be very slow recovery of plants in the areas that are burned; other areas from which ash is removed will be no different from the areas that are only burned. The two treatments will recover at the same rate to match untouched control plots. From the second model, the hypothesis is that areas burned and then cleaned of all ash will recover more quickly than areas burned, but otherwise untouched. The two hypotheses can be distinguished in one experiment.

The first experimental treatment is the fixed factor of fire, with three levels. The first level comprises areas experimentally burned (under a controlled regime to prevent spread of the fire). The second level consists

Table 12.7. Table of multipliers and mean square estimates for an unreplicated repeated-measures experimental design. The data conform to the model in Equation 12.9. The experimental treatments (factor A; $i = 1 \ldots a$) are three levels of burning, burning and removal of ash and control ($a = 3$). b plots of each are established and censussed at each of c times of sampling. A and C are fixed factors (times of sampling are times since the start of the experiment)

Source of variation	Table of multipliers			Degrees of freedom	Mean square estimates
	i	j	k		
Treatment = A	0	b	c	$a-1$	$\sigma_e^2 + c\sigma_{B(A)}^2 + bck_A^2$
Plots(treatment) = B(A)	1	1	c	$a(b-1)$	$\sigma_e^2 + c\sigma_{B(A)}^2$
Time of sampling = C	a	b	0	$c-1$	$\sigma_e^2 + \sigma_{CB(A)}^2 + abk_C^2$
Treatment × time = A × C	0	b	0	$(a-1)(c-1)$	$\sigma_e^2 + \sigma_{CB(A)}^2 + bk_{AC}^2$
Time × plots(treatment) = C × B(A)	1	1	0	$a(b-1)(c-1)$	$\sigma_e^2 + \sigma_{CB(A)}^2$
Total	–	–	–	$abc-1$	–
Residual = σ_{ijk}^2	1	1	1	–	σ_e^2

Note: Residual cannot be estimated

of burning and then removing the layer of ash from the plots. The third level is plots that are untouched so that recovery can be identified when it has occurred. It might be appropriate to consider a fourth treatment in which there is no burning, but the surface of the ground is treated to control for any other influences of disturbances due to removing ash. The desirability and feasibility of this will depend entirely on the nature of the plants and the ground. Such a control is not included here.

So, three series of replicated plots are established. In two, control burnings are done. In half of these plots, the ash from burning is blown away by a mixture of careful brushing and, perhaps, large fans. Obviously, the plots must be large enough to allow a controlled fire. The establishment of the treatments is very expensive. In each plot, the entire area can very quickly be sampled for percentage cover of plants and number of shrubs. So, at each chosen time after the start of the experiment, every plot is completely sampled. These are repeated measures.

The analysis is done according to the partitioning of variation in Table 12.7. The data, in fact, conform to the model:

$$X_{ijk} = \mu + A_i + B(A)_{j(i)} + C_k + AC_{ik} + CB(A)_{kj(i)} + e_{ijk} \qquad (12.9)$$

where X_{ijk} represents the data in plot j ($j = 1 \ldots b$) in treatment i ($i = 1 \ldots a$), at time k ($k = 1 \ldots c$). A_i, B_j and C_k are the effects, if they exist, of different treatments, plots within treatments and times of sampling, respectively. AC_{ik} and $CB(A)_{kj(i)}$ represent interactions among factors and e_{ijk} represents the individual deviation of plot j in treatment i at time k from the average of all theoretical plots that might instead have been sampled. e_{ijk} represents any source of variation in X_{ijk} that is not accounted for by the other terms in the model. Such errors or deviations cannot be measured because only a single plot (j) is repeatedly sampled and therefore there can be no separate estimation of e_{ijk} from other sources of variability.

The analysis is shown in Table 12.7. Note the similarity of Equation 12.9 to the model for a split-plot design in Equation 12.4 (which are identical in algebraic form because $l = 1$). The measurements for factor C in that design are, essentially, 'repeated' on the same plots. Anyway, the analysis of the two models is virtually identical (compare Table 12.7 with Table 12.3). It is, however, more difficult to identify what the e term represents here (Equation 12.9) than previously (Equation 12.4), where smaller spatial variation among experimental units in each plot was inevitable, but visible.

The repeated-measures analysis can proceed only if it can be assumed that there is no interaction among times and plots (i.e. that $\sigma^2_{\mathrm{CB(A)}} = 0$). Otherwise there is no possible test for differences among plots (B(A) in Table 12.7). There are, theoretically, tests for differences among treatments (A), times of sampling (C) and their interaction (A $\times$ C). These are, however, not interpretable if there are interactions among times and plots (as explained in Section 10.6). See also the discussion of split-plot designs in Section 12.3.

If there are interactions among times of sampling and plots within treatments, there can be no overall interpretation of interactions between times of sampling and differences among treatments. A much better approach is to be rigorous in the design of the experiment and to pursue one of two other options. First, if the potential interactions among times of sampling and plots are not important with respect to the stated hypothesis, set up enough plots so that separate, independent ones are sampled at each time. The data will then conform to the following model:

$$X_{ijk} = \mu + A_i + C_k + AC_{ik} + e_{j(ik)} \tag{12.10}$$

where X_{ijk}, A_i, C_k, AC_{ik} are as before (Equation 12.9) and $e_{j(ik)}$ represents the jth replicate plot ($j = 1 \ldots b$ plots) sampled in treatment i at time k.

There must now be a total of *bc* plots established for each of the *a* treatments, so that a different *b* can be sampled on each of the *c* occasions. This experiment has a straightforward, orthogonal two-factor design.

As a second approach, if interactions among times of sampling do not matter, a separate analysis can be done at any chosen times. Thus, at each of the *c* times of sampling, there is a set of data conforming to the model:

$$X_{ij} = \mu + A_i + B(A)_{j(i)} \quad \text{equivalent to} \quad X_{ij} = \mu + A_i + e_{j(i)} \qquad (12.11)$$

which is obviously a one-factor analysis. The F-ratios for detecting differences among treatments (i.e. among A_i values) could usefully have probabilities of Type I error adjusted to take into account the series of *c* tests being done in the whole experiment (as explained in Sections 6.9 and 8.6). The Bonferroni procedure would use $\alpha' = \alpha/c$ for each test, where α is the accepted probability of Type I error (say, $P = 0.05$) and *c* is the number of times for which analyses must be done.

A further approach, if the interactions with time are important, is to abandon analyses of differences among plots at each time. Instead, examine the temporal *trend* in each plot. For example, if the hypothesis can be stated along the lines of 'temporal trends in the measured variable will vary under the different experimental treatments', the replicate plots each provide a measure of the temporal trend. Thus, the set of *b* plots in each of the *a* treatments provides *b* replicates, independent estimates of the temporal trend in each of the *a* treatments. This is considered briefly in Chapter 13.

Yet again, there are multivariate approaches. In an ecological context, Gurevitch & Chester (1986) considered procedures to turn univariate, repeated-measures analyses into multivariate analyses. They and Winer *et al.* (1991) considered the possibility of using a modified F-ratio suggested by Geisser & Greenhouse (1958) to overcome the problem of non-independence among sequential, repeated measures on the same plots. There are several difficulties with this approach (see e.g. Boik, 1979).

Instead, Gurevitch & Chester (1986) suggested their multivariate analysis, as, in a different context, did Green (1993). The suggestion does, however, deal with only one of the problems: non-independence among times of sampling. The other problem is the lack of replication leading to the uncheckable assumption of no interaction between times of sampling and replicated plots in each treatment. This is not at all solved by the multivariate analysis of variance, which has as an assumption that the degree to which the data in a plot are correlated from time to time must be the same for all plots. This is the assumption of homogeneous covariances for all pairs of plots (see Johnson & Field, 1993). This is, essentially, the same assumption that there is no interaction between times of sampling and differences among plots – which cannot be evaluated because there is no independent measure of residual variation (e terms in Equation 12.9).

So, yet again, provision of proper independent replicates would seem a more compelling strategy than other devices suggested. Nevertheless, where interactions that must be assumed to be zero are realistically likely to be zero, there are many useful purposes to analyses of repeated measures. Excellent accounts of procedures are available in the text by Crowder & Hand (1990).

12.6 Asymmetrical controls: one factor

It is often appropriate to design experiments in which there are several treatments to be compared against a control, but for which the hypothesis causes no requirement for the non-control treatments to be compared with each other. Here is an example.

Many herbivorous animals include in their diet a range of species of plants. One model to explain such observations is that a varied diet is necessary because different plants provide various essential components (e.g. certain minerals or particular amino acids). The animals seek out these plants in order to guarantee an adequate provision of all their requirements.

From this, an appropriate hypothesis is that depriving the animal of a given component of the diet will cause a decline in rate of growth, or reproductive output, or survival, if the dietary component is crucial for this. The null hypothesis is that, because the varied diet is not crucial for the well-being of the animal, removal of a given component will make no difference to any of these variables.

Suppose the animal feeds by moving from one patch of food to another and that the various plants can easily be removed from areas of habitat (such a case was described by Kitting (1980)). The experiment then consists of a series of treatments, in each of which a particular component of the diet is removed so that growth, reproduction or survival can be compared with a control, where no change is made to the diet. For each component of diet manipulated, the null hypothesis is the one-tailed $\mu_X \geq \mu_C$, where μ_X represents the mean of the relevant variable (growth, reproduction, survival) when component X is removed and μ_C represents the mean in controls. There is, however, no purpose, given the hypothesis, in comparing the removals of the various components of diet. The experiment could be done as a series of separate comparisons of a control group of animals with a set deprived of a component of diet. It is, however, much more efficient and cost-effective to use one sample of controls.

This scenario leads to a one-factor analysis of variance of the a treatments which are one control and $i = 1 \ldots (a-1)$ manipulated diets to investigate $(a-1)$ species of plants. If there are significant differences among treatments in the single, fixed-factor analysis of variance, multiple contrasts are needed to determine which treatments differ. The only appropriate comparisons are of the control treatment against each of the others. Dunnett (1955) invented a test for comparing a single control with each of a series of treatments. The procedure is designed to control the experiment-wise rate of Type I error (as explained in Chapter 8) over the $(a-1)$ comparisons. Dunnett (1955) used t-tests to compare each pair of means and developed tables of adjusted probabilities for experiments of different sizes.

Miller (1981) recommended use of a step-wise procedure, for which the Student–Newman–Keuls (SNK) procedure (as described in Section 8.6.5.2) is appropriate. As pointed out by Day & Quinn (1989), there are no problems of excessive Type I error in groups of means in an SNK test when only two means are ever compared. So, test the mean of the control group, in turn, against the most different experimental group and then the next most different and so on. Once a comparison is not significant, no further testing is done, because no treatments with a mean even more similar to the control can be different from it.

12.7 Asymmetrical controls: fixed factorial designs

There are numerous ways in which the logic of an asymmetrical control should be used in the design of an experiment. The most obvious and common use in ecological experiments should be in studies of competition. These always require different experimentally manipulated treatments to be separately compared to controls.

The typical approach to a study of competitive interactions should arise from observations of patterns of size, growth or other variable that is correlated negatively with abundance or with density of other species that use the same resources. Alternatively, there may be some patterns of association of growth or reproductive output that is in some way related to availability of resources. These observations can all be explained by a model involving competitive interactions within a species or among the species that use the resources. Competition is a process leading to 'harm' to individuals when other individuals of the same or different species use the same resource(s) and the resource(s) are in short supply (Birch, 1957).

Sometimes competition is invoked simply because of ecological ideology. The 'thought-police' of ecological theory have dictated that competition must be important and therefore it is so. Under such a paradigm (Simberloff, 1980; Underwood & Denley, 1984), competition is invoked without there being clear patterns of observations that it is supposed to explain (Underwood, 1990). This inevitably leads to endless wrangling (Diamond & Gilpin, 1982; Connor & Simberloff, 1984, 1986). Sometimes, it can lead to poor experimentation and the apparent desire to find evidence – any evidence at all – to prop up the model of competition (Underwood, 1986).

Fortunately, however, competition is one of the ecological processes that has been investigated by numerous experiments (for reviews, see Connell, 1983; Schoener, 1983; Underwood, 1986, 1992b; Snaydon, 1991). Many of these are beset by problems of inadequate or no replication, inappropriate or no controls, fundamental problems of logic and resultant confounding (Underwood, 1986). The catalogue of problems will not be re-examined here (except one fundamental lack of logic which is re-iterated below in Section 12.8). Instead, the appropriate use of an asymmetrical experimental design is described (for a specific case, see Underwood, 1978).

As an example, it has been observed that animals of two species coexist in a habitat, at varying densities from place to place. They are known to eat the same foods. Where animals of one species (A) are relatively dense, or, where they are sparse and species B are numerous, individuals of species A are small, relative to their sizes where they and B are sparse. The model to explain this is that interspecific competition for food from B and intraspecific competition for food among A cause decreased rates of growth of individuals of A. There are many possible hypotheses to do with altering the supply of food and predicting what will happen and manipulating the densities of the two species to predict effects. Under many conditions in the field, manipulations of the resources are more difficult than manipulations of the animals. The resources may be intrinsically difficult to alter and it is often not clear how much of a manipulation might be appropriate to invoke a response. Also, at this stage, you have no observations about resources for which you need to propose explanatory models, so there is no basis for experimentation with the resources. First, you have to demonstrate that competitive interactions are, indeed, occurring.

So, consider the hypothesis about intraspecific competition that increases in numbers of A, where they are sparse, will cause decreased rates of growth of juveniles of A, resulting in adult sizes corresponding to those seen naturally where A are at larger densities. Careful preliminary observations will indicate exactly what sorts of difference to anticipate at various increased densities, if competition is, as proposed, responsible for variations in size. Such observations will provide the necessary information about the numbers of animals to add and the magnitudes of decreases in size that should result. The latter is necessary for any calculation of the power of the experiment (see Section 8.3).

This experiment would be a fixed, single-factor design. The control treatment would be independent replicated plots with natural small densities of A. The experimental treatments would be replicated plots with several pre-determined levels of increased density. All the plots would originally have been ones with small numbers of A, so that there are no influences due to the possibility that there are other variables related to patchy habitat, but not related to density, that may affect the outcome. Of course, there may be more complications because of the need to provide controls for enclosing animals (Dayton & Oliver, 1980) and for the disruption to the animals that are added to the experimental enclosures to increase densities (Underwood, 1986, 1988). These necessary controls will be ignored here in order to simplify the illustration of an asymmetrical design.

At the same time, we can hypothesize that because there is (under the model proposed to account for the original observations) interspecific competition for food, additions of numbers of B will cause decreased growth and therefore smaller sizes than in control areas where B are absent. Again, this could be done as a single factorial experiment with patches of small density of A serving as controls and similar patches, with pre-determined numbers of B added, as the experimental treatments.

There are, however, all sorts of reasons why examinations of the relative effects of intra- and interspecific competition need to be compared (Schoener, 1983; Underwood, 1986). Any proper interpretation of competitive interactions must be able to account for the effects of both kinds of process.

It makes sense therefore to examine the two hypotheses in the same experiment, so that interspecific increases in density and intraspecific increases in density can be examined separately, yet their interrelationship

Table 12.8. Analysis of an asymmetrical experiment to test the hypothesis about inter- and intraspecific competition. The design is illustrated in Figure 12.7. All data are analysed by a single-factor analysis of variance. Then, the four non-control treatments are analysed as a two-factor analysis, with increased density (treatments 2 and 4 versus 3 and 5) and species added (treatments 2 and 3 versus 4 and 5) as the two factors. There are n replicates of all treatments

Source of variation	Degrees of freedom		
Among all treatments[a]	4		
Controls versus others[b]		1	
Among others[c]		3	
Increased density[d]			1
Species added[d]			1
Inc. density × sp. added[d]			1
Residual[a]	$5(n-1)$		
Total[a]	$5n-1$		

[a] One-factor analysis of variance of all data.
[b] Sum of squares by subtraction of among others from among all treatments.
[c] Sum of squares by addition of those in[d].
[d] Two-factor analysis omitting controls (treatment 1).

is also investigated. If inter- and intraspecific influences are equivalent, there would be no statistical interaction in the appropriate analysis (see Table 12.8).

The point that causes asymmetry in the design is that there really can be only a single set of replicates of the control treatment. Thus, there cannot be orthogonality in the design to make two factors (increased density and species added – see Figure 12.7 and Table 12.8). Even if it were thought appropriate to double the number of replicates of the control treatment and then to allocate half of them to be the controls for intraspecific competition and half for interspecific competition, there is no satisfactory way to assign the replicates to the two 'controls'. Anyway, this would add to the cost of the experiment.

Hence, the asymmetrical analysis in Table 12.8 of the design in Figure 12.7. The differences among all five treatments can be calculated, as can the sum of squares representing differences among the four non-control treatments (2 to 5 in Figure 12.7). The latter can be partitioned into the sums of squares associated with each of the two factors and their interaction. The difference between the two can be shown to be a comparison of the control with the other four treatments. Because, in this experiment,

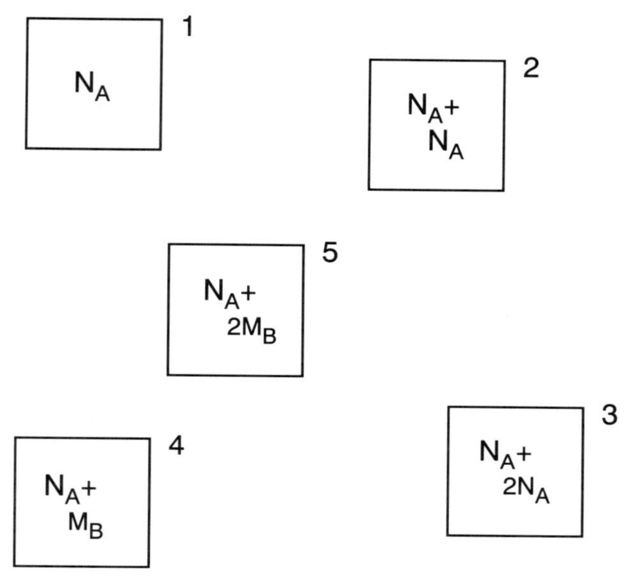

Figure 12.7. An asymmetrical design of two fixed factors in an experiment to test hypotheses about competition. One factor is increased density, the second is the species added (A or B). There is a single control (1) of species A at natural density, without species B. All treatments must be independently replicated *n* times. N_A, M_B are the natural densities of species A and B, respectively, where they are sparse. Where species A or B are relatively abundant their densities are $3N_A$ (as in the treatment $N_A + 2N_A$) and $2M_B$, respectively. Comparison of treatments 1, 2 and 3 tests for intraspecific competition. Comparison of treatments 1, 4 and 5 examines interspecific competition.

all factors are fixed, all sources of variation are tested by *F*-ratios of their mean squares against the residual mean square.

The analysis can be followed by appropriate multiple comparisons, where these are relevant. There are also further ways of partitioning the data (Winer *et al.*, 1991) and extensions to other situations, e.g. where there are more than two competing species, are possible (Underwood, 1986). The uses of these designs in analyses of competition have been extensively described by Underwood (1978, 1984, 1986, 1992b). They are, however, very useful in other types of experiment where two or more factors are orthogonal to one another, but there can be only a single control. For example, studies of the effects of and interactions between two types of pollutant, two drugs or other treatments that are to be applied in several concentrations (equivalent to density of species in the competition experiment), will require only a single group of replicated controls.

Note, however, in the case of studies of competition that there are important issues of the appropriate ways to compare the intensity of competition between members of the same species (the intraspecific component) and the interspecific effects of members of a different species. The symmetry or asymmetry of these two competitive interactions has some theoretical interest (Schoener, 1983; Fletcher & Creese, 1985), but is difficult to quantify where experiments are done with different densities of the two species as illustrated here (Underwood, 1986).

12.8 Problems with experiments on ecological competition

The use of asymmetrical (or, for that matter, any other) experimental designs in studies of competitive interactions raises an important issue about the logical basis for an experiment. This has been discussed in detail elsewhere (Underwood, 1986, 1992b), with other problems. The particular issue is, however, that of designing the experiment in such a way that the results and therefore the interpretation are entirely unconfounded.

It has sometimes been the case that experimental manipulations of animals and plants (to test the hypothesis that altering their numbers will, because of competition, change some aspect of their ecology) have confused two things. For example, Bertness (1981) found two species of hermit crabs coexisting in rock pools in Panama. He proposed that they competed for shells and tested hypotheses derived from this model. He introduced crabs into pools so that he had three treatments: 200 of species A alone in a pool, 200 of species B alone in a pool and 100 of each species together in a pool. These treatments were then replicated. The hypothesis was presumably that the presence of B was somehow detrimental to A, so that A alone would fare better (their mean value of any variable of interest would be greater) than when in a mixture with B. But there are now two differences between the two treatments. First is the planned difference between being alone or with B. Then there is the fact that the density of A differs between the two treatments. So, suppose that resources are in short supply and intraspecific competition for resources causes decreases in growth, reproduction, survivorship or some other variable. Suppose further that competition from species B is identical in intensity or effect. Under the model that competition for resources occurs, the hypothesis would be that addition of any A or any B to some number of A will cause an intraspecific or an interspecific decrease in some variable in A. If the experiment consists of halving the density of A and simultaneously replacing them with Bs, two things have altered. The density of A has gone down (thus decreasing the effect of intraspecific competition) and the density of B has gone up (thus increasing the effect of interspecific competition). There would then be *no* difference between the animals of species A in the control and those in the experimental treatment. The null hypothesis of no difference will be retained – despite the competition actually occurring. This type of experiment can be useful only if there is marked asymmetry in the effect of competition and B must cause more difference to A than extra A do.

Another variety of confounding is widespread in experiments on plants (see e.g. the review by Trenbath, 1974). There are numerous examples of what is called a 'replacement series' or 'reciprocal α design' (deBenedictus, 1974), where two species or varieties or strains are grown together in mixtures. Often, the treatments are 100% of species of variety A, then 90% A with 10% B, 80% A with 20% B, all the way to 100% B. Any comparison between any two treatments has two differences (a decrease in A and an increase in B), so sensible interpretation is usually impossible. Probably, these designs were originally used to determine in what combination of mixtures the maximal yield of fruit, flowers, seed or biomass can be harvested. The unconfounded design (sometimes called a 'mechanical diallele' design; Putwain & Harper, 1970) is as recommended above. There should be a control of some relevant number of species A and a series of treatments of that number of A with different numbers of B.

12.9 Asymmetrical analyses of random factors in environmental studies

Asymmetrical experimental designs are also important in environmental studies to detect or estimate the magnitude of anthropogenic disturbances against a noisy background. The reason for asymmetry is that usually, if not always, there is only one potentially impacted location. It has therefore been generally the case that only one control (i.e. undisturbed) location has been compared to it (as recommended by Stewart-Oaten *et al.*, 1986). Green (1979) in a very capable book provided the logical rationale for detection of an environmental impact. There must be some pattern of difference from before to after a planned disturbance in the relationship between the mean of whatever variable is measured in the disturbed location compared with that in the control. Bernstein & Zalinski (1983) and, later, Stewart-Oaten *et al.* (1986) described why replicated measures in time before and again after the disturbance are necessary. They also demonstrated procedures of analysis and, in the latter case, the importance of randomized temporal sampling. These concepts are now summarized as BACI designs (Before compared with After and Control compared with Impacted). Green's (1979) major contribution was to draw attention to, and insist on, a focus on the interaction between these two fixed factors. The hypothesis that a planned disturbance will lead to an environmental impact is tested using the null hypothesis that there will be no interaction.

This simple design, despite its widespread use, is illogical and uninterpretable in any situation where the variable of interest varies through time in a different way from place to place even when there is no disturbance. Stewart-Oaten *et al.* (1986) recognized this and explained why it was inappropriate to attempt to analyse such variables. If, for example, the mean abundance of some species varies differently through time in any two locations sampled, there are interactions between locations and time even when there are no disturbances. So, if one of the locations is disturbed by some human activity, it should come as no surprise that there is an interaction in the analysis of the data.

To overcome this problem, it is necessary to determine what happens in the disturbed (i.e. potentially impacted) location compared with what happens in several control, undisturbed locations. This has been developed as an experimental design by Underwood (1992a, 1993, 1994). The details and the possible complexities for interpretation do not have to be discussed again here. It is, however, important to demonstrate why an asymmetrical design when one factor is random is fundamentally different from the fixed case discussed above (in Section 12.7). It is also important for me to thank Dr K. R. Clarke (of the Plymouth Marine Laboratory) for his help in unravelling this topic.

To demonstrate what is going on, consider an overly simplified situation that is parallel to that for the fixed case in Section 12.7 and Table 12.8. A measure is to be made in one disturbed location and several randomly chosen control (i.e. undisturbed) locations. Unlike the requirements for an environmental survey, there will only be measurements at one time. To develop the analysis, consider first that there are, in fact, many disturbed and many control locations available. Several (b) are chosen at random from each set and each is sampled with n independent quadrats. The model of the analysis is a straightforward nested one (as in Chapter 9):

$$X_{ijk} = \mu + D_i + L(D)_{j(i)} + e_{k(ij)} \qquad (12.12)$$

where D_i is disturbed ($i = 1$) or control ($i = 2$) and $L(D)_{j(i)}$ represents the particular characteristics of location j ($j = 1 \ldots b$) that is either one of the

Table 12.9. Symmetrical and asymmetrical analysis of locations in an environmental survey

(a) Symmetrical, nested design. There are b randomly chosen control and b randomly chosen disturbed locations, each sampled with n replicates

Source of variation	Degrees of freedom	Mean square estimates	F-ratio versus
Disturbed versus control = D	1	$\sigma_e^2 + n\sigma_{L(D)}^2 + 2bnk_D^2$	Locations (D)
Locations (D)	$2(b-1)$	$\sigma_e^2 + n\sigma_{L(D)}^2$	Residual
Residual	$2b(n-1)$	σ_e^2	
Total	$2bn-1$		

(b) Asymmetrical design. There are b randomly chosen control locations and one disturbed location, each sampled with n replicates

Source of variation	Degrees of freedom	F-ratio versus
Disturbed versus control = D	1	Locations (D)
Locations (D)	$b-1$	Residual
Residual	$(b+1)(n-1)$	
Total	$n(b+1)-1$	

disturbed locations or one of the controls. $e_{k(ij)}$ represents variation due to replicate k in the jth location in either disturbed or control sets. Thus, locations are nested in either disturbed or control treatments and are a random factor (because other locations could have been sampled). The analysis of this sort of data is in Table 12.9a.

Now, suppose there are no replicate locations that are disturbed. There is, fortunately, only going to be one airport, marina, sewage outfall, harbour, whatever. There are still many control locations available, so a selection of these may be made at random. The model for the data is then exactly as in Equation 12.12 for data in control locations ($i = 2$), but $j = b = 1$ for the single disturbed location (when $i = 1$). The analysis remains as in Table 12.9b, but there are only nested locations in the control treatment.

Note how different this is from the fixed case in Table 12.8. Note also that there are few degrees of freedom for detecting differences between control and potentially impacted locations. This does not automatically mean that differences are unlikely to be detected. Environmental impacts *are* differences through time in the magnitudes of difference between control and impacted locations. This example was for illustrative purposes only. Full details of how to detect various patterns of interaction at various scales of time and space and some considerations of power are available in Underwood (1991b, 1992a, 1993, 1994).

Asymmetrical analyses are of undoubted use in various branches of ecology. All that is required in planning such experiments is clarity of thought and clear exposition of the hypotheses – including which factors are fixed or random and therefore how estimates provided by mean squares are derived.

13

Analyses involving relationships among variables

So far, the concepts considered involve only one variable. In many experiments, it is extremely useful to investigate hypotheses about more than one variable and to test specified patterns of relationship between variables. Multivariate procedures for this are too large a topic to consider here. Analyses of regressions are, however, well worth introducing because of their widespread use and relationships to the topics considered earlier. This is a large and complex topic and the various issues will be considered only briefly here, in the context of tests of ecological hypotheses.

13.1 Introduction to linear regression

Regression, in its simplest form, is the analysis of a functional relationship where one variable (Y) is dependent on or linearly related to the magnitude of another (X). The analysis can be used to test hypotheses about the relationship and to estimate the parameters of the equation linking the magnitude of Y to that of X. The hypotheses predict the form of the relationship (linear in the simplest case) and the equation linking Y with X (i.e. the slope and/or intercept (or elevation) of the linear relationship).

The underlying regression equation in a linear case is a straight line passing through the means of the two variables (μ_y, μ_x). The equation predicts how far a given value of Y is from its mean, taking into account the magnitude of X. The equation for Y as a linear function of X is:

$$Y_i = \alpha + \beta X_i \tag{13.1}$$

where α is the intercept (the magnitude of Y when $X = 0$) and β is the regression coefficient or slope of the line. To make the line pass through

μ_Y, μ_X requires the equation:

$$Y_i = \mu_Y + \beta(X_i - \bar{X}) \tag{13.2}$$

where β is the regression coefficient.

Usually, when it is thought that there is a linear functional relationship between two variables, we are proposing either that there is some unspecified slope (i.e. there is a relationship rather than no relationship; β is >0 or β is <0) or we hypothesize a particular relationship (β equals some specified value). In the first case, the hypothesis does not specify a particular value and the null hypothesis is H_0: $\beta = 0$.

So, if we observe that insects in various patches of flowers move different distances to feed on pollen, we might propose the model that the distance moved is less where there are more flowers. Assume for a moment that the relationship might be linear. We can predict, as the hypothesis, that the slope of the regression of mean distance moved (Y) or number of flowers per unit area in each plot (X) will be negative. Less distance will be moved where there are more flowers. The hypothesis is that $\beta < 0$. The null hypothesis in this case is the one-tailed one: H_0: $\beta \geq 0$.

In the second case, there is a particular value specified for β. In this situation, the statistical null hypothesis must be the actual hypothesis. For example, after observing that rats eat different amounts of food, we might discover in the literature a common explanation that rats eat a particular proportion (say, one tenth) of their body weight, per day. As a result, we might suggest this model applies to our rats and therefore we propose that the regression of amount eaten (Y) on body weight (X) should have a slope of $\beta = 0.1$.

In this case, we must use this as the statistical null hypothesis (as discussed in Chapter 5), H_0: $\beta = 0.1$. This will have all the problems of logic and power associated with such goodness-of-fit null hypotheses (as discussed in Chapter 5).

These null hypotheses have all the same features as those described for analyses of means of variables in Chapters 5 and 6. It will therefore come as no surprise that appropriate tests for null hypotheses about regressions tend to be similar in form to those used for means.

The underlying model (in Equation 13.2) is examined by taking a sample of n elements, for each of which you measure both variables. Now you have n values of the paired data (X_i, Y_i; $i = 1 \ldots n$) – the magnitudes of the two variables. These data are used to estimate the parameters of the regression equation (μ_y, β). It is going to be assumed here that the values of X_i are fixed, in the sense that they are determined by the experimenter. They are created as, say, pre-determined distances along a transect or chosen densities of putatively competitive snails or plants, or temperatures in a laboratory study.

The alternative is that they are not fixed. The X and Y values are recorded at whatever magnitude in samples. Thus, in an experiment to test the hypothesis that the height (Y) of some shrub is a linear function of the concentration of phosphorus in the soil (X) data are gathered in some study-area. Random shrubs are located and measured. The soil around the roots is sampled and its concentration of phosphorus determined. This sounds easy, but there are numerous difficulties of sloppy definition of phosphorus in this example, but let's ignore that. We now have paired data Y_i and X_i values, but the X values are themselves sampled, subject to error and not fixed. This case poses problems, briefly discussed below (Section 13.3.3).

So, for the fixed-X case, we know $\bar{X}$ as a precise estimate of the mean of X (we chose the set of values of X) and we have data that conform to the following linear model:

$$Y_i = \bar{Y} + b(X_i - \bar{X}) + e_i \tag{13.3}$$

where $\bar{Y}$ is the sample estimate of the mean of Y (i.e. of μ_Y), b is the sample estimate of the regression coefficient β and e_i represents sampling error (i.e. the data for Y at any value of X are sampled with error). The nature of the error, or deviation (e_i) terms will be considered in Section 13.3. The purpose of the analysis is to calculate b and to use it in tests of null hypotheses about β.

Of various methods used to analyse regression, least-squares regression is considered here. It determines the straight line that best fits the data, by being, on average, as close as possible to all the data points. This is illustrated in Figure 13.1.

As has occurred elsewhere (see the discussion of the location parameter in Section 3.6), once the line of best fit is found, the error terms will sum to zero, so they are squared and summed and the analysis is to determine the value of b that minimizes this sum. Hence, the procedure is called 'least squares'.

$$\sum_{i=1}^{n} e_i^2 = \sum_{i=1}^{n} [(Y_i - \bar{Y}) - b(X_i - \bar{X})]^2 \tag{13.4}$$

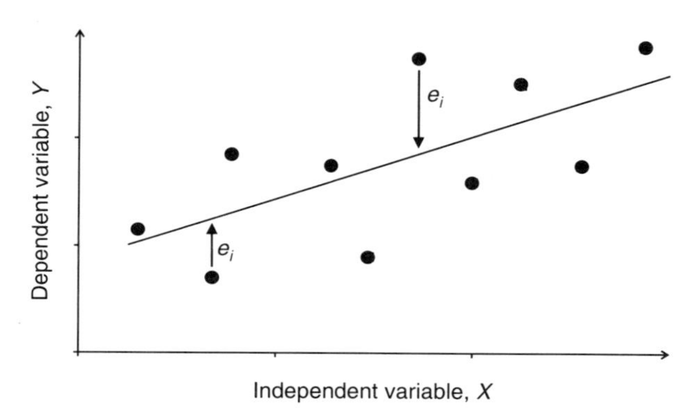

Figure 13.1. A linear, least-squares regression. The data are paired values of X (the independent variable) and Y (the dependent variable), collected as $n = 10$ pairs, X_i, Y_i. X is considered fixed, i.e. chosen as a fixed factor and measured without error. The least-squares regression line is the one that minimizes, on average, the squared vertical deviations of Y_i values from the line. These vertical deviations (e_i values) are illustrated as arrows.

This is minimized by differentiating the sum of squared deviations with respect to b and setting the differential to zero, which produces:

$$b = \frac{\left[\left(\sum_{i=1}^{n} X_i Y_i\right) - \frac{\left(\sum_{i=1}^{n} X_i\right)\left(\sum_{i=1}^{n} Y_i\right)}{n}\right]}{\left[\left(\sum_{i=1}^{n} X_i^2\right) - \frac{\left(\sum_{i=1}^{n} X_i\right)^2}{n}\right]} \tag{13.5}$$

13.2 Tests of null hypotheses about regressions

Whether we have a general hypothesis in mind ('there is a relationship between Y and X') with the null hypothesis $\beta = 0$, or whether there is a more specific relationship ($\beta = 0.1$ as in the second example in Section 13.1), we need a test.

If the assumptions discussed in the next section are true, the following t-test can be used:

$$t = \frac{b - H}{\mathrm{SE}_b} \tag{13.6}$$

where b is the calculated (least-squares) sample estimate of β, H is the value of β specified in the null hypothesis and SE_b is the standard error of b:

Table 13.1. Data and analysis of a linear regression between two variables. The dependent variable (Y) is the number of bites taken by fish in ten minutes; the independent variable (X) is the size of fish (cm). It has been proposed that larger fish feed faster (i.e. $\beta > 0$) and therefore the null hypothesis is H_0: $\beta \leq 0$. Data are from $n = 11$ fish

Fish no.	Size, X (cm)	Mean no. of bites, Y (in 6 separate minutes)
1	11.0	4.50
2	9.0	4.84
3	10.8	4.67
4	7.4	4.16
5	16.0	5.84
6	14.9	5.67
7	13.8	4.84
8	8.6	4.33
9	9.4	4.67
10	11.5	5.16
11	12.3	4.84

$$\sum_{i=1}^{n} X = 124.7 \quad \sum_{i=1}^{n} Y = 53.52$$

$$\sum_{i=1}^{n} X^2 = 1487.71 \quad \sum_{i=1}^{n} Y^2 = 263.079 \quad \sum_{i=1}^{n} XY = 619.003$$

$$\sum_{i=1}^{n} x^2 = 74.065 \quad \sum_{i=1}^{n} y^2 = 2.680 \quad \sum_{i=1}^{n} xy = 12.281$$

$$\bar{X} = 11.34 \quad \bar{Y} = 4.87$$

$$b = \frac{\sum_{i=1}^{n} xy}{\sum_{i=1}^{n} x^2} = \frac{12.281}{74.065} = 0.166$$

$$a = \bar{Y} - b\bar{X} = 4.87 - 0.166(11.34) = 2.99$$

For any value of X, estimated value of Y from regression is:

$$\hat{Y}_i = a + bX_i = 2.986 + 0.166X_i$$

$$t = \frac{b - 0}{\mathrm{SE}_b} = \frac{0.166}{\sqrt{\left\{\dfrac{\left[\sum_{i=1}^{n} y^2 - \dfrac{\left(\sum_{i=1}^{n} xy\right)^2}{\sum_{i=1}^{n} x^2}\right]}{(n-2)\left(\sum_{i=1}^{n} x^2\right)}\right\}}} = \frac{0.166}{\sqrt{\left(\dfrac{2.680 - \dfrac{12.281^2}{74.065}}{9(74.065)}\right)}} = 5.34$$

9 degrees of freedom, $P < 0.001$

$$\mathrm{SE}_b = \sqrt{\left\{ \left[\sum_{i=1}^{n} y_i^2 - \frac{\left(\sum_{i=1}^{n} xy_i \right)^2}{\left(\sum_{i=1}^{n} x_i^2 \right)} \right] \Big/ \left[(n-2) \left(\sum_{i=1}^{n} x_i^2 \right) \right] \right\}} \tag{13.7}$$

where $\sum_{i=1}^{n} y_i^2 = \sum_{i=1}^{n} (Y_i - \bar{Y})^2$, $\sum_{i=1}^{n} xy_i = \sum_{i=1}^{n} (X_i - \bar{X})(Y_i - \bar{Y})$ and $\sum_{i=1}^{n} x_i^2 = \sum_{i=1}^{n} (X_i - \bar{X})^2$.

The probability of getting the observed value of t if the null hypothesis is true can be found in a table of the t-distribution, using $(n-2)$ degrees of freedom, where n is the number of replicate data points sampled. A worked example is provided in Table 13.1.

It will be convenient to simplify Equation 13.5 to its shorthand version:

$$b = \frac{\sum_{i=1}^{n} xy_i}{\sum_{i=1}^{n} x_i^2} \tag{13.8}$$

This estimate of slope of the line (i.e. b from Equation 13.5) allows calculation of a, the intercept in the traditional equation for a straight line:

$$a = \bar{Y} - b\bar{X} \tag{13.9}$$

We now have $Y_i = a + bX_i$ to allow the line to be plotted easily. None of the algebra has any underlying assumptions – except that there is a linear relationship (as required in the model and hypothesis being investigated) and that the data are independently sampled (see also Section 13.3.1 below).

13.3 Assumptions underlying regression

There are several important assumptions underlying the preceding test for regression. Some of these are entirely under the control of the experimenter, because good experimental design will solve the potential problems before they occur. Other assumptions relate to properties of the data being analysed. Therefore, there is nothing that can be done to solve the problems, except by care in interpretation after the data are analysed.

Most of the assumptions of regression analyses are the same as those discussed in great detail in Chapter 7, under analysis of variance. It would be sensible to re-read the relevant sections there.

13.3.1 Independence of data at each X

The most widely ignored assumption in regressions in ecology is that data (Y values) are independently sampled at each value of X. For example, analyses of rate of growth of individual organisms are often done with measurements of the same individuals at successive times in an experiment. These data are not independent; quite often the size at one time is a very specific function of previous size. Consequently, the sizes measured at each time are not independent – they are determined by the previous size.

The solution to the general problem of analyses of temporal data taken repeatedly from the same sample units (individuals, plots, quadrats, etc.) is always to design the experiment so that independence is achieved. In the case of growth of animals and plants, if n individuals were to be measured at each of t times, start the experiment with nt individuals. At each time of sampling, measure n of them and then ignore these for the remainder of the experiment. In this way, n independently sampled replicates will be available each time.

This is not strictly true if the sampling or measurement of the first n individuals affects the subsequent measurements on the others. Thus, if animals are removed from a pen so that they can be measured, the procedure may alarm the other animals. If the fright affects their subsequent growth, future measurements are not independent estimates of size – they are affected by previous measurements. Vigilance is always necessary.

There are numerous ways in which data may not be independently sampled for regression analyses (see the more detailed discussion in Chapter 7). The most common non-independence in regressions in ecology is, however, repeated measurements of the same units.

Where there is positive correlation among data sampled at different values of the independent variable (X), the resulting estimate of b is too small. b (from Equation 13.8) is an underestimate or is biased downwards. The most widespread procedure for testing for non-independence in regressions is the Durbin–Watson test (Durbin & Watson, 1950, 1951).

The procedure involves calculating the regression and then the residual deviations from it. These are then used to determine the degree of correlation between the residuals at different intervals, or lags, through the data. If the degree of correlation through time is known, or can be estimated, it is possible to correct for it in simple regressions (see e.g. Doran & Guise, 1984). Further details are not considered here.

Correlation or non-independence crops up in numerous ways in ecology. Apart from the cases of temporal sampling, there are others. For example, it was fashionable some years ago to seek for density-dependent processes in abundances of populations by analysis using regressions. The procedure was to plot the estimated abundance of a population in one year as X and the estimated abundance in a subsequent year as Y. If the slope of the regression was negative, this supposedly indicated density dependence in the populations. Density-dependent mortality (e.g. due to intraspecific competition) would cause numbers in the population to decline. Thus, numbers in any subsequent year would be reduced more (would become relatively smaller) when numbers in a previously sampled year were large.

Unfortunately, the procedure is fraught with numerous difficulties because it is usually done with data sequentially plotted as an independent and then again as a dependent variable. Except for the very first census, all estimates of abundance in one year (year i) are used twice. Each is a dependent variable (Y_{i+1}) to plot against the estimate from the previous year (X_i) Each is then used as a dependent variable (X_{i+1}) for the plot of the following year (Y_{i+2}). This had major problems of underestimation of slopes of regression – leading to excessive numbers of apparent negative slope and apparent density–dependence (Maelzer, 1970; St Amant, 1970).

A new version of this problem has been re-invented more recently. Because of the popularity of 'supply-side' ecological ideas (see e.g. Underwood & Fairweather, 1989), it has been of recent interest to examine early recruitment of animals at some period after their colonization of, or settlement in, the habitats occupied by adults. For example, Connell (1985) and Davis (1988) examined settlement of marine invertebrates and their subsequent survival after the first few weeks. They wished to test the hypothesis that numbers of survivors (recruits) would be related to numbers of original settlers. They used regressions of number of surviving recruits (Y) on number of initial settlers (X) to test the null hypothesis of no relationship.

This is inappropriate because the number of recruits in any replicate (Y_i) can be no greater than the number of settlers (X_i). If no mortality occurs, $Y_i = X_i$ for all i. If any animals die, Y_i must be less than X_i. The null hypothesis of no relationship is clearly inappropriate (Ito, 1972; Holm, 1990; McGuinness & Davis, 1989).

Much more thought about the hypotheses and the data is needed before one undertakes regression analyses about ecological processes.

13.3.2 Homogeneity of variances at each X

The assumption requires that the data (Y values) collected at each value of X have the same variance. In many ecological situations, this is extremely unlikely. For example, in studies of the growth of animals, as individuals grow, there will be increasing variance with time because small individuals are fairly similar in size and have small variance. As time goes by, there are differences among the individuals in the rate of growth (because of genetic, behavioural and other differences). So, as the mean size increases, so does the variance. Any regression analysis of growth (i.e. with size as Y and time as X) will have heterogeneous variances.

As another example, consider the situation in an experiment on predation by birds on insects. In a set of experimental plots (for which the models and hypotheses are not presented here), the proportion of marked beetles eaten by birds is recorded at several times. The plots are sampled independently (as they must be), so that different plots are sampled each time. Further, for the sake of illustration, assume that birds tend to keep visiting a plot in which they have already found beetles.

At the start of the experiment, birds have found few plots. Few beetles have been eaten in few plots, so the mean proportion eaten per plot is quite small, but the variance among plots is also small (Figure 13.2). As time elapses, the mean proportion of beetles eaten increases because birds have had time to eat more of them. The birds, however, also find more of the plots. Consequently, the mean and variance among plots both increase. Finally, when enough time has passed for the birds to have

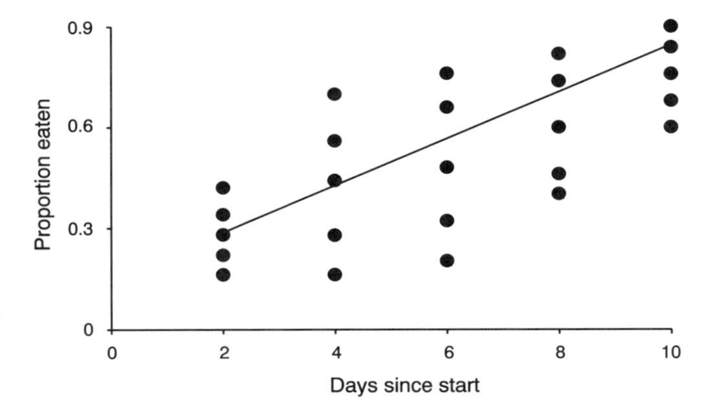

Figure 13.2. Proportions of beetles eaten by birds in plots sampled at different times. As time goes by, more beetles are eaten, but birds gradually find all the plots. Variance among plots increases then decreases.

eaten most of the beetles, they will have done so in most of the plots, so the variation among plots is decreased again. In Figure 13.2, the variance is greater at the intervening times of sampling. So, any regression to determine the rate of predation will be subject to potentially heterogeneous variances.

Dealing with heterogeneity can be difficult. Tests for heterogeneity may be found in Goldfeld & Quandt (1965). If you have several estimates of Y at each value of X, any test for heterogeneity of variance can be used (see Section 7.16). You can estimate the variance of Y from the replicated sampling at each value of X.

If variances are heterogeneous, the regressions can sometimes be done using a weighted regression. The principle is to weight the data so that those with larger variance are given less 'weight' or influence in the analysis, relative to data with small variance. Procedures for this are described by Green & Margerison (1977), Steel & Torrie (1980) and Doran & Guise (1984).

Also, if a simple procedure for weighting the data cannot be found, it will be necessary to transform the data to remove the heterogeneity of variances. The principles of this are described in Section 7.18, with respect to transformations in analysis of variance. There are, however, severe problems with transformations because they are scale dependent (see Box & Cox, 1964), so that the regressions are not estimated with the same precision if the scale is changed.

Furthermore, the relationship between Y and X will no longer be the same after transformation, which has important implications for the null hypothesis. For example, consider the analysis of the linear relationship between weight of food consumed (Y) and weight of animal (X). If the data have increasing variances at increasing values of X, transformation to logarithms can be considered as a device to remove heterogeneity.

The relationship between $\log(Y)$ and $\log(X)$ is no longer linear, so a linear regression cannot be used. This causes a change to the null hypothesis and to the form of regression analysed. Much professional statistical advice is needed where regression analyses do not have homogeneous variances.

13.3.3 X values are not fixed

The linear regression analysed using least-squares procedures is only valid where X values are fixed, i.e. measured without error. If both Y

and X values are measurable without error, the regression relationship is *functional* in the terminology of Kendall & Stuart (1977). In many areas of biology and ecology, we deal with situations where X is fixed (as in fixed-factor analysis in Chapter 8) and measured without error, but Y is randomly variable. These are handled by the least-squares procedures described above. Thus, in experiments, we can create or control X. It might be the initial density of plants or the distance from shelter or the amount of time elapsing or the number of contacts with conspecifics, etc. Careful thought during the design of such an experiment will often ensure that X is fixed and made error free.

In other circumstances, however, X is a randomly sampled variable or cannot be measured without error. Thus, X is no longer fixed.

Under these circumstances, the estimate of slope of the regression from a least-squares analysis is biased towards zero (i.e. smaller in absolute value than the true regression coefficient). This will tend to decrease the chances of detecting a significant regression (i.e. will cause excessive Type II errors – failure to detect a departure from the null hypothesis of no relationship between Y and X). The review by McArdle (1988) is an excellent introduction to the issues and a synthesis of available procedures. There is no room here to discuss the details.

13.3.4 Normality of errors in Y

The final assumption in regressions is that the data, Y values, have normal distributions at each value of X. To detect departures from this requires many data to be sampled at each value of X. This means that the assumption is rarely testable in practice. If variances are homogeneous, the residuals from the regression line (i.e. the difference between every value of Y and the value predicted by the regression line) should be normally distributed. Pooling them together and examining their frequency distribution could be done.

Also, remember that means of samples are much more likely to be normally distributed than the individual replicates (see Chapter 5). Therefore, if data are unlikely to be normally distributed, take several replicate measures of Y at each X and analyse the regression of sample means ($\bar{Y}$ values) instead of Y. The sample means are much more likely to be normally distributed than are the individual values. Otherwise, there are non-parametric solutions (see e.g. Sokal & Rohlf, 1981).

Table 13.2. Analysis of variance of a linear regression of Y_i on X_i, sampled with n data

Source of variation	Sum of squares	Degrees of freedom	Mean square	F-ratio
Regression = Reduction in variability due to regression	$SS_R = \frac{\left(\sum_{i=1}^{n} xy_i\right)^2}{\left(\sum_{i=1}^{n} x_i^2\right)}$	1	$SS_R/1$	MS_R/MS_D
Deviations from regression $\left(\sum_{i=1}^{n} e_i^2\text{; see Equation 13.4}\right)$	$SS_D = \left(\sum_{i=1}^{n} y_i^2\right) - \frac{\left(\sum_{i=1}^{n} xy_i\right)^2}{\left(\sum_{i=1}^{n} x_i^2\right)}$	$n-2$	$SS_D/(n-2)$	
Total variation	$\sum_{i=1}^{n} y_i^2$	$n-1$		

For definitions of $\sum_{i=1}^{n} x_i^2$, $\sum_{i=1}^{n} xy_i$, $\sum_{i=1}^{n} y_i^2$, see Equations 13.7 and 13.8.

13.4 Analysis of variance and linear regression

Instead of the above *t*-test, a regression can be handled by analysis of variance. An analysis of linear regression of paired data from two sampled variables (Y_i, X_i; $i = 1 \ldots n$) calculates the total variation of the Y values from their mean and then the variation of the Y values from the fitted regression line. This effectively partitions the total variability in Y (from the mean $\bar{Y}$) into two components or sources: the deviations around the fitted regression (as in Figure 13.1) and the remainder. The remainder is therefore the reduction in the variability due to fitting the regression. Obviously, if the deviations from regression are large (not much smaller than the total variability about $\bar{Y}$), there is no relationship with X. If, however, the variability of deviations is small and the reduction due to fitting the regression is large, then the regression is a reasonably good fit to the data.

The total variation in Y values was identified under Equation 13.7. The deviations from the fitted regression were shown as $\sum_{i=1}^{n} e_i^2$ in Equation 13.4, which can be expanded and simplified to give the sum of squared deviations in SS_D in the analysis of variance in Table 13.2. The F-ratio for testing whether a regression is significant is a test of the null hypothesis that $\beta = H$ (where H is some hypothesized value) (see e.g. Snedecor & Cochran, 1989; Winer *et al.*, 1991). This is the same test as the *t*-test described in Section 13.1. If the *t*-value calculated there is squared, you get F, as in Table 13.2. The same assumptions as described in Section 13.3, of course, also underlie the use of the F-ratio.

The advantage of considering the simple, linear regression this way is that it leads to straightforward extensions of the algebra and tests to cover more complex circumstances. Two of these (partial and polynomial regressions) are introduced briefly below (Sections 13.6 and 13.7).

13.5 How good is the regression?

It is often quite important to know how well the calculated regression fits the data. For example, suppose we have observed, from one place to another, different numbers of flowers per plant in a particular species. There seem to be fewer where there are more plants, so we propose the model that number of flowers is a negative function of intraspecific density. Imagine that such a relationship would be linear, so we hypothesize a negative linear regression between number of flowers and density. The null hypothesis is $\beta \geq 0$.

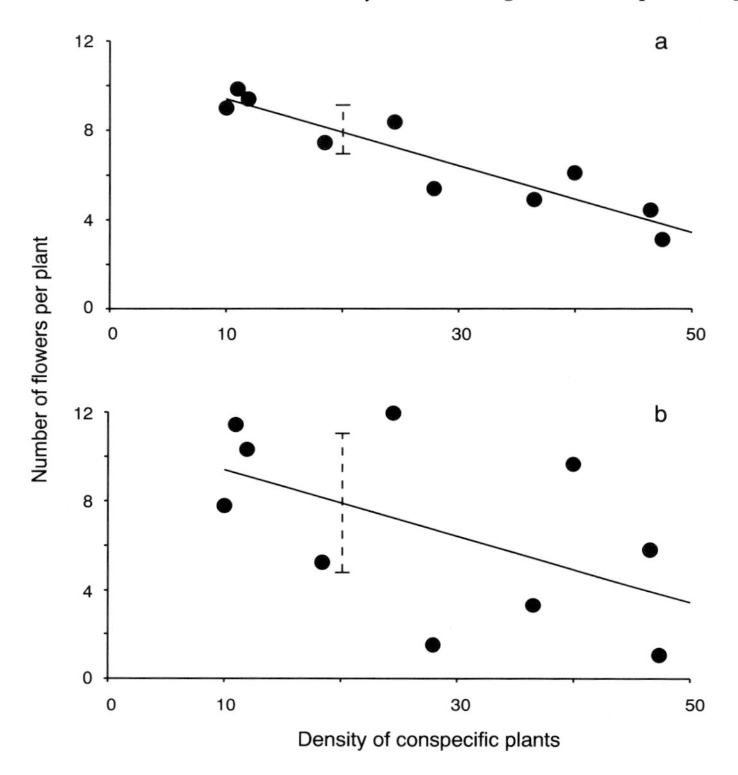

Figure 13.3. Two regressions with the same intercept and slope. In the first case, the regression is a good fit – the data are close to the estimated line ($r^2 = 0.87$). In (b), the data are more variable and are, on average, further from the line ($r^2 = 0.29$). Obviously, future predictions are more accurate in the first case. The approximate ranges of variability in the data at a density of 20 are shown as the dashed lines in each graph.

Suppose that we reject the null hypothesis (as in Figure 13.3) and now wish to know how many flowers per plant are likely to occur at a particular density. For either of the examples in Figure 13.3, the predicted number of flowers for a density of 20 per unit area is 8. In the first case, this is fairly precise; the data suggest a range of about ±1. In the second case, the estimate is not very precise at all; the data suggest a possible range of about ±3.5.

It is therefore helpful to estimate the precision of the prediction, i.e. a measure of how close the regression line is to the sampled data. An obvious way to do this is to estimate the deviations of the data from the line, as in Equation 13.4. This measure, on its own, is not useful because it is without reference to any standard measure. Therefore, it is customary to consider it with reference to the total variation in Y, as in Equation 13.7 and see Table 13.3.

If no regression relationship exists ($\beta = 0$), Y is μ_Y at all values of X (consult Equation 13.2 and set $\beta = 0$: X disappears). So, if there is no regression, the line of best fit is $\bar{Y}$. Therefore, the deviations from regression are the same as the total variation in Y. Where an absolute regression relationship exists, the Y values must all lie on the regression line. All deviations are zero. Any regression between Y and X must therefore reduce the proportion of variability in Y that is measured by the deviations of Y values from the regression line. The decrease in variation due to fitting the regression is therefore some measure of how good the regression is as a description of the data.

The reduction in variation due to fitting the regression can be expressed as a proportion of the total variation in Y. This proportion is known as the *coefficient of determination* and is called r^2:

$$r^2 = \frac{\sum_{i=1}^{n}(Y_i - \bar{Y})^2 - \sum_{i=1}^{n}[Y_i - \bar{Y} - b(X_i - \bar{X})]^2}{\sum_{i=1}^{n}(Y_i - \bar{Y})^2} = 1 - \frac{\sum_{i=1}^{n} e_i^2}{\sum_{i=1}^{n} y_i^2} \tag{13.10}$$

where $\sum_{i=1}^{n} e_i^2$ is as in Equation 13.4 and $\sum_{i=1}^{n} y_i^2$ is as under Equation 13.7. r^2 can also be calculated as the squared value of the correlation coefficient, r, between Y and X. This measures the goodness-of-fit of the regression to the data. If the data fit perfectly, there are no deviations and $r^2 = 1$. If there is no relationship, $\beta = 0$ and the deviations from the line $b = 0$ are the same as from the line $\bar{Y}$. Therefore, the deviations ($\sum_{i=1}^{n} e_i^2$ values) equal the variation in Y ($\sum_{i=1}^{n} y_i^2$ values) and $r^2 = 0$.

It is advisable, therefore, to describe a regression in terms of its equation, its significance with respect to the relevant null hypothesis and its goodness-of-fit (as measured by r^2). It is also possible to define the standard error or confidence interval for any predicted value (see e.g. Snedecor & Cochran, 1989).

Finally, as one word of warning – extrapolations or predictions outside the range of the regression are extremely dangerous things. In Figure 13.4, for example, are data representing proportional survival of animals through time in an experiment. They are not at all linear but, measured over some short time interval, may seem to be. Predictions about the proportion surviving during that interval are reliable. Predictions outside it

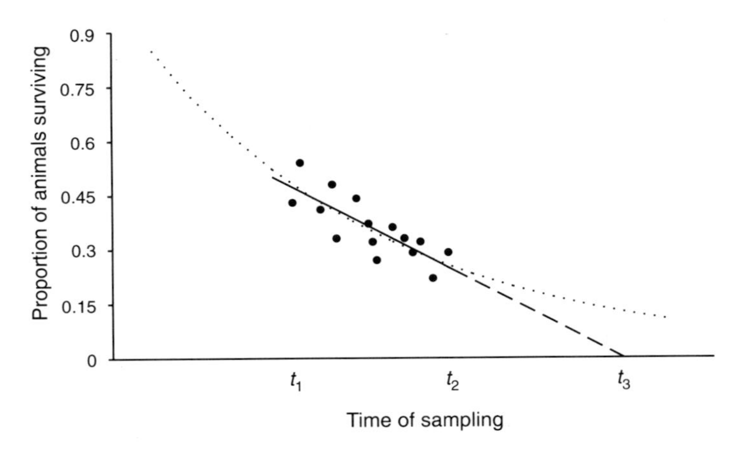

Figure 13.4. Predictions from a linear regression are only appropriate for the range over which it has been sampled (here from t_1 to t_2 as shown by the continuous line). The real relationship (the proportion of animals surviving through a season) is non-linear (the dotted line). The proportion predicted, as an extrapolation from the regression, to survive to time t_3 (which is outside the period observed) is shown by the dashed line. It is obviously wrong.

are clearly wrong (the linear regression over a short, early period would predict *no* survivors after a long period).

Interpolation to make predictions about values of the dependent variable (Y) from a regression equation and a chosen value of the independent variable (X) are reasonable, particularly if the confidence limit is calculated around the regression. Extrapolation outside the range of the independent variable is an unwise procedure. If it is necessary, you should propose the hypothesis that the calculated regression is appropriate over the range including the value of X for which you wish to extrapolate. This should then be tested by making the prediction and collecting data to determine whether it is realistic. If not, no mistaken action will have been taken on the basis of a false prediction and evidence will be available that the regression does not hold outside the observed range. This would normally result in trying to construct a more realistic regression equation that applies over the entire range of observations including the new ones. This is an application of the procedure advocated in Chapter 2.

13.6 Multiple regressions

There is no reason to limit regressions to situations where there is only one independent variable. For example, the growth of plants might be a

Table 13.3. Multiple regressions of growth as increase in height of plants (Y_i) with density of plants (X_{1i}) per 100 cm^2 and concentration of soluble phosphorus (X_{2i} as ppm). There are $n = 10$ samples

Y_i	X_{1i}	X_{2i}
1.1	18	34
2.4	16	43
3.8	10	47
5.4	6	67
2.3	21	56
3.7	4	53
5.1	3	61
2.0	12	28
3.4	13	36
3.9	7	38

$$\sum_{i=1}^{n} X_1 = 110 \quad \sum_{i=1}^{n} X_2 = 463 \quad \sum_{i=1}^{n} Y = 33.10$$

$$\sum_{i=1}^{n} X_1^2 = 1544 \quad \sum_{i=1}^{n} X_2^2 = 22893 \quad \sum_{i=1}^{n} Y^2 = 126.33$$

$$\sum_{i=1}^{n} x_1^2 = 334 \quad \sum_{i=1}^{n} x_2^2 = 1456.1 \quad \sum_{i=1}^{n} y^2 = 16.77$$

$$\sum_{i=1}^{n} X_1 Y = 302.50 \quad \sum_{i=1}^{n} X_2 Y = 1643.60 \quad \sum_{i=1}^{n} X_1 X_2 = 4813$$

$$\sum_{i=1}^{n} x_1 y = -61.60 \quad \sum_{i=1}^{n} x_2 y = 111.07 \quad \sum_{i=1}^{n} x_1 x_2 = -280$$

$$Y_i = \bar{Y} + b_1(X_{1i} - \bar{X}) + b_2(X_{2i} - \bar{X}_2) \quad \text{or} \quad Y_i = a + b_1 X_{1i} + b_2 X_{2i}$$

$$b_1 = -0.144$$

$$b_2 = 0.049$$

$$a = \bar{Y} - b_1\bar{X}_1 - b_2\bar{X}_2 = 2.63$$

function of the concentration of phosphorus in the soil and the density of surrounding plants. In this case, the appropriate regression model is a *multiple* regression – Y is affected by two independent variables.

$$Y_i = \mu + \beta_1(X_{1i} - \bar{X}_1) + \beta_2(X_{2i} - \bar{X}_2) + e_i \qquad (13.11)$$

where μ is the mean of the population being sampled; Y_i values are sampled values; X_{1i} and X_{2i} are the corresponding values of concentration of phosphorus and density of plants, respectively. These have means $\bar{X}_1$ and $\bar{X}_2$; the regression coefficients of these two variables are β_1 and β_2 and e_i represents error from the underlying model.

Table 13.4. Tests of importance of independent variables in a multiple regression. Y is the increase in height of plants, X_1 is the density of plants, X_2 is the concentration of phosphorus as in Table 13.3

(a) Linear regression on X_1 alone

$Y_i = 5.34 - 0.18X_{1i}$

Source of variation	Sum of squares	Degrees of freedom	Mean square	F-ratio
Regression	$SS_{X_1} = 11.36$	1	11.36	16.71 $P < 0.01$
Deviations	5.41	8	0.68	
Total	16.77	9		

(b) Linear regression on X_2 alone

$Y_i = -0.22 + 0.08X_{2i}$

Source of variation	Sum of squares	Degrees of freedom	Mean square	F-ratio
Regression	$SS_{X_2} = 8.47$	1	8.47	8.14 $P < 0.01$
Deviations	8.30	8	1.04	
Total	16.77	9		

(c) Multiple regression (as in Table 13.3)

Source of variation	Sum of squares	Degrees of freedom	Mean square	F-ratio
Regression on X_1 and X_2 = SS_R	14.25	2	7.13	19.81 $P < 0.01$
Effect of X_1 after X_2 = $SS_R - SS_{X_2}$	5.78	1	5.78	16.06 $P < 0.01$
Effect of X_2 after X_1 = $SS_R - SS_{X_1}$	2.89	1	2.89	8.03 $P < 0.01$
Deviations = SS_D	2.52	7	0.36	–
Total	16.77	9	–	–

Correlation: X_1 and X_2 $r = -0.40$ $P > 0.05$, 8 degrees of freedom
X_1 and Y $r = -0.82$ $P < 0.05$, 8 degrees of freedom
X_2 and Y $r = 0.71$ $P < 0.05$, 8 degrees of freedom

To estimate the regression coefficients from the data, the same procedure of differentiation as in Section 13.1 results in the following equations:

$$b_1 = \frac{\left(\sum_{i=1}^{n} x_{2i}^2\right)\left(\sum_{i=1}^{n} x_{1i}y_i\right) - \left(\sum_{i=1}^{n} x_{1i}x_{2i}\right)\left(\sum_{i=1}^{n} x_{2i}y_i\right)}{\left(\sum_{i=1}^{n} x_{1i}^2\right)\left(\sum_{i=1}^{n} x_{2i}^2\right) - \left(\sum_{i=1}^{n} x_{1i}x_{2i}\right)^2} \tag{13.12}$$

$$b_2 = \frac{\left(\sum_{i=1}^{n} x_{1i}^2\right)\left(\sum_{i=1}^{n} x_{2i}y_i\right) - \left(\sum_{i=1}^{n} x_{1i}x_{2i}\right)\left(\sum_{i=1}^{n} x_{1i}y_i\right)}{\left(\sum_{i=1}^{n} x_{1i}^2\right)\left(\sum_{i=1}^{n} x_{2i}^2\right) - \left(\sum_{i=1}^{n} x_{1i}x_{2i}\right)^2} \tag{13.13}$$

where x_{1i}^2, x_{2i}^2, $x_{1i}x_{2i}$, $x_{1i}y_i$, $x_{2i}y_i$ are all abbreviated as indicated in Equations 13.7 and 13.8. An example is calculated in Table 13.3.

Now it is possible to fit the multiple regression using both variables. If it is significant, it may be of interest to determine which of the two variables is (or whether both of them are) driving the values of Y. This is easy to accomplish by first fitting each variable on its own, i.e. ignoring the other one. So, the linear regressions of Y on X_1 and on X_2 are calculated and then the differences between how well the data fit each of these and how well they fit the multiple regression are compared. The mechanics are shown in Table 13.4.

If a good fit is obtained with, say, X_1 alone, there is little residual deviation to be removed by fitting the regression on X_2. In such a case, most of the reduction in variability due to the partial regression is due to X_1. Consequently the effort of fitting X_2 after (or over and above X_1) would be small and non-significant.

In the example in Tables 13.3 and 13.4, the growth of plants is significantly positively related to the amount of phosphorus in the soil. An increase of 1 ppm of phosphorus increases the growth of plants of 0.08 cm (as in Table 13.4b). At the same time, there is a significant negative effect of density of plants. An increase of 1 plant per 100 cm^2 causes a decrease in growth of plants of −0.18 cm (as in Table 13.4a).

When considered together, there is a significant relationship with the combination of variables (Table 13.4), but each variable is important. An increase of 1 ppm of phosphorus causes an increase of 0.049 cm in the height of plants, on average at any density. An increase of 1 plant per 100 cm^2 causes a decrease in growth of plants of −0.144 cm at any concentration of phosphorus. Both are significant relationships (as in Table 13.4c).

There is, however, an extremely important assumption underlying the use and interpretation of individual variables in multiple regressions. If the various explanatory variables (the X values) are to be interpreted separately, there must be no *collinearity* or correlation between any of the independent variables. In the example here, if the density of plants is, itself, a function of the concentration of phosphorus so that X_1 is a function of X_2, the analysis becomes futile. There is first a statistical and then a logical problem. If the two variables are correlated, much of the influence on Y of X_1 includes the effects of X_2, and vice versa. As a result, the multiple regression on X_1 and X_2 may well be significant. Removing one variable still leaves a significant regression on the other, but the sum of squares for the regression on X_1 will be very similar to that in the partial regression. Similarly, the sum of squares for the regression on X_2 will be similar to that for the partial regression. As a result, neither of them will be significant when tested after the regression on the other. You finish up with a significant regression on two variables, but neither is significant!

The logical dilemma is self-induced. If two variables are tightly correlated, there is no way to decide which of them, on its own, matters. Neither has any effect independent of the other. There is a 'holistic' or 'Gestalt' response of growth of plants to the combinations of the two variables. This is all you can know and all you *need* to know. Attempts to separate the effects of two confounded variables by observation alone are always silly (whoever does them!).

The solution is, of course, to propose models of separable effects of the two variables, from which you make hypotheses about their separate, individual influences and any interactions between them. Then you design experiments to manipulate the densities of plants at chosen concentrations of phosphorus and to manipulate the concentrations at chosen densities. Such an experiment would unravel any independent effects of each factor and their interaction. It is completely irrational to ignore the need for such experimental evaluations of complex hypotheses.

13.7 Polynomial regressions

Relationships between two variables are unlikely to be linear – there are many other possibilities. For example, the rate of growth of plants in areas with different availabilities of nitrogen could well be quadratically

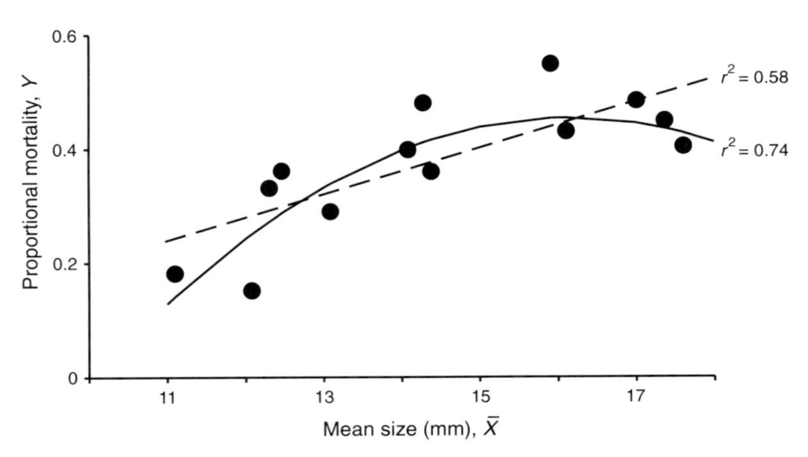

Figure 13.5. Quadratic regression of proportion of snails eaten by birds (Y) on their mean size ($\bar{X}$) and mean size squared ($\bar{X}^2$). Data and analyses are in Table 13.5. The dashed line is the linear regression and the continuous line the quadratic regression. Goodness-of-fit is measured by the coefficient of determination (r^2), as shown for each regression.

related to concentration of nutrients. The plants do not grow well where there is inadequate nitrogen, nor where there is an excessive amount. They do well in a 'Goldilocks' fashion, i.e. where things are 'just right'. Similarly, the proportional mortality of prey of many invertebrate predators is a function of their size relative to the predator. It is, however, a concave downward curved relationship because predators eat few small prey and cannot eat many large prey.

Under these circumstances, a linear fitted regression is not a good description of the relationship (Figure 13.5). It may not be known (or guessable) what is the exact relationship between Y and X. A polynomial fit to the data is often a useful first step in trying to find a useable model. Polynomial analyses are the same as multiple regressions, except that the terms in the model are all different powers of the same variable.

So, proportional mortality of prey in a given period of time (Y) may be described as a quadratic function of their mean size (X):

$$Y = a + b_1X + b_2X^2 \tag{13.14}$$

The analysis proceeds exactly like that for a multiple regression, with $X_1 = X$ and $X_2 = X^2$. An example is given in Table 13.5.

The same principles apply to any higher powers of polynomial. You can, theoretically, fit polynomials up to the power of $(n - 1)$ for a set of

Table 13.5. Quadratic regression of proportional mortality (Y), over a month, of snails of different mean sizes (X) being eaten by birds. Data are as in Figure 13.5

Proportional mortality	Mean size (mm)	Proportional mortality	Mean size (mm)
0.18	11.1	0.36	14.4
0.15	12.1	0.55	15.9
0.33	12.3	0.43	16.1
0.36	12.5	0.48	17.0
0.29	13.1	0.45	17.4
0.40	14.1	0.40	17.6
0.48	14.3		

Linear regression: $Y = -0.21 + 0.0405X$

Quadratic regression: $Y = -2.75 + 0.3969X - 0.0013X^2$

Source of variation	Sum of squares	Degrees of freedom	Mean square	F-ratio
Linear regression on X	$SS_{RL} = 0.094$	1	0.094	15.67 $P < 0.01$
Deviations	$SS_{DL} = 0.067$	11	0.006	
Quadratic regression on X, X^2	$SS_{RQ} = 0.119$	2	0.059	14.75 $P < 0.01$
Residual	$SS_{DQ} = 0.042$	10		0.004
Total		12		
Addition of X^2	$SS_{DL} - SS_{DQ} = 0.025$	1	0.025	6.25[a] $P < 0.01$

[a] F = Mean square addition of X^2/Mean square deviations from quadratic regression.

n data. As each extra polynomial (power of X) is included in the model, it can be analysed as in Table 13.5.

Polynomial regressions are also very useful for determining relationships among means after an analysis of variance. For many experimental treatments, the levels of a factor are not separate categories (plus or minus predators, different caging treatments, patches of habitat, etc.). Instead they are levels of a continuous variable (density of a competitor, temperature, etc.). For such factors, it is often appropriate that the relevant hypothesis is not about differences among treatments, but is rather about the relationships between the variable measured in the experiment and the levels of the experimental treatment (see also the last part of Section 8.6.5.4).

Consider an example like that in Table 13.5. Suppose it has been observed that the numbers of some species being attacked by predators vary from place to place and the prey also vary in size from one place to another. A model to explain the rate of predation would then be that the predators are more effective at finding and eating prey of some sizes (note there are many alternatives to this model, but it will illustrate the point). We can propose the hypothesis that if sets of plots are prepared, so that we place the same size of prey in each of a set and different sizes in the different sets, the rate of predation in the plots will vary because there are differences in the sizes of prey among sets of plots. We can test this by choosing, say, four different sizes of prey (e.g. 13, 14, 15 or 16 mm mean size for the case in Table 13.5) and having several independent replicate plots each with a chosen number of prey of the chosen size. An analysis of variance (one factor with four fixed levels of sizes and replicate plots) might indicate that there are differences in proportional mortality among sizes. There is, however, no particular reason why there should be any difference between the adjacent sizes in their proportional mortality. If you examine Figure 13.5, you will see little difference in mortality across each pair in that range of sizes.

Attempts to identify differences using multiple comparison procedures (as in Chapter 8) will fail to find any pattern. These are inappropriate tests (see also Dawkins, 1981; Perry, 1986). It would be better to analyse the data as a polynomial regression (unless the shape of the relationship had been more specifically hypothesized in advance), using the fixed sizes as 'X' values and the mean proportional mortality as 'Y' values. There are three degrees of freedom for comparing the four treatments. One of these can be attributed to the linear regression, one to the quadratic relationship and the third to the cubic relationship between mortality and size. In this case, the X values are equally spaced, but that is not an essential feature of polynomial analyses. The details are not considered further here, but are straightforward extensions of the analysis in Table 13.5.

There is no point in attempting to separate means of a series of levels of a factor if their pattern of difference is an overall relationship with little difference between adjacent treatments. Regression analysis of the means is far superior because it tests hypotheses about the relationship, not about differences among treatments.

Note the two different uses of polynomial regression in this example. First, it was observed that rate of predation seemed to vary with size of prey. The model proposed to explain this was that there really was a relationship between mortality and size, leading to the hypothesis that if prey of different sizes were observed over a period of time, their rate of predation would vary in some systematic manner with size. The null hypothesis was therefore no relationship and a regression analysis of rate of mortality on size was appropriate. This was tested for the existence of the claimed *pattern* in the data.

The second component was the model that the observed pattern was really due to some response by the predators to the sizes of the prey. The model proposed to account for the observed pattern was that predators (by their feeding behaviour in response to prey of different sizes) accounted for the pattern. Hence the hypothesis and experimental manipulation. The manipulation controlled for differences in microhabitat (eliminating models to do with certain sizes of prey happening to be in microhabitats where predators are most active), local density of prey (eliminating models to do with prey of certain sizes happening to be in larger density, creating patches of intense mortality when predators find them because of density-, not size-, dependent foraging by the predators), etc. The polynomial regression analysis of the mean mortalities of prey in the experimental plots tests hypotheses about the process causing the pattern.

The framework of analysis of variance of regressions can be extended, using the same principles and procedures, to any combination of independent variables (i.e. any complexity of partial regression) and to any combinations of powers of variables. Alternatively, it can be used to examine any linear model constructed to describe the relationship between variables in a regression. It is, however, worth remembering that the more terms, powers, parameters or variables put into a regression, the more replication (or the larger the sample size) that may be necessary to ensure adequate power in tests of the appropriateness of the model.

13.8 Other, non-linear regressions

There are many and varied other regression models. For example, there are many possible complex models of relationship of Y on one or more variables. There are two obvious reasons for wanting to fit complex curves. First, the underlying logical model being examined may be a complex relationship – such as predicting that growth of animals follows some physiological model leading to a particular form of asymptotic curve. Alternatively, there may be a complex non-linear relationship proposed between the density of a population and the age at which reproduction occurs.

The second reason for wanting to test hypotheses about complex regressions is to be able to fit the most appropriate curve (i.e. regression model) to a set of data so that it can be used for predicting future values of Y from chosen values of X.

The general procedures for fitting complex curves are well explained in such texts as that by Snedecor & Cochran (1989) and modern statistical packages can do the asymptotic iterations of the calculations without difficulty. Consequently, the important issue to remember here is that the assumptions underlying linear regressions (independence of data, homogeneity of variables, X being measured without error, normality of errors at each X) also apply to more complex models. The same problems will eventuate if these assumptions are ignored. Considerably more care is needed in the analysis and interpretation of complex regressions. Above all, the models and hypotheses leading to complex regression equations must be very clearly thought out so that meaningful conclusions can be drawn from any analysis of the data. Analyses of regression are not considered further here.

13.9 Introduction to analysis of covariance

Often, despite the best intentions of experimenters, initial allocations of experimental units to treatments are not equally representative. For example, frogs of the same initial weights may be used in experiments to examine their rates of predation on various insect species. Despite the effort to match the sizes of the frogs, the animals may not be equal in some other important variable, such as breadth of head. So, if the rate of ingestion of insects depends on the sizes of frogs' heads, differences from one container to another will, at least in part, be due to the initial sizes of frogs and not the hypothesized differences because the species of insects differ.

One way to control for such differences, if you can anticipate what variables matter, is to balance the samples by ensuring that equally representative samples of measures of those variables are allocated to each treatment. As with the frogs, this cannot always be done. Analysis of covariance is a procedure to help control, retrospectively, such variables, provided their influence can be estimated and, as discussed below (Section 13.13), is equal in every treatment.

Consider the hypothetical example in Figure 13.6. It has been observed that in areas where frogs forage, survival differs between insects of two species (A and B) even though frogs attempt to catch either sort of insect with equal frequency and energy. The model put forward to explain these results is that one of the insects (A) is better able to avoid the frogs' lunges. From this one would predict a similar average rate of attack by frogs on the two species, but a different average proportion of successful attacks. The null hypothesis for the latter part of this is that the mean proportion of successful attacks on A will be smaller than or equal to the mean proportion of successful attacks on B.

This is tested by allocating one sample of frogs to containers with insects of species A and another sample to containers with B. Because previous froggy work has revealed differences in rates of consumption of insects according to the weight of frogs, great care should be taken to allocate frogs to the two treatments equitably according to weight. At the end of the experiment, the data look like one of the sets in Figure 13.6 – where an alert assistant has noticed during the experiment that the frogs with big mouths got more insects than did those with small mouths (there may well be an important moral in this tale).

In Figure 13.6a and b, the null hypothesis is true – the frogs do not have differential success attacking the two species. In the other cases, A is less successfully lunged for and the data are consistent with the hypothesis. In the cases illustrated, the relationship of successful attacks to size of mouth was evident in Figure 13.6b and d. In cases shown in Figure 13.6a and c, there was no relationship. In an analysis of variance (or for that matter, a t-test) of these data, $\bar{Y}$ values would not differ significantly in Figure 13.6a and b and should differ in Figure 13.6c and d.

There is, however, a problem in cases where there is a positive relationship between proportion of successful attacks and size of mouth, but frogs of different sizes are inadvertently assigned to the two treatments. In this case, where larger frogs (on average) are present, there is a difference in $\bar{Y}$ values between the two treatments (Figure 13.6e). In such cases, analysis of variance will reveal a significant difference between the two treatments,

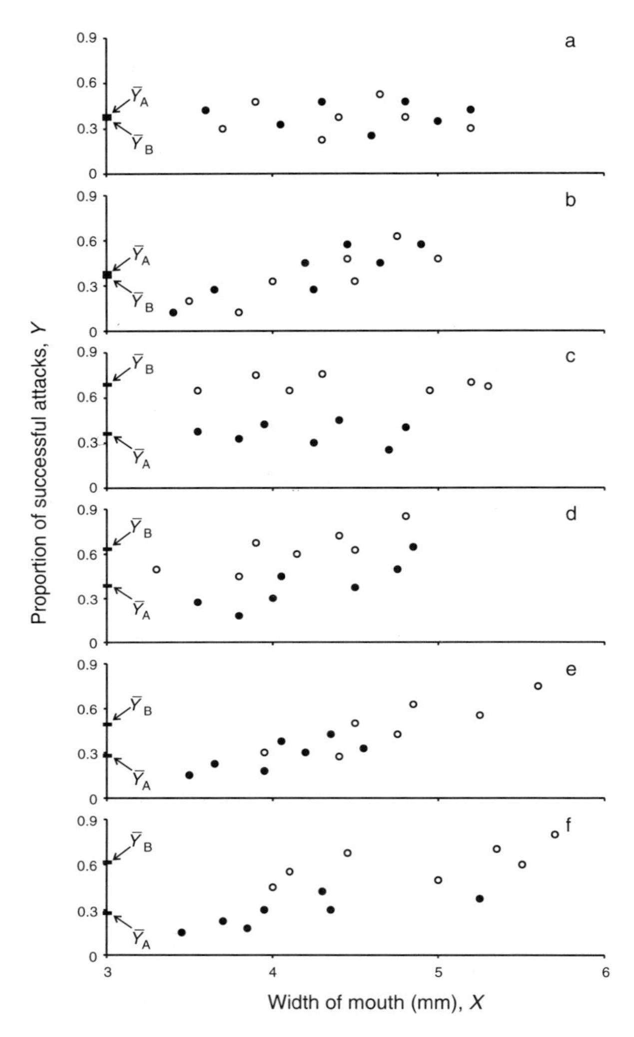

Figure 13.6. Effect of a covariate (X), the width of mouth of frogs, on a variable (Y), the proportion of successful attacks on insects. In the experiment, frogs are placed with insects of species A (filled circles) or B (empty circles). $\bar{Y}_A$ and $\bar{Y}_B$ are the mean proportions of successful attacks on each species. In (a) and (b), there is no difference in the rate of success on either species; in (c) and (d) frogs catch more of B than of A. In (b) and (d), there is a positive relationship between proportional success and size of mouth of the frog. In (e) and (f) there are, due to bad luck, differences in the sizes of frogs in the two treatments. As a result, in both cases, there is a difference between the mean proportional success ($\bar{Y}_A$ and $\bar{Y}_B$) for the two species, but in (e), this is entirely attributable to the covariate, size of mouth (X). In (f), there really is a difference in catchability of the insects.

but it is entirely attributable to the different sizes of frogs – not the difference between treatments (types of insects) per se.

Contrast the conditions in Figure 13.6e and f. In the former (as above), differences between treatments are entirely attributable to differences in sizes of the frogs. In the second case, there is also a difference in the success of attacking the two species of insects. In both cases, there is a relationship of Y to the variate (size of frog). We need to be able to separate the effects of differences among treatments from the differences due to the covariate of size. This is the purpose of analysis of covariance.

The other use of analysis of covariance is when we know, before we plan an experiment, that the variable about which we are hypothesizing is correlated with some other variable. If we can estimate the relationship between the two, we can use the extra information gained about the covariate to increase the precision with which we estimate the difference among treatments. For example, in Figure 13.6, in those cases where there is a clear regression of percentage success on size, the variation within treatments is large because of the scatter of Y values due to the spread of X values. We could, theoretically, reduce the variance among Y values by choosing to use in the experiment only a small range of sizes (X values) of frogs. This would drastically reduce the scatter of Y values in each treatment. It would, however, also reduce the generality of any findings, which would apply only to that restricted range of sizes of frogs. Also there is no logical structure in the construction of the original hypothesis that might suggest how to choose a relevant range of sizes. So, instead, analysis of covariance can be used to reduce the variation among measurements by taking account of the relationship with a relevant and measurable covariate.

The principle, for either purpose, is to use information about the relationship of the variable (Y) to the covariate (X) in order to estimate what would be the values of the variable, in each treatment, if all measurements were done at the same value of X. In other words, the procedure uses the regression relationship between Y and X to adjust the data (Y) to the values they would have been if all replicate frogs had an identical size of mouth (a particular value of X). Then, as the analysis of covariance, the adjusted values of Y are analysed. This allows a test of the null hypothesis of no differences among Y, having removed any scatter due to X.

13.10 The underlying models for covariance

First, it is necessary to estimate the appropriate regression relationship between the variable being analysed (Y) and a specified covariate (X) or

Table 13.6. Algebra of analysis of covariance in a one-factor experiment (treatments $1 \ldots i \ldots a$ in the fixed factor A, each replicated n times) with a single covariate (X), to which the data are linearly related. Y_{ij} and X_{ij} represent the paired data for the variable Y and covariate X for replicate j in treatment i

Treatment							
1		...	i		...	a	
Variable	Covariate		Variable	Covariate		Variable	Covariate
Y_{11}	X_{11}		Y_{i1}	X_{i1}		Y_{a1}	X_{a1}
Y_{12}	X_{12}		Y_{i2}	X_{i2}		Y_{a2}	X_{a2}
$\vdots$	$\vdots$		$\vdots$	$\vdots$		$\vdots$	$\vdots$
Y_{1j}	X_{1j}		Y_{ij}	X_{ij}		Y_{aj}	X_{aj}
$\vdots$	$\vdots$		$\vdots$	$\vdots$		$\vdots$	$\vdots$
Y_{1n}	X_{1n}		Y_{in}	X_{in}		Y_{an}	X_{an}

series of covariates (see Section 13.15.1). Then it is necessary to use this relationship to adjust the data to a chosen value of the covariate to analyse them. This procedure involves fitting, in turn, a series of three models for regression and then comparing the deviations from each of these models.

Consider a general experiment with one fixed factor (A), with levels $1 \ldots i \ldots a$. As a specific example, this could be an experimental test for differences among the mean growths of animals fed on one of three diets. Their rates of growth are simply recorded as their final sizes after a designed period of feeding. Each experimental treatment has n replicates. There is a known covariate (X). In this introduction, the variable measured is considered to be linearly related to the covariate. In the case of the animals in the experiment on growth, this might be demonstrated by the rate of growth (i.e. final size after a fixed period) being a negative linear function of original size, i.e. slightly larger animals grow slightly slower. This may not be unreasonable in real experiments and will serve as a specific example here.

So, the data consist of the n paired sets of original size and final size of animals in each of the a treatments (as in Tables 13.6 and 13.7). The relevant null hypothesis is that there is no difference among the means of the a treatments.

13.10.1 Model 1: Regression in each treatment

The data in each treatment, considered separately, can obviously be

Table 13.7. Example of a single-factor analysis of covariance for an experiment on $a = 3$ diets (a fixed factor), each replicated with $n = 8$ animals. Data are Y = final size, X = initial size; for sums of squares, see text and Table 13.6. Data are plotted in Figure 13.9

	Diet					
	1		2		3	
Replicate	X	Y	X	Y	X	Y
1	10.10	20.0	10.10	18.7	9.85	15.2
2	11.15	19.6	10.45	18.1	10.05	14.4
3	10.75	19.3	11.20	18.3	10.75	14.2
4	11.30	19.2	11.65	17.7	11.25	14.5
5	11.95	19.5	12.40	18.3	11.85	13.7
6	12.15	19.1	13.00	18.0	11.45	13.8
7	12.45	19.6	13.35	17.4	13.25	13.4
8	13.35	18.8	13.95	17.5	13.60	12.8

examined as a regression (see Equation 13.1):

$$Y_{ij} = \mu_i + \beta_i(X_{ij} - \bar{X}_i) + e_{ij} \tag{13.15}$$

This is the familiar linear regression for a sampled set of data. μ_i, β_i, $\bar{X}_i$ belong to treatment i. Then, using the same structure as in the single-factor analysis of variance (Section 7.4), if we assume that treatments differ in an additive manner, we can reconsider the means of each treatment (μ_i values) as being:

$$\mu_i = \mu + A_i \tag{13.16}$$

where μ is the overall mean of all populations in all treatments and A_i is the additive effect of treatment i if the null hypothesis is false (Section 7.4). Thus, the data in any treatment can be described as:

$$Y_{ij} = \mu + A_i + \beta_i(X_{ij} - \bar{X}_i) + e_{ij} \tag{13.17}$$

which combines the logic of analysis of variance and the structure of regression. This is the first model – to fit the linear regressions separately in each experimental treatment.

13.10.2 Model 2: A common regression in each treatment

The second model is the attempt to fit a *common* regression to the data in each treatment. This requires a regression to be calculated that has, over all treatments, the smallest squared deviations (as in the least-squares

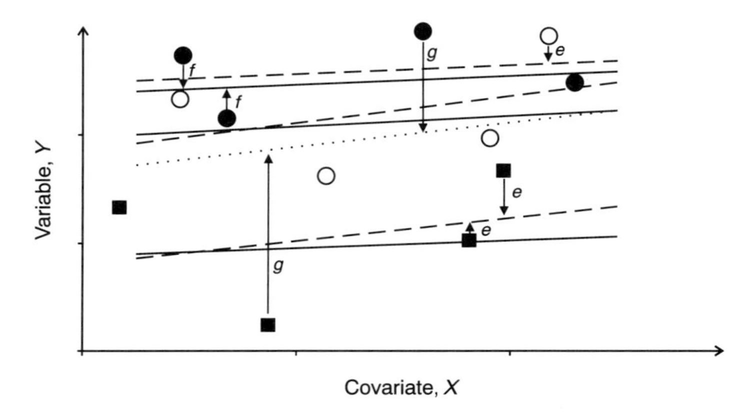

Figure 13.7. Example of data, regressions and deviations in an analysis of covariance on samples from $a = 3$ experimental treatments, sampled with $n = 4$ replicates of the variable (Y) and a linearly related covariate (X). Filled circles, empty circles and filled squares are treatments 1, 2 and 3, respectively. Dashed lines are the regressions under model 1 (Equation 13.17) with deviations e. Continuous lines are the common regression fitted separately to each sample, as under model 2 in Equation 13.18, with deviations f. The dotted line is the total regression on all data combined, under model 3 (Equation 13.27), with deviations g.

regression in Section 13.1), but has the same slope in every treatment. This is illustrated in Figure 13.7. The equation for the data when fitted to this regression is:

$$Y_{ij} = \mu + A_i + \beta_c(X_{ij} - \bar{X}_i) + f_{ij} \tag{13.18}$$

where the data are fitted separately in each treatment (using $\mu + A_i$ and $\bar{X}_i$) but must all have the same, common slope β_c. Individual deviations from this are denoted by f_{ij}.

Note the difference between this model and the previous one (Equation 13.17). These two models would be the same if every treatment had the same regression relation between Y and X (so that all β_i equalled β_c).

The important thing about this model is that there must be a common regression relationship in all the treatments to ensure that the adjustments made to the data to remove the effects of the covariate (X) are equivalent in every treatment. Problems when the treatments do not have the same regression relationships are considered later (Section 13.14).

As will be seen by examination of Equations 13.17 and 13.18, we can construct a test for whether regressions are not similar by use of the deviations from the two models. The summed squared deviations from each model are:

$$\sum_{i=1}^{a}\sum_{j=1}^{n} e_{ij}^2 \quad \text{and} \quad \sum_{i=1}^{a}\sum_{j=1}^{n} f_{ij}^2 \tag{13.19}$$

The deviations are squared (for reasons explained before, Section 3.8) and summed over all a treatments. Obviously, the null hypothesis that all regressions are the same and are therefore equal to the common regression is:

$$\text{H}_0: \quad \beta_1 = \beta_2 = \ldots = \beta_i = \ldots = \beta_a = \beta_c \tag{13.20}$$

By fitting the data first to their own regression, calculated from one sample at a time, the deviations from model 1 will usually be smaller than those from model 2. The parallel lines for every sample, i.e. the common regression, will not be as close a fit. Even slight differences among the slopes of the samples will usually ensure that model 2 is not as good a fit as model 1. The only circumstance in which the fit to model 2 can be as close as the separate fits in model 1 is when all the regressions really are the same as the common one (β_i values all equal β_c). So, when the null hypothesis (Equation 13.20) is true, except for sampling error in each treatment:

$$\sum_{i=1}^{a}\sum_{j=1}^{n} e_{ij}^2 = \sum_{i=1}^{a}\sum_{j=1}^{n} f_{ij}^2 \tag{13.21}$$

When there are differences in slopes among treatments and the null hypothesis is not true, the common regression cannot be as a good a fit as the individual ones (Figure 13.7). So:

$$\sum_{i=1}^{a}\sum_{j=1}^{n} e_{ij}^2 < \sum_{i=1}^{a}\sum_{j=1}^{n} f_{ij}^2 \tag{13.22}$$

This forms the basis for the test for heterogeneity of regressions described below (Section 13.13.3).

The deviations are calculated for each of these models by replacing the terms in the model by our estimates of them. So, model 1 (Equation 13.17) becomes:

$$Y_{ij} = \bar{Y}_i + b_i(X_{ij} - \bar{X}_i) + e_{ij} \tag{13.23}$$

where $\bar{Y}_i$ is the sampled mean of the variable (Y) in treatment i (and estimates μ_i or $\mu + A_i$ in Equations 13.15, 13.16 and 13.17). b_i is the regression coefficient in treatment i calculated as in Section 13.1 and Table 13.1.

Table 13.8. Calculations of a single-factor analysis of covariance for data as in Tables 13.1 and 13.2. See text for further details

	x_i^2	xy_i	y_i^2	Deviations		
				Sum of squares	Degrees of freedom	Mean square
Model 1 (Individual regressions)						
Sample						
1						
⋮						
i	$\sum_{j=1}^{n}(X_{ij}-\bar{X}_i)^2$	$\sum_{j=1}^{n}(X_{ij}-\bar{X}_i)(Y_{ij}-\bar{Y}_i)$	$\sum_{j=1}^{n}(Y_{ij}-\bar{Y}_i)^2$	$SS_i = y_i^2 - \frac{xy_i}{x_i^2}$	$n-2$	$SS_i/(n-2)$
	$=\sum_{j=1}^{n}X_{ij}^2 - \frac{\left(\sum_{j=1}^{n}X_{ij}\right)^2}{n}$	$=\sum_{j=1}^{n}X_{ij}Y_{ij} - \frac{\left(\sum_{j=1}^{n}X_{ij}\right)\left(\sum_{j=1}^{n}Y_{ij}\right)}{n}$	$=\sum_{j=1}^{n}Y_{ij}^2 - \frac{\left(\sum_{j=1}^{n}Y_{ij}\right)^2}{n}$			
⋮						
a						
Summed				$SS_S = \sum_{i=1}^{a} SS_i$	$a(n-2)$	MS_{sum}
Model 2 (Common regression)						
	$x_c^2 = \sum_{i=1}^{a} x_i^2$	$xy_c = \sum_{i=1}^{a} xy_i$	$y_c^2 = \sum_{i=1}^{a} y_i^2$	$SS_C = y_c^2 - \frac{xy_c^2}{x_c^2}$	$a(n-1)-1$	MS_{com}
Differences among slopes				$SS_C - SS_S$	$a-1$	MS_{slopes}

Model 3 (Total regression)					
$x_T^2 = \sum_{i=1}^{a}\sum_{j=1}^{n}(X_{ij} - \bar{X})^2 = \sum_{i=1}^{a}\sum_{j=1}^{n}X_{ij}^2 - \frac{\left(\sum_{i=1}^{a}\sum_{j=1}^{n}X_{ij}\right)^2}{an}$	$xy_T = \sum_{i=1}^{a}\sum_{j=1}^{n}(X_{ij} - \bar{X})(Y_{ij} - \bar{Y}) = \sum_{i=1}^{a}\sum_{j=1}^{n}X_{ij}Y_{ij} - \frac{\left(\sum_{i=1}^{a}\sum_{j=1}^{n}X_{ij}\right)\left(\sum_{i=1}^{a}\sum_{j=1}^{n}Y_{ij}\right)}{an}$	$y_T^2 = \sum_{i=1}^{a}\sum_{j=1}^{n}(Y_{ij} - \bar{Y})^2 = \sum_{i=1}^{a}\sum_{j=1}^{n}Y_{ij}^2 - \frac{\left(\sum_{i=1}^{a}\sum_{j=1}^{n}Y_{ij}\right)^2}{an}$	$SS_T = y_T^2 - \frac{(xy_T)^2}{x_T^2}$	$an - 2$	MS_{tot}
Differences among adjusted means			$SS_T - SS_C$	$a - 1$	MS_{adj}

H_0: No differences among slopes is tested by $F = \frac{MS_{slopes}}{MS_{sum}}$ with $(a - 1)$ and $a(n - 2)$ degrees of freedom.

H_0: No differences among adjusted means is tested by $F = \frac{MS_{adj}}{MS_{com}}$ with $(a - 1)$ and $a(n - 1) - 1$ degrees of freedom.

Hence, the deviations can be estimated as:

$$\sum_{i=1}^{a}\sum_{j=1}^{n}[Y_{ij} - \bar{Y}_i - b_i(X_{ij} - \bar{X}_i)]^2 \quad (13.24)$$

Deviations in model 2 are calculated from the following sampled estimate of Equation 13.18:

$$Y_{ij} = \bar{Y}_i + b_c(X_{ij} - \bar{X}_i) + f_{ij} \quad (13.25)$$

where $\bar{Y}_i$ is as in Equation 13.23 and b_c is the common regression fitted separately to each treatment. The calculation of b_c is given later in Equations 13.34 and 13.35.

13.10.3 Model 3: The total regression, all data combined

If there are no differences among experimental treatments, means of all treatments are equal (i.e. $\mu_1 = \mu_2 = \ldots = \mu_i = \ldots = \mu_a = \mu$). This occurs when the following null hypotheses are true:

$$\begin{aligned} H_0:&\quad \mu_1 = \mu_2 = \ldots = \mu_i = \ldots = \mu_a = \mu \\ H_0:&\quad A_1 = A_2 = \ldots = A_i = \ldots = A_a = 0 \end{aligned} \quad (13.26)$$

Under this scenario, it should make no difference from which treatment the data come, so they can all be put together. Now:

$$Y_{ij} = \mu + \beta_T(X_{ij} - \bar{X}) + g_{ij} \quad (13.27)$$

where β_T is the total regression (all data combined), using $\bar{X}$, the average of all values of the covariate. The deviations from this model (g_{ij} values) are illustrated in Figure 13.7.

The deviations from this model (Equation 13.27) should, under most circumstances, be larger than those under model 2, because the line of best fit for all the data is probably not very close to some of the sets of data (for example, in Figure 13.7). Generally, therefore, the deviations under model 3 can only be the same magnitude as under model 2 if there are no differences among treatments, so that all A_i terms in Equation 13.26 are zero and the regressions in each treatment (β_i values) are all equal to the common regression (i.e. equal to β_c). The latter condition was examined by comparison of deviations from models 1 and 2 (Section 13.10.2). The former condition is, of course, the null hypothesis specified in Equations 13.26. Therefore, this null hypothesis can be considered as:

$$\sum_{i=1}^{a}\sum_{j=1}^{n} f_{ij}^2 = \sum_{i=1}^{a}\sum_{j=1}^{n} g_{ij}^2 \quad (13.28)$$

except for sampling error.

If there are differences among the treatments, model 3 cannot be as good a fit as model 2. Therefore:

$$\sum_{i=1}^{a}\sum_{j=1}^{n} f_{ij}^2 < \sum_{i=1}^{a}\sum_{j=1}^{n} g_{ij}^2 \quad (13.29)$$

As with models 1 and 2, the deviations are calculated using our estimates of the parameters of model 3. Thus, Equation 13.27 is replaced by:

$$Y_{ij} = \bar{Y} + b_{\mathrm{T}}(X_{ij} - \bar{X}) + g_{ij} \quad (13.30)$$

where $\bar{Y}$ is the sampled mean of all values of the variable from all treatments combined and b_{T} is the regression of all values of Y_{ij} on X_{ij} from all the treatments combined (as in Section 13.4 and Table 13.8). Thus, the deviations from model 3 are estimated as:

$$\sum_{i=1}^{a}\sum_{j=1}^{n} [Y_{ij} - \bar{Y} - b_{\mathrm{T}}(X_{ij} - \bar{X})]^2 \quad (13.31)$$

This is the basis for an analysis of covariance. It can be shown that the statistical test constructed to test the null hypothesis in Equations 13.26 is entirely equivalent to a test of data adjusted to remove any influence of the covariate X. It is, however, more usual in an analysis of covariance to adjust the data to their values at $\bar{X}$ rather than zero. It makes no difference what value of X is used, because the treatments are all being adjusted using a common slope. Thus, the data are on parallel regressions and are being adjusted using the same slope. At any value of X, the adjusted values of Y have the same differences among treatments. It is identical with a test for differences among the intercepts or elevations of the regressions for each treatment. The intercept of a regression is the value of the dependent variable Y when the covariate or independent variable X is zero.

This is explained in Figure 13.8. Two sets of data have sufficiently similar slopes of regression for a common slope (b_c) to apply to each of them. They do, however, differ in intercept (in the first case) because there are differences between the two treatments. The data (Y values) are 'adjusted' by predicting what value they would have if the value

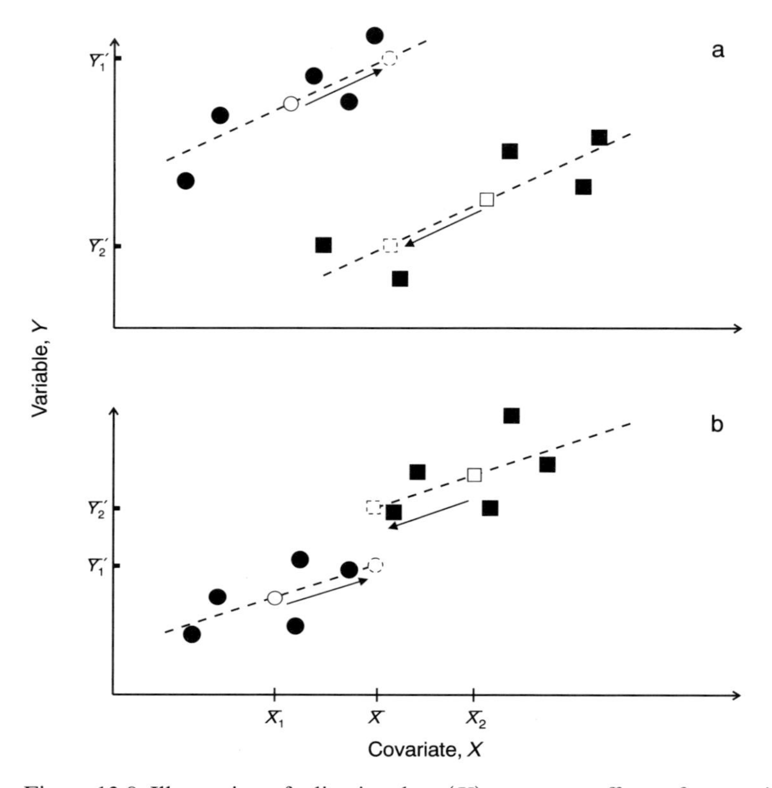

Figure 13.8. Illustration of adjusting data (Y) to remove effects of a covariate (X), by calculating the predicted values of each Y at $\bar{X}$, using the common regression in each sample. Dashed lines indicate the common regression (with slope b_c) fitted in each of the two treatments (1, filled circles; 2, filled squares). Data are solid symbols, with means as empty symbols at the mean value of the covariate in each treatment. Dashed open symbols are the same data adjusted to $X = \bar{X}$ (as shown by the arrows). In (a), treatments differ; adjusted means ($\bar{Y}_1'$, $\bar{Y}_2'$) differ; in (b) treatments (and adjusted means) do not differ.

of all the covariates were $\bar{X}$. These adjusted values (i.e. values in every treatment calculated as what they would be if every value of the covariate was $\bar{X}$) are then analysed by the analysis of covariance (which turns out to be the same as the comparison of models 2 and 3 above).

In the second case (Figure 13.8b), the variables (Y) do differ in the two samples, but, in fact, there are no differences between the two treatments. The null hypothesis in Equations 13.26 is true. The difference between the two sets of Y values is due to the differences between the two sets of X values that happen to have finished up in the two treatments. There is no systematic difference due to treatments. When the Y values are adjusted by predicting what they would be in each treatment (if all values of X_{ij} were equal to $\bar{X}$), they no longer differ significantly.

So, provided there is sufficient similarity of slopes (β_i values) among the treatments that a common slope (β_c) is a realistic fit, the data can be adjusted to the values for $X = \bar{X}$. The adjustment is to move the mean value of the variable in every treatment (i.e. $\bar{Y}_i$) from its value at the mean of the covariate in its treatment ($\bar{X}_i$) to its predicted value at the mean of the covariates in all treatments ($\bar{X}$). This is done using the common regression in every treatment (b_c) so that the adjustment is identical in every treatment. Adjusted means can then be compared as if there were no influence of the covariate because the covariate has been 'set' to the same value in every treatment. Adjustment is therefore done by calculating the adjusted means as:

$$\bar{Y}_i' = \bar{Y}_i - b_c(\bar{X}_i - \bar{X}) \tag{13.32}$$

where $\bar{Y}_i'$ is the adjusted mean in treatment i and $\bar{Y}_i$, b_c, $\bar{X}_i$ and $\bar{X}$ are as before.

Justification to fit a common slope is provided by testing and failing to reject the null hypothesis in Equation 13.20. If the common slope is fitted, the adjusted data can be compared by testing the null hypothesis in Equations 13.26. Such a comparison is an analysis of covariance.

13.11 The procedures: making adjustments

The calculations for models 1 and 3 are the same as those for any linear regression (Section 13.4). The deviations are calculated from the regression in each treatment and from the regression combining all the data together. These calculations are shown in Table 13.8.

For model 2, the common regression is needed. This is calculated using differentiation, as for the linear regression in Section 13.1. By re-arrangement of Equation 13.18 and by replacing $(\mu + A_i)$ by its estimate $\bar{Y}_i$ for each treatment and β_c by its estimate b_c, the following is the equation for the summed, squared deviations from model 2 (see Equation 13.25):

$$\sum_{i=1}^{a}\sum_{j=1}^{n} f_{ij}^2 = \sum_{i=1}^{a}\sum_{j=1}^{n} [Y_{ij} - \bar{Y}_i - b_c(X_{ij} - \bar{X}_i)]^2 \tag{13.33}$$

Table 13.9. Example of calculations of single-factor linear analysis of covariance using data in Table 13.7 and formulae in Table 13.8. Estimated regression coefficients (*b* values) are also shown

Treatment	x_i^2	xy_i	y_i^2	b	Sum of squares	Degrees of freedom	Mean square
1	7.46	−1.85	0.95	−0.25	0.49	6	0.08
2	13.56	−3.29	1.38	−0.24	0.58	6	0.10
3	13.05	−6.50	3.82	−0.50	0.58	6	0.10
Summed					1.66	18	0.09
Common	34.06	−11.64	6.15	−0.34	2.17	20	0.11
Differences among slopes					0.51	2	0.26
Total	35.15	−5.51	131.35	−0.16	130.49	22	5.93
Differences among adjusted means					128.32	2	64.16

H_0: Homogeneity of deviations from individual regressions,

$$\text{Cochran's test } C = \frac{0.10}{0.28} = 0.36,\ P > 0.05$$

H_0: Homogeneity of individual regressions,

$$F_{\text{slopes}} = \frac{0.26}{0.09} = 2.89,\ P > 0.05;\ 2 \text{ and } 18 \text{ degrees of freedom}$$

H_0: No differences among adjusted means,

$$F_{\text{adj}} = \frac{64.16}{0.11} = 583.27,\ P < 0.01;\ 2 \text{ and } 20 \text{ degrees of freedom}$$

The least-squares formula for calculating b_c is:

$$b_c = \frac{\sum_{i=1}^{a}\left[\sum_{j=1}^{n}(X_{ij} - \bar{X}_i)(Y_{ij} - \bar{Y}_i)\right]}{\sum_{i=1}^{a}\left[\sum_{j=1}^{n}(X_{ij} - \bar{X}_i)^2\right]} \qquad (13.34)$$

Using the abbreviated notation in Equation 13.8, this is:

$$b_c = \frac{\sum_{i=1}^{a}\sum_{j=1}^{n} xy_i}{\sum_{i=1}^{a}\sum_{j=1}^{n} x_i^2} \qquad (13.35)$$

This is shown in Table 13.8 and an example is given in Table 13.9 and Figure 13.9. The deviations from the regressions in all three models are therefore calculated. The first procedure is therefore to test whether there is satisfactory homogeneity of slopes to allow fitting a common regression to all the treatments. This test is a comparison of the deviations from models 1 and 2 (see above in Section 13.10, Equations 13.21 and 13.22).

The deviations from model 1 (i.e. $\sum\sum e_{ij}^2$) are estimated in Table 13.8 as MS_{sum}, the mean square calculated from the sum of squares of

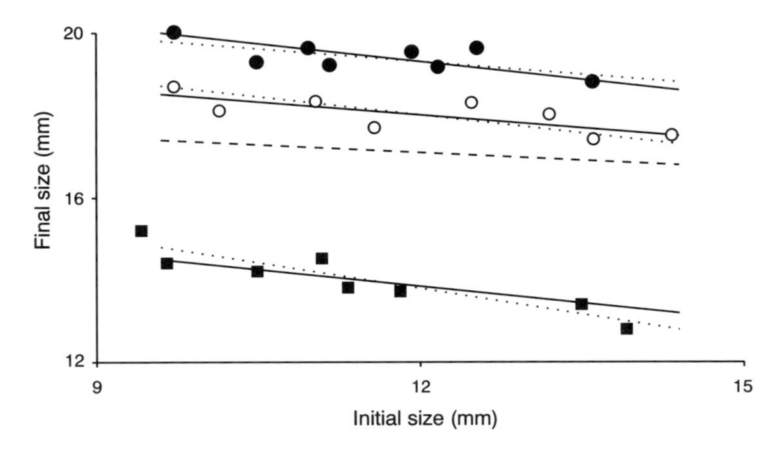

Figure 13.9. Data from the worked example in Tables 13.7 and 13.9. Animals were fed on one of three experimental diets (1, filled circles; 2, empty circles; 3, filled squares). There were $n = 8$ replicates in each diet. Final size at the end of the experiment is regressed on the initial size (the covariate) of each individual. Individual regressions (dotted lines), the common regression in each treatment (continuous lines) and the total regression (dashed line) are shown.

deviations from all the individual regressions. The deviations from model 2 are also estimated in Table 13.8. The derivation of an appropriate test is given in numerous texts (e.g. Snedecor & Cochran, 1989; Winer *et al.*, 1991). To test the null hypothesis of no difference among regressions (as in Equation 13.21), the following calculation is done:

$$\text{MS}_{\text{slopes}} = \text{Mean square for differences among slopes}$$
$$= \frac{\text{SS}_{\text{com}} - \text{SS}_{\text{sum}}}{(a-1)} \tag{13.36}$$

i.e. it is the mean square for the difference between deviations from model 2 and the deviations from model 1. If the null hypothesis is true and all treatments have similar regression slopes, this mean square should be similar to that for deviations from model 1. Thus, the null hypothesis is:

$$\text{H}_0\text{:} \quad \frac{\text{MS}_{\text{slopes}}}{\text{MS}_{\text{sum}}} = 1$$
$$\text{and} \quad \text{H}_\text{A}\text{:} \quad \frac{\text{MS}_{\text{slopes}}}{\text{MS}_{\text{sum}}} > 1 \tag{13.37}$$

with $(a-1)$ and $a(n-2)$ degrees of freedom (as in Table 13.8). If the regressions are all identical, these two mean squares should be the same, except for sampling error. If the regressions have different slopes, the ratio of these mean squares will be large because model 2 does not fit the data as closely as does model 1.

This null hypothesis can be tested using a one-tailed F-ratio, provided certain assumptions are met by the data (as in Section 13.13). If the calculated value of F is significantly large, the null hypothesis should be rejected. The relationships between the variable Y and the covariate X therefore differ among treatments. The ecological interpretation of this is discussed in Section 13.12 and what to do about it is discussed in Section 13.14.

What matters here is that significant heterogeneity of regressions prevents an analysis of covariance being done. That analysis can be done only if there is a common regression for adjusting the data in every treatment.

Note also that this test for heterogeneity of regressions is useful in its own right. If, for any reason, you wish to compare a series of linear regressions to test the null hypothesis that they have the same slope (i.e. the null hypothesis in Equation 13.20), this test will do the job, provided its underlying assumptions are met (see Section 13.13).

If the test for heterogeneity produces a non-significant outcome, the analysis of covariance can be done. The deviations from the total regression are shown in Table 13.8. A test for the null hypothesis that there are no differences among samples (i.e. among treatments, as in Equations 13.26) is provided by comparing the deviations from models 2 and 3. The actual formulation of the test is, again, not derived here, but, if the adjusted values (with all X_{ij} values moved to $\bar{X}$) in every treatment are identical, the null hypotheses in Equations 13.26 are true. The test is:

$$\mathrm{MS}_{\mathrm{adj}} = \frac{\mathrm{SS}_{\mathrm{total}} - \mathrm{SS}_{\mathrm{com}}}{(a-1)} \tag{13.38}$$

i.e. it is the mean square for the difference between deviations from model 3 and the deviations from model 2. If the null hypothesis of no differences among treatments is true, this mean square should be equal to that for deviations from model 2. Thus, the null hypothesis is:

$$\begin{aligned} \mathrm{H_0}: \quad & \frac{\mathrm{MS}_{\mathrm{adj}}}{\mathrm{MS}_{\mathrm{com}}} = 1 \\ \text{and} \quad \mathrm{H_A}: \quad & \frac{\mathrm{MS}_{\mathrm{adj}}}{\mathrm{MS}_{\mathrm{com}}} > 1 \end{aligned} \tag{13.39}$$

with $(a-1)$ and $(a(n-1)-1)$ degrees of freedom (as in Table 13.8).

Again, if certain assumptions are met by the data, this can be evaluated by a one-tailed F-ratio. If rejected, the interpretation is that the treatments differ due to some component or effect over and above any differences due to the values of the covariate (X_{ij} values). This has the same interpretation of significant differences among treatments as in an analysis of variance. It is also identical with a test of whether the intercepts (or elevations) of a set of regression relationships differ. The intercept of a linear regression is the value of Y when $X = 0$. This is obviously the same as the value of Y adjusted to its value, from the regression, when X is zero, i.e. it is an adjusted value. The only difference from the test for adjusted means in the analysis of covariance is that test uses values adjusted to $\bar{X}$, not zero. The procedure is outlined in Table 13.10 and a worked example is given in Table 13.9, using the data from Table 13.7.

So far, the analysis has been described for the case where data are balanced so that the size of sample is the same in each treatment. This is not necessary in the case of a single factor, although, as has been the case elsewhere, the analysis is more robust with respect to the underlying assumptions when samples are balanced (see Section 7.10). If data are not balanced, the formulae to be used can be found in various texts. Complete details are available in Huitema's (1980) excellent book.

Table 13.10. Sequence of procedures in analysis of covariance

1. Plot the data. Examine data for conformity to assumption of normality.
2. Fit model 1 by calculating deviations from individual regressions in each treatment.
3. Test null hypothesis of homogeneity of deviations from individual regressions. If heterogeneous, STOP.
4. Fit model 2 by calculating deviations from the common regression separately in each treatment.
5. Test null hypothesis of homogeneity of regressions in all treatments. If heterogeneous, STOP. Examine regressions (i.e. slopes) to determine how and why treatments have caused different slopes.
6. Fit model 3 by calculating deviations from the total regression, all treatments combined.
7. Do analysis of covariance to test null hypothesis of no difference among treatments.
8. If significantly different, do relevant multiple comparisons.

13.12 Interpretation of the analysis

There are several components to an interpretation of an analysis of covariance. First is the issue of what to make of the situation where there is heterogeneity of regression coefficients, or slopes. On the one hand, this is a great nuisance because it means that the treatments cannot be compared after a uniform procedure of adjustment. There is no common slope to use equitably in each treatment. This is discussed below (Sections 13.13 and 13.14).

On the other hand, information that regressions differ is, itself, very informative. Consider a study of herbivory by insects on a plant at different heights up a hillside. It has been proposed that damage to the plants will be greater where there are more insects and will be greater at lower levels because there are more insects lower down. To test this hypothesis, the percentage area of damage to the leaves is measured in a sample of leaves at each height and the numbers of insects are recorded for each leaf. A positive relationship between percentage damage and number of insects has been proposed in the hypothesis. This is not likely to be linear in the real world, but pretend it is for the purpose of discussion. Suppose preliminary testing shows that this regression relationship is not uniform among the treatments – there is heterogeneity of regressions from one height to another (as in Figure 13.10e, below). The slope of the regression is smaller higher on the hill.

In this case, the interpretation is that individual insects are more damaging in the lower habitats. So, although we still do not yet know if the average amount of damage to the plants is different from one height to another (the original hypothesis of interest), we do know that the intensity of herbivory per capita is greater at the lower levels. Apart from seeking other experimental designs to pursue the original aims and hypotheses (see Section 13.14), we now have good observations about the effect of the interaction between insects and plants in different habitats along a gradient. This is an important advance – it tells us that the original models about herbivory that led to the hypothesis being tested are over-simplified. Insects are not a constant source of damage from one part of a hillside to another.

The next step would sensibly be not to by-pass the result in a headlong pursuit of the original aims, but to reconstruct the logic of the study. Now, models of herbivory must be proposed that explain the differential effects at different heights. Are some parts of the gradient more amenable, by climate or absence of predators, to grazing activities of the insects? Do the insects themselves vary from one part of a hillside to another? Are they, perhaps, genetically or historically different? Do they carry pathogens that damage the plants more in some places? Do the plants vary, so that they are, for example, more stressed due to other environmental factors and therefore more susceptible to damage in lower parts of the habitat?

All of this requires considerable ecological ingenuity to unravel. Sets of competing models and hypotheses need to be carefully constructed. These will also almost certainly require much more complex experimentation involving transplantation of insects and plants separately up and down the hillside (with appropriate controls and spatial replication). All in all, this has the makings of a big project!

The interpretation has therefore not suffered at all from the discovery that regressions did not have the same slope. It is vital information in its own right. It is quite disappointing to see the common ecological response. In many papers, the fact that slopes of regressions differ is ignored. This leads to totally illogical and erroneous analyses of covariance. It also leads to loss of important ecological insights into the processes involved. The *reasons* for heterogeneity of regressions are usually just as interesting and worthy of attention as the existence of differences among adjusted means!

Should it be necessary to determine which regressions differ (i.e. to identify the particular pattern of differences that would represent the particular alternative to the null hypothesis), there are procedures for multiple comparisons of the slopes. These have been excellently summarized by Huitema (1980).

Now, what happens if it is possible to complete the analysis of covariance? Suppose the regressions were similar and there are significant differences among adjusted means of the treatments. As in the case of analysis of variance (Section 8.6), multiple comparisons are necessary to sort out the specific alternative hypothesis. Again, the procedures, in both *a priori* and *a posteriori* forms, are described in full by Huitema (1980). The generally most useful test for *a posteriori* comparisons (see Section 8.6.5) is that due to Tukey (1949a) and Bryant & Paulson (1976).

BPT (Bryant, Paulson and Tukey) tests, or other equivalent procedures, are more complex than the multiple comparisons after an analysis of variance. This is because the construction of standard errors for each pair-wise comparison of adjusted treatment means involves differences among values of the covariate (X_{ij} and $\bar{X}_i$ values), in addition to the variance of the variable itself (Y_{ij} values).

Alternatively, if there were *a priori* planned comparisons (see Section 8.6.4) an appropriate procedure is to use Dunn–Bonferroni tests (Dunn, 1961; see the account in Huitema, 1980).

Finally, if there are no differences among the treatments in the mean values of the covariate ($\bar{X}_i$ values are all similar), there is little point in attempting an analysis of covariance. If there is little difference in means of the covariate, adjusting the values of the variable (Y_{ij}) to their values at a chosen magnitude of X can make very little difference. The Y values are already centred at the same average value ($\bar{X}_i$) in each treatment! Some authorities consider that a useful preliminary would be an analysis of variance of the covariates from the different treatments. If they are significantly different, on average, among treatments, the covariate could cause problems for any interpretation of an analysis of the Y variable. Then, an analysis of covariance is worth a try.

13.13 The assumptions needed for an analysis of covariance

There are numerous assumptions underlying an analysis of covariance, because it combines those needed for regressions and those for analysis of variance.

13.13.1 Assumptions in regressions (see Section 13.3)

The data (the variable Y) must be independently sampled in each treatment and at each value of the covariate (X). Further, the data in each regression should be normally distributed (a weak assumption that is less important than the others) and should have the same variance at every value of the covariate (i.e. at every value of X), as described in Section 13.3. These are the assumptions of tests and analyses of the regression relationships between the variable (Y) and covariate (X).

13.13.2 Assumptions in analysis of variance (see Sections 7.13 to 7.16)

Analysis of covariance requires independence of data among the treatments, as discussed extensively in Section 7.14. As considered there, this requires considerable skill in designing the experiment to ensure that data really are independently sampled.

Then there is the assumption of homogeneity of variances for all treatments, as required in analyses of variance (Section 7.16). In the case of analysis of covariance, this assumption is that the scatter of data around each regression should be the same. Thus, what is needed is homogeneity of variances of deviations from the regressions in all of the treatments. In other words, there should be a similar degree of goodness-of-fit of data in each treatment to the regression fitted separately in each treatment (as in model 1, above).

This assumption can be tested by any of the procedures used in analyses of variance (Section 7.16). The variance of deviations from a regression is estimated as the deviations mean square, with $(n - 2)$ degrees of freedom for a sample of size n. So, the deviations mean squares from a set of regressions can be tested for heterogeneity by a Cochran's test (if n is the same in every treatment) or some other test if n differs. The use of Cochran's test was explained in Section 7.16 and is illustrated in the analysis of covariance in Table 13.9.

One problem in analysis of covariance is that where data are heterogeneous with respect to deviations from the individual regressions, transformations are not readily useful. Transforming the data in the various samples in an attempt to cause them to have similar deviations from the regressions in each treatment would cause the regressions themselves to differ among treatments. Any transformation that caused homogeneity of deviations from regressions in each treatment must differentially alter the values of Y at each value of X. The alteration cannot be the same in every treatment. Therefore, the regressions of Y on X can no longer themselves be homogeneous.

If there is heterogeneity of variances of deviations, you need a great deal of help from professional statisticians. You might, however, stop to think about the reasons for heterogeneity. Look at Section 7.16 for some advice about the biology and the mechanics of heterogeneity of variances! Whatever happens – do not ignore the assumption of homogeneity of variances of deviations from regressions.

13.13.3 Assumptions specific to an analysis of covariance

13.13.3.1 Homogeneity of regressions

In addition to the previous assumptions, there are some necessary for covariance to test hypotheses about means in different treatments. The most crucial is that there is homogeneity of regressions among treatments, i.e. the slopes are reasonably parallel. This is most obvious by examining the rationale for the analysis presented as a series of three models in Section 13.10. If there is no common slope, data cannot be adjusted according to a common relationship. As a result, any outcome is possible in the analysis of covariance. To put this another way, the procedure cannot adjust the variable to remove any influence of the covariate if the adjustment depends on the value of the covariate! Homogeneity of regressions is a critical necessity of an analysis of covariance.

Apart from interpreting the heterogeneity of regressions as an outcome of the experiment (as explained in Section 13.12), there are ways to proceed when regressions are heterogeneous. These are considered in a preliminary way in Section 13.14.

13.13.3.2 Independence of treatment and covariate

The next assumption underlying an analysis of covariance is that the covariate is, itself, not affected by the treatment. This has nothing to do with *statistical* properties of the data or their analysis, but has everything to do with the logic of certain types of experiment.

Consider the situation outlined in Section 13.12. The study is of herbivory by insects on leaves. It has been hypothesized that the average amount of damage to leaves will be less higher up a hillside. Suppose that insects are, in contrast to the situation discussed before and illustrated in Figure 13.10, similar in their destructiveness to plants at all heights on the hill.

There is, however, a greater density of insects at the lower sites. Consequently, the null hypothesis (no difference in mean percentage damage from one height to another) should be rejected. It would be in an analysis of variance because the covariate (density of insects) would be ignored.

An analysis of covariance would, however, remove the influence of density and make a comparison of average damage at a fixed density at

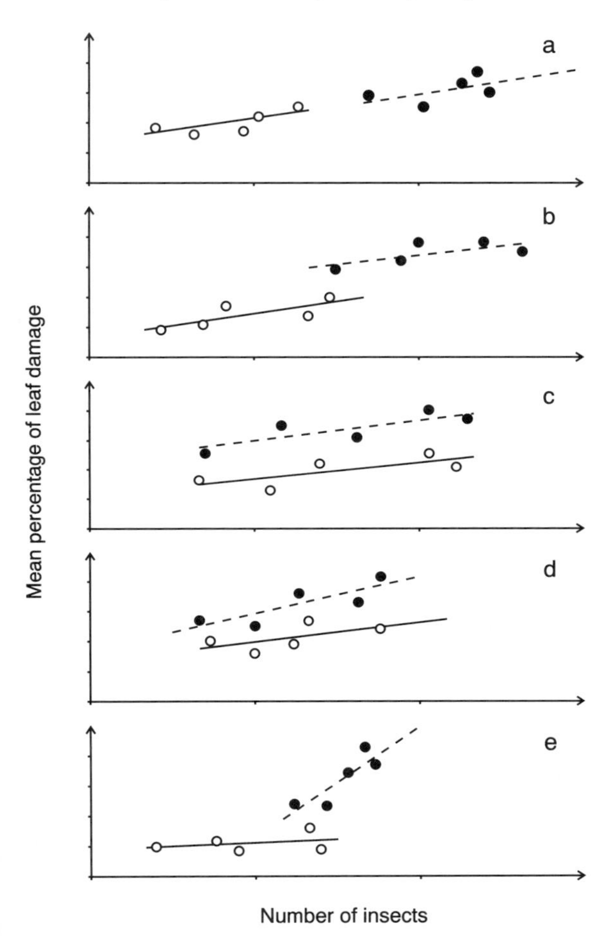

Figure 13.10. Different patterns in data under different models about ecological processes. Data are samples of percentage damage to leaves at two heights on a hillside, sampled in five replicate patches. In each case, the mean number of insects in the patch was recorded as a covariate. The models are described in the text (Sections 13.12 and 13.13). (a) Damage differs between heights; numbers of insects also differ, but they are equally effective at both heights; (b) and (c) plants respond to insects differently at each height, so damage differs because of different attributes or responses by plants. Numbers of insects may differ (b) or not (c) between the two heights. In (d), insects vary in their effectiveness at the two heights, so regressions differ. In (e), insects are as in (d), but also differ in densities between heights. In all cases, filled circles and dashed lines are data from the lower height; open circles and continuous lines are the upper height.

all heights (i.e. $\bar{X}$). This should show no differences among samples (because, as stated above, the insects have similar effects per capita). Now, the real difference in average damage per capita from one height to another has been removed by the analysis. You would conclude that the null hypothesis should be retained!

The problem comes about because of different, but often confused, logical models that might be the reason for doing an experiment. The issue is whether or not the covariate is part of the process or mechanism being investigated. In the case just considered, the logic would be along the lines of the following:

Observation: Apparently different amounts of damage to leaves at different heights.

Model 1: This has been seen because it is so. Even though there is variation in amount of damage from place to place at a given height, there really is a difference from one height to another.

Hypothesis: If the amount of damage is measured carefully in representative samples at each height, there will be differences.

In this case, herbivores are not part of the model. All that matters is that there should be differences in $\bar{Y}$ from one treatment to another – including differences that might be due to different densities of herbivores. So, analysis of variance is appropriate. Unknown (yet) to you, there are differences in densities from height to height, but these *cause* the observations. You are attempting to demonstrate that differences exist – so you should *not* be attempting to eliminate the influences of insects because these would also remove the average differences in amounts of herbivory.

In contrast, here is a second (and more complex) model that might explain the observations:

Model 2: Damage is due solely to insects. The numbers of insects are smaller at higher levels on the hillside and therefore there is more damage on average lower than higher on the hillside.

Hypothesis: If the amount of damage and the number of insects are carefully measured in representative samples at each height, there will be differences in both variables (Y and X) from height to height. But, if the amount of damage is adjusted to remove the influence of the numbers of insects, there will be no difference in adjusted average damage.

Now, it is predicted that analyses of variance of Y and, separately, of X will show differences among heights. Analysis of covariance should remove the differences among mean amounts of damage – because the differences were supposedly attributable to the effects of the covariate. This is illustrated in Figure 13.10a.

The third possibility is that outlined originally in Section 13.12.

Model 3: Different amounts of damage are due to a combination of possibly different numbers of insects at each height and some response by the plants that causes more damage lower than higher on the hillside for the same amount of herbivory.

Hypothesis: If the amount of damage and the number of insects are carefully measured in representative samples at each height, there will be differences in average damage (Y) from height to height. But, if the amount of damage is adjusted to remove the influence of the numbers of insects, there will still be the hypothesized difference in adjusted average damage.

Under this hypothesis, there will be differences among heights in analyses of variance of Y but there may or may not be for X (depending on the range of X and the relationship between Y and X). Analysis of covariance should, however, demonstrate differences among heights over and above the influence of numbers of insects. This is illustrated in Figure 13.10b and c.

The distinction between these latter two models is why you should use an analysis of covariance. The distinction between these two models and the first one cannot be made by use of an analysis of covariance. It is inappropriate there because the differences in magnitude of the covariate could be the cause of the observed differences in amount of damage. The cause, or process, has not been specified in the model, so removing it from the data would be silly.

Finally, on this topic, a further model is:

Model 4: Different amounts of damage are due to the insects being more effective (i.e. more damaging) at lower heights, even though they occur in similar numbers.

Hypothesis: If the amount of damage and the number of insects are carefully measured in representative samples at each height, there will be differences in mean damage (Y) from height to height. The relationship between amount of damage and number of insects will be greater in the lower parts of a hillside.

This was the case discussed in Section 13.12 and illustrated in Figure 13.10d. There should be no difference in an analysis of variance of numbers of insects (the covariate alone), but the regressions at different heights should be heterogeneous, making the analysis of covariance impossible, but furnishing the interpretable outcome to the test of this hypothesis. The heterogeneity of the regressions would be examined by the preliminary test of the slopes.

Of course there could be a further possibility (model 5) that the insects are in different densities (the analysis of variance of the covariate alone is significant), but also have differential effects per capita (i.e. a combination of models 2 and 4 above). This is shown in Figure 13.10e.

There are, of course, circumstances where an experiment has been planned which involves a covariate, but it is not anticipated that the covariate will be influenced by the experimental treatments. For example, consider an experiment on frequency of visitation by pollinating insects to plots of ground that have different amounts of shade. Why one might do this (i.e. the observations, model and hypothesis) is not given in detail here, but the notion is that more insects are found in shady sites because they are less likely to be seen by their predators. In the experiment, some plots are shaded, others are open. The number of visits per plot is the variable of interest, but the amount of nectar per flower varies from place to place, so is recorded as a covariate. The analysis would be an analysis of covariance of adjusted mean number of visits per plot, having adjusted the data for both treatments to the average amount of nectar (assuming that number of visits turns out to be linearly, or in other ways simply related to amount of nectar per plant). There is no expectation that the covariate will differ between the two treatments. The purpose is to eliminate any chance differences that happen to show up in the patches used for the experiment.

It would, however, be worth analysing the nectar data in their own right. If shade, for example, causes the plants to increase their production of nectar, that would explain the results obtained for the insects. The analysis of covariance would show no difference between treatments (i.e. the model about predation *is* wrong). The unexpected discovery of different amounts of nectar caused by shade would, however, lead to an appropriate model being proposed to explain the original observations.

The crucial thing is to be very clear which model(s) is (are) being examined, so that a correct and logically sound interpretation can be made after the appropriate analysis. This is much more important than considering some of the reasons for different interpretations to be artefacts caused by problems with the assumptions of the analyses! If you don't know what the models and hypotheses are, no amount of analysis is going to help. Any of the models about processes (models 2 to 5) could be an explanation for differences in the attacks by insects on plants at different heights.

To distinguish among these needs careful experimental design and analysis. They are proposed only, however, if it is *known* (i.e. not guessed) that there really is a pattern of different amounts of damage at different heights. Hence, model 1 needs to be evaluated. To save time, the logical sequence could be foreshortened by collecting data for the other models (i.e. sampling insects at the same time), but it is important not to lose sight of the nature and order of the relevant models.

13.13.3.3 Other assumptions

The analysis of regressions in its strictest form assumes that the independent variable (i.e. the covariate in this context) is fixed and measured without error. This was discussed in Section 13.3.3. In the case of analysis of covariance, this does not matter at all, provided that the relationship between Y and X is the same in each sample and X values are largely over the same range in every sample. An algebraic proof of this result was given by Winer *et al.* (1991) and the same conclusion was reached by Scheffé (1959). For Y and X to have the same relationship in each sample requires there to be homogeneity of regressions (β_i values all the same) and the deviations from the regressions must have the same variance in every treatment. These two assumptions should be tested (as described above) before doing the analysis of covariance.

Finally, of course, this introduction has been about linear regressions. It is assumed that the relationship between Y and X is linear. If it is not, a more complex regression model is needed (Section 13.15.2).

13.14 Alternatives when regressions differ

13.14.1 A two-factor scenario

As discussed in Sections 13.12 and 13.13, testing the hypothesis of no difference among the adjusted means of several treatments cannot proceed if the regressions in the treatments are heterogeneous. Under these circumstances, there is the possibility of stopping, satisfied with the result that the effect of the treatments is to alter the relationship between the variable and covariate.

Often, however, this is unsatisfactory. Instead, there are two possible procedures to continue with the original aims of the experiment. The most obvious is to use some of the data in a two-factor experiment, using parts of the range of the covariate as levels of a second factor.

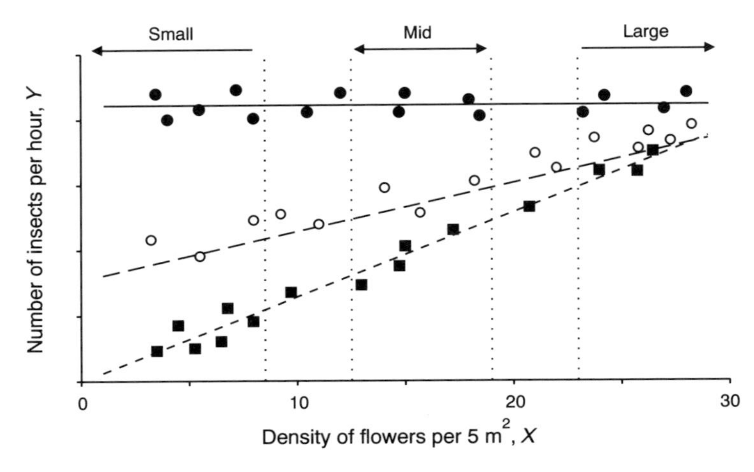

Figure 13.11. Hypothetical data from an experiment to test the null hypothesis of no difference in the mean number of insects visiting patches of flowers in three habitats (say, open fields (filled circles), woodland (empty circles) and at the edges of roads (filled squares)). The density of flowers in each patch is measured as a covariate. The relationship between number of insects and number of flowers varies from one habitat to another. The data can be divided into those replicate patches with a small, medium or large density of flowers and examined by a two-factor analysis of variance. For further details, see the text.

This is illustrated in Figure 13.11. In an experiment with $a = 3$ treatments, replicated $n = 15$ times, the regression between Y and X varied among the treatments. But a graph of the data indicated that there were interesting patterns of difference among the treatments that should still be explored. In the case illustrated, there appear to be no differences when densities of flowers (X) are large and clear differences among treatments when densities are small.

This graph is very similar to a graph of data from an orthogonal, two-factor experiment where there is interaction among treatments (see Chapter 10). This is, of course, exactly what is happening here. The differences among the levels of the experimental treatment (the mean numbers of insects in the three habitats) vary according to the magnitude of the covariate. If the covariate (density of flowers) had been considered as a second experimental factor, the data could have been analysed as a two-factor analysis of variance, with several levels of the factor represented by the covariate. This is shown in Figure 13.11. If the density of flowers (the covariate) is divided into discrete bands or ranges, representing small, medium and large densities, these are equivalent to three fixed levels of the factor density of flowers. Wherever possible, the decision on how to divide the values of the covariate into ranges or bands should be made on some *a priori* basis. Nevertheless, you could proceed even where there is no reason other than expediency for dividing up the covariate.

In the experiment illustrated, by good luck sufficient replicates happen to fall in each of these ranges for every treatment. Only three replicates are in the small range of density for treatment 2 and only three are in the large range for treatment 3. Thus, data can be sub-sampled out of those available to make an orthogonal, balanced set with three levels of the experimental treatment (the habitats) and three levels of the second factor (density of plants), all nine combinations sampled with $n = 3$. This allows a two-factor analysis of variance with two fixed factors. The interaction will be highly significant and multiple comparisons can be used to reveal the precise patterns of differences in the means in each habitat.

Sometimes, a better alternative will be to design a new experiment using the preliminary findings to guide the choice of level of the factor representing the covariate. This would usually allow the levels of the new factor (i.e. different ranges of the span of the covariate) in each to be very small, reducing the variance in each treatment. Where the range of the values of the covariate is large, there is usually greater variance among the values of the variable (Y) in every combination of treatments because the values are spread out by their relationship to X. By making the range of values of X very small, as could be done in a planned experiment, the variance among replicates in each combination of treatments would probably be reduced. Thus, the residual mean square in the analysis would be small and the experiment therefore as powerful as possible.

Also, by planning the experiment to include the covariate as a second factor, you could avoid problems of heterogeneity of variances. In the example, the regression for treatment 3 is steeper than that for treatment 1. Over the same range of values of X, the values of Y will therefore have a greater range (and variance) in treatment 3 than in treatment 1. This will often lead to heterogeneity of variances. Thus, keeping the ranges of X small in the various levels of the covariate factor is a sensible plan. It can be achieved in a designed experiment.

13.14.2 The Johnson–Neyman technique

As an alternative to the above two-factor analysis, Johnson & Neyman (1936) described a test for the case of two levels of treatment and one covariate. The formulae and interpretation were described by Huitema

(1980). The procedure is straightforward when there are only two samples (levels of the treatment) being compared. It is often much more suitable than the two-factor analysis described above because it allows differences in Y to be examined at every value of X – there is no arbitrary division of X into discrete ranges or levels. It also uses all of the data – no data points have to be omitted to create balanced samples.

It does, however, become messy when there are more than two levels of a treatment. Several procedures may be appropriate (see Polthoff, 1964). In essence, these consist of taking, in turn, each possible pair of regressions and analysing the differences between the two samples of Y. The probabilities are adjusted for each pair to ensure that Type I error stays as specified, despite the numerous comparisons being made. If you want to do these tests, read Polthoff (1964) or Huitema (1980).

Otherwise do as suggested above. Bite the bullet and accept the heterogeneity of regressions. Construct biological models that take this discovery into account and plan new experiments that will allow you to test the relevant hypotheses without analyses of covariance.

13.14.3 Comparisons of regressions

As mentioned earlier, there are situations when the object of an analysis is to compare regressions. This is appropriate when it has been hypothesized that regressions differ among treatments in an experiment. For example, the plots of final size (Y) against initial size (covariate X) are not likely to have similar slopes for animals grown at different densities, if rate of growth is affected by density. In such a situation, it would normally have been hypothesized that the regressions will differ due to the effects of density. Then, the null hypothesis (no differences among slopes – see Equation 13.21) would be the focus of attention. This is analysable by the test for heterogeneity of regressions as described in Section 13.12 and illustrated as the Cochran's test in Table 13.9. Multiple comparisons to determine which regressions differ are described by Huitema (1980).

13.15 Extensions of analysis of covariance to other designs

The one-factor, linear analysis of covariance with one covariate can be extended in several directions for more complex hypotheses. There could be more than one covariate (i.e. data could be partially regressed on more than one independent variable) or non-linear regression between the variable and the covariates. Or there could be a multifactorial experimental design with one or more covariates. Finally, there could be complex combinations of these. The principles, as they pertain to the design of experiments are introduced briefly.

13.15.1 More than one covariate

There is no great complexity in having multiple regressions (see Section 13.6) on more than one covariate – provided all the relevant assumptions are likely to have been met. All of the assumptions of regressions and analyses of variance, plus the ones described for covariance must all be considered. In addition, to have a realistic chance of interpreting the analysis correctly, there should be independence between the covariates (see the discussion of this for multiple regressions in Section 13.6).

Here is an example where analysis of covariance using two covariates would be appropriate. It is hypothesized that the number of plants attacked by weevils varies from one area to another because the plants in different areas sprouted at different times during spring. It is proposed that plants sprouting at different times have different chemistry, so that their susceptibility to attack differs. To test this, an experiment will be done in areas where weevils are abundant. Replicated experimental plots will have seeds sown at intervals to produce a set of plots that sprout in each of, say, four weekly intervals. Thus, there is one fixed experimental factor with four levels (the four intervals). At the end of a pre-determined period since sprouting, the number of plants attacked by weevils will be recorded and analysed to test the null hypothesis of no difference among intervals.

There are, however, two uncontrollable influences that should be taken into account as covariates. First, the number of plants sprouting in each plot will not be constant. It is important to remove any influence of different numbers of plants from being responsible for differences among the intervals. Of course, this might more simply be done by recording the proportion of plants attacked, but that is not the basis for the observations being explained. The model is about *numbers* of plants.

So, assume for a moment that the number of plants attacked is a simple linear function of the density of plants in the experimental plots. This is possibly realistic and using density of plants per plot as a covariate would remove any influence of variations in densities from the comparison of the intervals.

At the same time, the numbers of weevils per plot will usually also influence the number of plants attacked. Again, assume that there is a linear relationship between number attacked and density of weevils. So, again, using density of weevils as a covariate would be helpful.

The data to be collected are now the number of plants attacked and the density of plants per plot at the end of the experiment (i.e. the variable Y and the first covariate X_1). In addition, the density of weevils per plot, perhaps averaged from several times of sampling during the experiment, is also recorded (as covariate X_2).

For this analysis to work, there should be no influence of the covariates on each other. If weevils do not kill the plants, X_2 is unlikely to influence X_1. The number of weevils present may, however, be influenced by the number of plants there. Let us assume this is not the case, so that the two covariates are independent of one another. It would, in a real experiment, be worth checking for correlation between observed values of densities of plants and weevils (X_1 and X_2).

Thus, the procedures are a fairly straightforward extension of the case for a single covariate. The computations are described by, for example, Huitema (1980) and Winer *et al.* (1991) and are not considered further here. Again, there need to be preliminary tests on the deviations from regressions in each treatment to determine whether they are sufficiently homogeneous to allow you to proceed. There also needs to be a test for satisfactory homogeneity of regression planes for each treatment so that common partial regressions can be used. These are the same preliminaries as used before.

13.15.2 Non-linear relationships

There is no reason for regressions between a variable and a covariate to be linear. Non-linear regression models can be analysed in exactly the same way as linear ones (consult Section 13.7 for further discussion of a polynomial regression). Again, the details aren't considered further – but all the provisos about being careful about the logic of the hypothesis and the assumptions of the tests still apply!

13.15.3 More than one experimental factor

There is no reason, in principle, why analysis of covariance should not be extended to more complex designs than the single experimental factor. A typical algebraic model for a replicated two-factor experiment (with factors A, levels $i = 1 \ldots a$ and B, levels $j = 1 \ldots b$) and one covariate (X) would be:

$$\text{Model 1:} \quad Y_{ijk} = \mu + A_i + B_j + AB_{ij} + \beta_{ij}(X_{ijk} - \bar{X}_{ij}) + e_{ijk}$$

$$\text{Model 2:} \quad Y_{ijk} = \mu + A_i + B_j + AB_{ij} + \beta_c(X_{ijk} - \bar{X}_{ij}) + f_{ijk} \qquad (13.40)$$

$$\text{Model 3:} \quad Y_{ijk} = \mu + \beta_T(X_{ijk} - \bar{X}) + g_{ijk}$$

with notation analogous to that used before (in Equations 13.17, 13.18 and 13.27), with Y_{ijk} representing the kth replicate in the combination of level i of factor A and j of factor B. A_i, B_j and AB_{ij} represent the effects of factors A and B and their interaction and β_{ij} is the regression of Y on X in the combination of levels i of A and j of B. β_c and β_T represent the common and total regressions, respectively.

The calculations necessary for this or any other model are not difficult. There are, however, some problems with how to partition complex models, so be cautious and get help before embarking on the analysis. Consult Winer *et al.* (1991) and Huitema (1980) for explicit directions for certain experimental designs.

14
Conclusions: where to from here?

14.1 Be logical, be eco-logical

The major purpose of the information provided here is to help you to get good professional help. Use the information to help to plan and to consider the relevant parts of the biology so that your discussions with statistical advisers include the biological reasons why the requirements and assumptions of statistical procedures may be problematic. Use available accounts of experiments and reviews of experimental procedures as tools to evaluate designs – not just as methods of identifying ways of doing experiments. Use other accounts (such as the text by Scheiner & Gurevitch (1993), which appeared during the final writing of this one). Constantly probe better ways of doing experiments.

Do not be frustrated that this book has not discussed such topics as how to deal with unbalanced sets of data. Apart from the serious recommendation that we try to avoid having unbalanced data in the first place, the issues are complex and beyond this discussion. You will need help, but the help you get will be improved if you are aware of the issues.

The preceding sections were chosen to be an introduction to the major themes of modern experimental design as needed by ecologists. The issues discussed are the most general in ecological experimentation. Spatial scales of patterns and processes vary. Intensities of effects and outcomes of processes vary from place to place and time to time. Numerous processes operate at the same time, but not in clearly parallel ways. We must develop theories and testable hypotheses to encompass the complexities caused by these three features of ecological systems. We must therefore routinely handle variation at several simultaneous spatial scales, requiring experiments that include randomly chosen replication in hierarchical designs. We must also be able to deal with the many and varied interactions among processes and between processes and spatial or temporal variability across a habitat. Hence the focus on designs that lead inexorably to analyses of variance and the effort expended here on these structures.

The use of any experimental framework and any procedures for analysis is, however, entirely dependent on the thought processes before an experiment and the careful interpretation after it. The three components – prior specification and formulation of testable hypotheses, the design, planning and implementation of the experiment and the analysis and interpretation of the results – must be unified by the logical structure of the research programme. This is why I found it necessary to discuss logical and philosophical frameworks for experimentation as part of a discussion of how to do experiments.

It is often the case that some statisticians and some ecologists argue that testing hypotheses is not a sensible course of action. Instead, they recommend describing ecological and biological systems by careful construction of the appropriate parametric models and confidence intervals around the estimates of parameters. This procedure does not make at all clear how the decision is made about what to measure in the first place.

Nevertheless, this argument against testing hypotheses revolves around the potential confusion that a statistical test of a statistical null hypothesis is somehow the same as a test of a logical hypothesis. Thus, it is always possible to assert that rejecting a null hypothesis (because a test is statistically significant) is the same thing as declaring some observed phenomenon to be ecologically important. Conversely, failing to reject a null hypothesis is sometimes confused with the assertion that the hypothesized process has no ecological or biological importance. These are important criticisms.

Critics of hypothesis testing are also correct that the probabilities used to reject or retain a null hypothesis in a statistical test are determined by the size of the samples, in conjunction with the choice of probability of Type I error and the intrinsic variation in the variable being sampled. Hence, the focus on power in various parts of this book (Sections 5.12, 6.7, 8.3, 8.4, 9.12, 10.11, 10.14).

These criticisms are good ones – if experimental data are not gathered thoughtfully, analysed carefully and interpreted sensibly in relation to a coherent research programme. The type of research programme outlined here (Chapter 2) requires that hypotheses should be tested only if their logical relationships to observations, models and theories are clear. Such relationships must underpin any attempt to interpret descriptive studies just as much as experimental tests of hypotheses. Ecology can improve only if the processes of thought precede inquiry and are made explicit – thus the attempt here to discuss experimental design in the context of explicit statements of hypotheses about models to account for observations.

Note, however, that the experimental framework advocated here *requires* that there is continual re-evaluation of models or theories, even when they have been supported by experiments. There can be no confusion of the results of a single experiment of arbitrary size (the problem for those who do not like hypothesis testing) if the use of the results (i.e. the decision to reject or retain the specified null hypothesis) is a sensible part of the evaluation of clearly stated models. Thus, in this framework, no single experimental result is adequate.

There are several other immediate consequences of fitting experimental tests of hypotheses properly into logically constructed ecological research programmes. These are discussed briefly in the next three sections.

14.2 Alternative models and hypotheses

Never be complacent about the conclusions reached so far. Remember that rejection of a null hypothesis means that the model from which it was derived has only been supported so far. This is not the same as it being *true*. Alternative models may still explain the results. Wherever possible, use your colleagues (advisers, students, friends) to provoke counter-explanations and theories. Most biological, particularly ecological, patterns have numerous, interactive causes, which vary in space and time. It is, for example, senseless to keep making claims for *one* process, such as predation, controlling the distribution or abundance of organisms if predation cannot also explain the variability in abundances from place to place. What is needed is a series of models about the things affecting predation (such as weather and competitive interactions among prey).

Constantly seek alternative models. Competing and conflicting hypotheses derived from a suite of models can be examined and contrasted. Then put as much effort as possible into the design of experiments to test the contrasting hypotheses from different models. Make sure that the requisite controls and replicates are available so that the experiment will test the stated hypothesis(es). Check such features of the design as will enable you to unconfound things, to make the data independent and to provide accurate and precise measures of differences or relationships. Never, ever, compromise because of the tired old excuse that it's too difficult/expensive/time-consuming. It is, in fact, difficult/expensive and time-consuming everywhere!

14.3 Pilot experiments: all experiments are preliminary

Another feature of biological experimentation that needs greater appreciation is that virtually all our work is preliminary. The hallmark of progressive ideas is that they progress! Even Newton was wrong and his ideas lasted a while.

Accepting that research projects are largely preliminary means that much more information can be used to plan the next phase of the study – to challenge the ideas more forcefully or to keep probing our theories in the light of the various experimental results already obtained. It would be very unusual to be able to identify all the possible sources of confusion in advance. We should not be shocked by the discovery that something we didn't know turned up and made our experiment uninterpretable or irrelevant or made our theory wrong. Things will keep going wrong, making interpretations difficult and making it necessary to investigate new and different hypotheses and to investigate the same hypotheses in different ways. We need to recognize the challenge of putting the hard work into sorting out what to do about the problems, in terms of further tests of the new and changing hypotheses, rather than into building monuments to our current ideas.

14.4 Repeated experimentation

From the foregoing consideration of the need for constant probing and re-evaluation of theories and models, it is surprising how few ecological studies involve repeated experimentation. The experiments don't have to be the *same*, but hypotheses need to be tested over and over again. Sometimes, direct repeats of the same experiment are required to re-test an hypothesis. Sometimes, some components of an experiment should be structured to provide a new, valid test of some previous hypothesis that may already have been tested originally in a completely different manner.

There are broadly three situations where doing experiments more than once (or in more than one place at a time) will be useful. Consider the observation that density of a plant varies from one place to another and in apparent negative correlation with the density of a species known to eat the plant. There are two classes of model to explain the observations by invoking grazing. First is the model that explains variations from place to place because grazers reduce the numbers of plants and grazing is a major process whatever else is going on. This leads to the hypothesis that, although there are all sorts of processes leading to different densities of plants from place to place, removal of grazers will generally lead to increased density of plants. The appropriate design of an experiment is then a nested one (as in Chapter 9). Grazers are removed from some patches of habitat and are not removed from others. The densities of plants per quadrat are then determined in n replicate quadrats per patch. For a moment, ignore controls. The patches are now nested within the treatments. Any difference due to grazing is over and above the differences among patches (i.e. due to any number of other processes).

In the second case, the model is that the process of grazing is important, but is, itself, affected by other processes. Thus, the result of grazing varies from place to place. In this case, it is hypothesized that removal of grazers will increase the numbers of plants, but the difference from patches with grazers will vary from place to place. This is the two-factor experiment discussed as a mixed model in Chapter 10. The grazing treatment is a fixed factor; places are randomly chosen to represent the variation observed across the study-area. A significant interaction would be followed by multiple comparisons to examine the grazing treatments in each place. One outcome of this experiment might well be that removal of grazers made little difference in some places. Under these circumstances, the importance of grazing would have to be gauged from the proportion of places in which removal of grazers made a difference to the number of plants.

Neither of these is actually *repeated* experimentation. In the first example, the repeated application of treatments to different patches is *required* to ensure the experiment is replicated and that differences between the treatments are unconfounded from any natural differences (i.e. unrelated to the treatment) between patches. In the second case, the experiment *must* be done in several places because the hypothesis requires that the treatments differ from place to place. Unless there is replication, the hypothesis cannot be examined at all!

In the third, but much less usual situation, an experiment is done more than once to provide confirmation of results of a previous, perhaps small experiment or to clarify what happened during earlier experiments or to expand the domain of validity of some theory supported by previous experiments. In this situation, some useful procedures will help to integrate the results. What is required is a so-called *meta-analysis* of a series of experiments, to combine the outcomes into one interpretable whole.

How to integrate results of several experiments has been discussed by Mead (1988). There has also been some recent (but rather confused) discussion of field ecological experiments by McKone & Lively (1993). This was confused by the odd use of the term 'nested' to describe a fully orthogonal experiment and a series of *a priori* planned comparisons of experimental treatments in several places (see also the discussion by Bennington & Thayne, 1994; Greenwood, 1994; Lively & McKone, 1994). A much more complete account of how to compare and combine series of experimental results in a formal meta-analysis was provided by Gurevitch *et al.* (1992) in their analysis of experiments on competition. This provides considerable insight and practical advice.

There are, however, two other techniques that help to allow evaluation of the repeated experimentation needed in tests of ecological theory. Suppose two small, independent experiments have been done to test an hypothesis about some process, for example predation and its influence on a rare species. Each experiment was done with limited numbers of prey (to avoid adding ecological experiments to the ever-expanding list of proximate causes of extinction of rare species). Neither experiment had large power. In each case, the probability of getting the observed result if the relevant null hypothesis were true was not significant, but quite small ($P < 0.10$ and $P < 0.06$, respectively). Can we combine these independent results? In doing so, we want to use the independent probabilities in a meta-analysis. How likely is the null hypothesis to represent reality if two separate tests give rather unlikely answers? We can combine the results using Fisher's (1935) formula:

$$C = -2\sum_{i=1}^{k} \log_e P_i \qquad (14.1)$$

where P_i is the probability associated with test i and there are k independent tests or experiments. C is distributed as χ^2 with $2k$ degrees of freedom. In the example above $C = 10.23$ with 4 degrees of freedom, which is significant ($P < 0.05$).

A second method for combining experimental results in more complex situations was used by Underwood & Chapman (1992). We had done a series of experimental manipulations of microhabitat for intertidal snails. In each experiment, there were the same six treatments. Each time we did the experiment, we were unable to determine which treatments differed. So, from a series of 12 experiments over several years, we ranked the order of means in each experiment and then examined the concordance of order of results. The number of times each treatment appeared first, second, etc., in order of magnitude of mean was then analysed by the procedure described by Anderson (1959). The results (Underwood & Chapman, 1992) were strikingly clear-cut. Thus, although no single experiment was ever going to be unambiguous, the set of repeated experiments was remarkably straightforward to interpret.

Although there are numerous problems involved in comparing experiments from one time or place to another, the available procedures should be much more widely used. This would allow separate studies to serve as independent reinforcement of accepted ideas and would reveal when previously exposed notions are incorrect.

Note also that because ecological processes tend to be very varied in time and place, a series of small experiments in different habitats, seasons, etc., will usually be more informative than one large one done once in one place.

14.5 Criticisms and the growth of knowledge

This section's title is taken from the book edited by Lakatos & Musgrave (1974). It encapsulates a major theme by which ecology can grow and advance: criticism.

We need much more explicit criticism of ecological and biological ideas, models, theories. We need much more consistent criticism of the experimental data used to support these ideas. Only by critical evaluation can we proceed to throw out those components of our thinking that are clearly wrong. Criticism of ideas is crucial. Criticism of our experimental results is the first step.

Above all, be self-critical. If some experimental test of an hypothesis is problematic because of difficulties with controls, independence, replication and other aspects of design, the person most responsible for explaining this is the person who did the experiment. Then an argument can be advanced to explain why the results and interpretation should be accepted, despite the problem. This argument will rest on ancillary evidence, inductive notions based on experience, analogy, etc. Some of it may be compelling. All of it needs to be aired. Otherwise, readers, referees, editors, etc., elsewhere in the world or working on different problems are entitled (and have a duty) to reject the findings.

If the reasoning that may support the shaky experimental results is explicitly discussed there will be two major advances. First, everything possible will have been done to clarify the issues for those who read the work so that they are well informed to make a decision about agreeing or disagreeing with its conclusions. Second and much more importantly, the processes, components of systems, issues, etc., will have been clearly identified to form the basis of a proper evaluation of whether anyone should agree with the conclusion. The information provided (the observations), the processes invoked (the models) and the assumptions made are all the raw material for designing better experiments. Criticism in the design of experiments and of their interpretation is the most useful tool we have. My hope is that this book will help to keep this tool sharp, flexible and practical.

References

Anderson, R. L., 1959. Use of contingency tables in the analysis of consumer preference studies. *Biometrics*, **15**: 582–590.

Andrew, N. L. & B. D. Mapstone, 1987. Sampling and the description of spatial pattern in marine ecology. *Annual Review of Oceanography and Marine Biology*, **25**: 39–90.

Andrewartha, H. G. & L. C. Birch, 1954. *The distribution and abundance of animals*. University of Chicago Press, Chicago.

Bacon, F., 1620. *Novum organum*, 1889 edition. Clarendon Press, Oxford.

Bartlett, M. S., 1937. Some examples of statistical methods of research in agriculture and applied biology. *Journal of the Royal Statistical Society Supplement*, **4**: 137–170.

Bartlett, M. S., 1947. The use of transformations. *Biometrics*, **3**: 39–52.

Bennington, C. C. & W. V. Thayne, 1994. Uses and misuses of mixed model analysis of variance in ecological studies. *Ecology*, **75**: 717–722.

Bernstein, B. B. & J. Zalinski, 1983. An optimum sampling design and power tests for environmental biologists. *Journal of Environmental Management*, **16**: 335–343.

Bertness, M. D., 1981. Competitive dynamics of a tropical hermit crab assemblage. *Ecology*, **62**: 751–761.

Birch, L. C., 1957. The meanings of competition. *American Naturalist*, **91**: 5–18.

Boik, R. J., 1979. Interactions, partial interactions, and interaction contrasts in the analysis of variance. *Psychological Bulletin*, **86**: 1084–1089.

Box, G. E. P., 1953. Non-normality and tests on variances. *Biometrika*, **40**: 318–335.

Box, G. E. P., 1954a. Some theorems on quadratic forms applied in the study of analysis of variance problems. 1. Effect of inequality of variance in the one-way classification. *Annals of Mathematical Statistics*, **25**: 290–302.

Box, G. E. P., 1954b. Some theorems on quadratic forms applied in the study of analysis of variance problems. 2. Effect of inequality of variance and correlation of errors in the two-way classification. *Annals of Mathematical Statistics*, **25**: 484–498.

Box, G. E. P. & D. R. Cox, 1964. An analysis of transformations. *Journal of the Royal Statistical Society, Series B*, **26**: 211–243.

Box, G. E. P. & G. M. Jenkins, 1976. *Time series analysis: forecasting and control*. Holden Day Inc., San Francisco.

Bryant, J. L. & A. S. Paulson, 1976. An extension of Tukey's method of multiple comparisons to experimental designs with random concomitant variables. *Biometrika*, **63**: 631–638.

Burdick, R. K. & F. A. Graybill, 1992. *Confidence intervals on variance components*. Marcel Dekker, New York.

Chalmers, A. F., 1979. *What is this thing called science?* Queensland University Press, Brisbane.

Chapman, M. G., 1986. Assessment of some controls in experimental transplants of intertidal gastropods. *Journal of Experimental Marine Biology and Ecology*, **103**: 181–201.

Cliff, A. D. & J. K. Ord, 1973. *Spatial autocorrelation*. Pion Ltd, London.

Cochran, W. G., 1947. Some consequences when the assumptions for the analysis of variance are not satisfied. *Biometrics*, **3**: 22–38.

Cochran, W. G., 1951. Testing a linear relation among variances. *Biometrics*, **7**: 17–32.

Cochran, W. G. & G. Cox, 1957. *Experimental designs*, second edition. Wiley, New York.

Cohen, J., 1977. *Statistical power analysis for the behavioral sciences*. Academic Press, New York.

Collins, H. M., 1985. *Changing order: replication and induction in scientific practice*. Sage, London.

Colquhoun, D., 1971. *Lectures on biostatistics*. Clarendon Press, Oxford.

Connell, J. H., 1970. A predator–prey system in the marine intertidal region. I. *Balanus glandula* and several predatory species of *Thais*. *Ecological Monographs*, **40**: 49–78.

Connell, J. H., 1975. Some mechanisms producing structure in natural communities: a model and evidence from field experiments. In *Ecology and evolution of communities*, edited by M. S. Cody & J. M. Diamond, Harvard University Press, Cambridge, MA, pp. 460–490.

Connell, J. H., 1983. On the prevalence and relative importance of interspecific competition: evidence from field experiments. *American Naturalist*, **122**: 661–696.

Connell, J. H., 1985. The consequences of variation in initial settlement versus post-settlement mortality in rocky intertidal communities. *Journal of Experimental Marine Biology and Ecology*, **93**: 11–46.

Connell, J. H. & W. P. Sousa, 1983. On the evidence needed to judge ecological stability or persistence. *American Naturalist*, **121**: 789–824.

Connor, E. F., 1986. Time series analysis of the fossil record. In *Patterns and processes in the history of life. Dahlem konferenzen*, edited by D. M. Raup & D. Jablonski, Springer-Verlag, Berlin, pp. 119–147.

Connor, E. F. & D. Simberloff, 1984. Neutral models of species' co-occurrences patterns. In *Ecological communities: conceptual issues and the evidence*, edited by D. R. Strong, D. Simberloff, L. G. Abele & A. B. Thistle, Princeton University Press, Princeton, NJ, pp. 316–331.

Connor, E. F. & D. Simberloff, 1986. Competition, scientific method and null models in ecology. *American Scientist*, **75**: 155–162.

Conover, W. J., 1980. *Practical nonparametric statistics*. Wiley & Sons, New York.

Cornfield, J. & J. W. Tukey, 1956. Average values of mean squares in factorials. *Annals of Mathematical Statistics*, **27**: 907–949.

Crawley, M. J., 1993. *GLIM for ecologists*. Blackwell Scientific Publications, Oxford.

Crowder, M. J. & D.J. Hand, 1990. *Analysis of repeated measures*. Chapman & Hall, London.

Cubit, J. D., 1984. Herbivory and the seasonal abundance of algae on a high intertidal rocky shore. *Ecology*, **65**: 1904–1917.

Davis, A. R., 1988. Effects of variation in initial settlement on distribution and abundance of *Podoclavella moluccensis* Sluiter. *Journal of Experimental Marine Biology and Ecology*, **117**: 157–168.

Dawkins, H. C., 1981. The misuse of *t*-tests, LSD and multiple-range tests. *Bulletin of the British Ecological Society*, **12**: 112–115.

Day, R. W. & G. P. Quinn, 1989. Comparisons of treatments after an analysis of variance. *Ecological Monographs*, **59**: 433–463.

Dayton, P. K., 1979. Ecology: a science or a religion? In *Ecological processes in coastal and marine systems*, edited by R. J. Livingstone, Plenum Press, New York, pp. 3–18.

Dayton, P. K. & J. S. Oliver, 1980. An evaluation of experimental analyses of population and community patterns in benthic marine environments. In *Marine benthic dynamics*, edited by K. R. Tenore & B. C. Coull, University of South Carolina Press, Columbia, pp. 93–120.

DeBenedictus, P. A., 1974. Interspecific competition between tadpoles of *Rana pipiens* and *Rana sylvatica*: an experimental field study. *Ecological Monographs*, **44**: 129–141.

Diamond, J. M., 1986. Overview: laboratory experiments, field experiments and natural experiments. In *Community ecology*, edited by J. M. Diamond & T. J. Case, Harper & Row, New York, pp. 3–22.

Diamond, J. M. & M. E. Gilpin, 1982. Examination of the 'null' model of Connor and Simberloff for species co-occurrences on islands. *Oecologia*, **52**: 64–74.

Doran, H. E. & J. W. B. Guise, 1984. *Single equation methods in econometrics: applied regression analysis*. University of New England Press, Armidale, New South Wales.

Duncan, D. B., 1955. Multiple range and multiple *F* tests. *Biometrics*, **11**: 1–42.

Dunn, O. J., 1961. Multiple comparisons among means. *Journal of the American Statistical Association*, **56**: 52–64.

Dunnett, C. W., 1955. A multiple comparison procedure for comparing several treatments with a control. *Journal of the American Statistical Association*, **50**: 1096–1121.

Durbin, J. & G. S. Watson, 1950. Testing for serial correlation in least-squares regression. *Biometrika*, **37**: 409–428.

Durbin, J. & G. S. Watson, 1951. Testing for serial correlation in least-squares regression. 2. *Biometrika*, **38**: 159–178.

Einot, I. & K. R. Gabriel, 1975. A study of the powers of several methods of multiple comparisons. *Journal of the American Statistical Association*, **70**: 574–583.

Eisenhart, C., 1947. The assumptions underlying the analysis of variance. *Biometrics*, **3**: 1–21.

Fairweather, P. G., 1991. Statistical power and design requirements for environmental monitoring. *Australian Journal of Marine and Freshwater Research*, **42**: 555–568.

Fairweather, P. G. & A. J. Underwood, 1983. The apparent diet of predators and biases due to different handling times of their prey. *Oecologia*, **56**: 169–179.

Federer, W. T., 1955. *Experimental design*. Macmillan, New York.

Feyerabend, P. K., 1975. *Against method*. New Left Books, London.

Fisher, R. A., 1928. On a distribution yielding the error functions of several well-known statistics. *Proceedings of the International Mathematical Congress, Toronto, 1924*, **2**: 805–813.
Fisher, R. A., 1935. *The design of experiments*. Oliver & Boyd, Edinburgh.
Fisher, R. A., 1955. Statistical methods and scientific induction. *Journal of the Royal Statistical Society, Series B*, **13**: 69–78.
Fisher, R. A. & F. Yates, 1953. *Statistical tables for biological, agricultural, and medical research*, fourth edition. Oliver & Boyd, Edinburgh.
Fleiss, J. L., 1969. Estimating the magnitude of experimental effects. *Psychological Bulletin*, **72**: 273–276.
Fletcher, W. J. & R. G. Creese, 1985. Competitive interactions between co-occurring herbivorous gastropods. *Marine Biology*, **86**: 183–192.
Frank, P. W., 1975. Latitudinal variation in life history features of black turban snail *Tegula funebralis* (Prosobranchia: Trochidae). *Marine Biology*, **31**: 181–192.
Gaylor, D. W. & F. N. Hopper, 1969. Estimating the degrees of freedom for linear combinations of mean squares by Satterthwaite's formula. *Technometrics*, **11**: 691–706.
Geary, R.C., 1936. Moments of the ratio of the mean deviation to the standard deviation for normal samples. *Biometrika*, **28**: 295–305.
Geisser, S. & S. W. Greenhouse, 1958. An extension of Box's results on the use of the *F* distribution in multivariate analysis. *Annals of Mathematical Statistics*, **29**: 885–891.
Goldfeld, S. M. & R. E. Quandt, 1965. Some tests for heteroscedasticity. *Journal of the American Statistical Association*, **60**: 539–547.
Gossett, W. S. (Student), 1908. On the probable error of the mean. *Biometrika*, **6**: 1–25.
Gottman, J. M., 1981. *Time series analysis: a comprehensive introduction for social scientists*. Cambridge University Press, Cambridge.
Graybill, F. A. & R. A. Hultquist, 1961. Theorems concerning Eisenhart's model II. *Annals of Mathematical Statistics*, **32**: 261–269.
Green, J. R. & D. Margerison, 1977. *Statistical treatment of experimental data*. Elsevier/North-Holland, New York.
Green, R. H., 1979. *Sampling design and statistical methods for environmental biologists*. Wiley, Chichester.
Green, R. H., 1993. Application of repeated measures designs in environmental impact and monitoring studies. *Australian Journal of Ecology*, **18**: 81–98.
Green, R. H. & K. D. Hobson, 1970. Spatial and temporal structure in a temperate intertidal community, with special emphasis on *Gemma gemma* (Pelecypoda: Mollusca). *Ecology*, **51**: 999–1011.
Greenwood, J. J. D., 1994. Statistical analysis of experiments conducted at multiple sites. *Oikos*, **69**: 334.
Greig-Smith, P., 1964. *Quantitative plant ecology*, second edition. Butterworth, London.
Gurevitch, J. & S. T. Chester, 1986. Analysis of repeated measures experiments. *Ecology*, **67**: 251–255.
Gurevitch, J., L. L. Morrow, A. Wallace & J. S. Walsh, 1992. A meta-analysis of competition in field experiments. *American Naturalist*, **140**: 539–572.
Hairston, N. G., 1989. *Ecological experiments: purpose, design and execution*. Cambridge University Press, Cambridge.
Hanson, N. R., 1959. *Patterns of discovery*. Cambridge University Press, Cambridge.

Harper, J. L., 1977. *Population biology of plants.* Academic Press, London.

Hartley, H. O., 1950. The maximum *F*-ratio as a short-cut test for heterogeneity of variance. *Biometrika*, **37**: 308–312.

Hartley, H. O., 1955. Some recent developments in analysis of variance. *Communications in Pure and Applied Mathematics*, **8**: 47–72.

Hawking, S. W., 1988. *A brief history of time.* Bantam, London.

Hocutt, M., 1979. *The elements of logical analysis and inference.* Winthrop, Cambridge, MA.

Hollander, M. & D. A. Wolfe, 1973. *Nonparametric statistical methods.* Wiley, New York.

Holm, E. R., 1990. Effects of density-dependent mortality on the relationship between recruitment and larval settlement. *Marine Ecology Progress Series*, **60**: 141–146.

Huitema, B. E., 1980. *The analysis of covariance and alternatives.* Wiley Interscience, New York.

Hurlbert, S. J., 1984. Pseudoreplication and the design of ecological field experiments. *Ecological Monographs*, **54**: 187–211.

Ito, Y., 1972. On the methods for determining density-dependence by means of regression. *Oecologia*, **10**: 347–372.

Johnson, C. R. & C. A. Field, 1993. Using fixed-effects model multivariate analysis of variance in marine biology and ecology. *Annual Review of Oceanography and Marine Biology*, **31**: 177–222.

Johnson, P. O. & J. Neyman, 1936. Tests of certain linear hypotheses and their application to some educational problems. *Statistical Research Memoirs*, **1**: 57–93.

Kendall, M. G. & A. Stuart, 1977. *The advanced theory of statistics.* Griffin and Co., London.

Kennelly, S. J. & A. J. Underwood, 1984. Underwater microscopic sampling of a sublittoral kelp community. *Journal of Experimental Marine Biology and Ecology*, **76**: 67–78.

Kennelly, S. J. & A. J. Underwood, 1985. Sampling of small invertebrates on natural hard substrata in a sublittoral kelp forest. *Journal of Experimental Marine Biology and Ecology*, **89**: 55–67.

Kershaw, K. A., 1957. The use of cover and frequency in the detection of pattern in plant communities. *Ecology*, **38**: 291–299.

Keuls, M., 1952. The use of studentized range in connection with an analysis of variance. *Euphytica*, **1**: 112–122.

Kitting, C. L., 1980. Herbivore-plant interactions of individual limpets maintaining a mixed diet of intertidal marine algae. *Ecological Monographs*, **50**: 527–550.

Koyre, A., 1968. *Metaphysics and measurement.* Chapman & Hall, London.

Kruskal, W. H. & W. A. Wallis, 1952. Use of ranks in one-criterion variance analysis. *Journal of the American Statistical Association*, **47**: 583–621.

Kuhn, T., 1970. *The structure of scientific revolutions*, second edition. University of Chicago Press, Chicago.

Lack, D., 1954. *The natural regulation of animal numbers.* Oxford University Press, New York.

Lakatos, I. & A. S. Musgrave (Editors), 1974. *Criticism and the growth of knowledge.* Cambridge University Press, Cambridge.

Legendre, P. & M.-J. Fortin, 1989. Spatial pattern and ecological analysis. *Vegetatio*, **80**: 107–138.

Lemmon, E. J., 1971. *Beginning logic*. Nelson, Surrey.
Levene, H., 1960. Robust tests for equality of variances. In *Contributions to probability and statistics*, edited by I. Olkin, S. G. Ghurye, W. Hoeffding, W. G. Madow & H. B. Mann, Stanford University Press, Stanford, CA, pp. 278–292.
Lively, C. M. & M. J. McKone, 1994. Choosing an appropriate ANOVA for experiments conducted at few sites. *Oikos*, **69**: 335.
Lund, R. E. & J. R. Lund, 1983. Probabilities and upper quantiles for the studentized range. *Applied Statistics*, **32**: 204–210.
Maelzer, D. A., 1970. The regression of $\log N_{n+1}$ on $\log N_n$ as a test of density dependence: an exercise with computer-constructed density-independent populations. *Ecology*, **51**: 810–822.
Martin, P. & P. Bateson, 1993. *Measuring behaviour*, second edition. Cambridge University Press, Cambridge.
Maxwell, S. E. & H. D. Delaney, 1990. *Designing experiments and analysing data: a model comparison perspective*. Wadsworth, Belmont, CA.
McArdle, B. H., 1988. The structural relationship: regression in biology. *Canadian Journal of Zoology*, **66**: 2329–2339.
McArdle, B. H. & K. J. Gaston, 1992. Comparing population variabilities. *Oikos*, **64**: 610–612.
McArdle, B. H., K. J. Gaston & J. H. Lawton, 1990. Variation in the size of animal populations: patterns, problems and artefacts. *Journal of Animal Ecology*, **59**: 439–454.
McGuinness, K. A. & A. R. Davis, 1989. Analysis and interpretation of the recruit-settler relationship. *Journal of Experimental Marine Biology and Ecology*, **134**: 197–202.
McKone, M. J. & C. M. Lively, 1993. Statistical analysis of experiments conducted at multiple sites. *Oikos*, **67**: 184–186.
McLean, R. A., W. L. Sanders & W. W. Stroup, 1991. A unified approach to mixed linear models. *American Statistician*, **45**: 54–64.
Mead, R., 1988. *The design of experiments: statistical principles for practical applications*. Cambridge University Press, Cambridge.
Mead, R. & R. N. Curnow, 1983. *Statistical methods in agricultural and experimental biology*. Chapman & Hall, London.
Medawar, P., 1969. *Induction and intuition in scientific thought*. Methuen, London.
Menge, B. A., 1976. Organization of the New England rocky intertidal community: role of predation, competition and environmental heterogeneity. *Ecological Monographs*, **46**: 335–393.
Merton, R. K., 1977. *The sociology of science: an episodic memoir*. Southern Illinois University Press, Carbondale, IL.
Mill, J. S., 1865. *A system of logic*, volume 2, sixth edition. Longman, Green & Co., London.
Miller, R. G., 1981. *Simultaneous statistical inference*. McGraw-Hill, New York.
Milliken, G. A. & D. E. Johnson, 1984. *Analysis of messy data*, volume 1: *Designed experiments*. Lifetime Learning, Belmont, CA.
Moran, M. J., 1985. Effects of prey density, prey size and predator size on rates of feeding by an intertidal predatory gastropod *Morula marginalba* Blainville (Muricidae) on several species of prey. *Journal of Experimental Marine Biology and Ecology*, **90**: 97–105.

Morrisey, D. J., A. J. Underwood, L. Howitt & J. S. Stark, 1992. Temporal variation in soft-sediment benthos. *Journal of Experimental Marine Biology and Ecology*, **164**: 233–245.

Mosteller, F. & C. Youtz, 1961. Tables of the Freeman–Tukey transformations for the binomial and Poisson distributions. *Biometrika*, **48**: 433–440.

Nelder, J. A., 1977. A reformulation of linear models. *Journal of the Royal Statistical Society, Series A*, **140**: 48–76.

Newman, D., 1939. The distribution of the range of samples from a normal population expressed in terms of an independent estimate of standard deviation. *Biometrika*, **31**: 20–30.

Paine, R. T., 1991. Between Scylla and Charybdis: do some kinds of criticism merit a response? *Oikos*, **62**: 90–92.

Pearson, E. S. & H. O. Hartley, 1958. *Biometrika tables for statisticians*, volume 1. Cambridge University Press, Cambridge.

Perry, J. N., 1986. Multiple-comparison procedures: a dissenting view. *Journal of Economic Entomology*, **79**: 1149–1155.

Peters, R. H., 1991. *A critique for ecology*. Cambridge University Press, Cambridge.

Pickett, S. A. & P. S. White (Editors), 1985. *The ecology of natural disturbances and patch dynamics*. Academic Press, New York.

Pielou, E. C., 1969. *An introduction to mathematical ecology*. Wiley Interscience, New York.

Pielou, E. C., 1974. *Population and community ecology: principles and methods*. Gordon and Breach, New York.

Polthoff, R. F., 1964. On the Johnson–Neyman technique and some extensions thereof. *Psychometrika*, **29**: 241–256.

Popper, K. R., 1968. *The logic of scientific discovery*. Hutchinson, London.

Putwain, P. D. & J. L. Harper, 1970. Studies on the dynamics of plant populations. 3. The influence of associated species on populations of *Rumex acetosa* L. and *R. acetosella* L. in grassland. *Journal of Ecology*, **58**: 251–264.

Ramsey, P. H., 1978. Power differences between pairwise multiple comparisons. *Journal of the American Statistical Association*, **73**: 479–485.

Rice, W. R., 1989. Analyzing tables of statistical tests. *Evolution*, **43**: 223–225.

Robinson, D. L., 1987. Estimation and use of variance components. *The Statistician*, **36**: 3–14.

Rosenthal, R. & R. L. Rosnow, 1985. *Contrast analysis: focussed comparisons in the analysis of variance*. Cambridge University Press, Cambridge.

Ryan, T. A., 1959. Multiple comparisons in psychological research. *Psychological Bulletin*, **56**: 26–47.

Ryan, T. A., 1960. Significance tests for multiple comparisons of proportions, variances, and other statistics. *Psychological Bulletin*, **57**: 318–328.

Satterthwaite, F. E., 1946. An approximate distribution of estimates of variance components. *Biometrics Bulletin*, **2**: 110–114.

Scheffé, H., 1959. *The analysis of variance*. Wiley, New York.

Scheiner, S. M. & J. Gurevitch, 1993. *Design and analysis of ecological experiments*. Chapman & Hall, New York.

Schoener, T. W., 1983. Field experiments on intraspecific competition. *American Naturalist*, **122**: 240–285.

Searle, S. R., G. Casella & C. E. McCulloch, 1992. *Variance components*. John Wiley & Sons, Inc., New York.

Shaw, R. G. & T. Mitchell-Olds, 1993. ANOVA for unbalanced data: an overview. *Ecology*, **74**: 1638–1645.

Siegel, S., 1953. *Nonparametric statistics for the behavioral sciences.* McGraw-Hill, New York.

Simberloff, D., 1980. A succession of paradigms in ecology : essentialism, materialism and probabilism. In *Conceptual issues in ecology*, edited by E. Saarinen, Reidel, Dordrecht, pp. 63–99.

Simpson, G. G., A. Roe & R. C. Lewontin, 1960. *Quantitative zoology*. Harcourt Brace, New York.

Snaydon, R. W., 1991. Replacement or additive designs for competition studies. *Journal of Applied Ecology*, **28**: 930–946.

Snedecor, G. W. & W. G. Cochran, 1989. *Statistical methods*, eighth edition. University of Iowa Press, Ames, IA.

Sokal, R. R. & F. J. Rohlf, 1981. *Biometry: the principles and practice of statistics in biological research.* W. H. Freeman, San Francisco.

Spearman, C., 1904. The proof and measurement of association between two things. *American Journal of Psychology*, **15**: 72–81.

St Amant, J. L. S., 1970. The detection of regulation in animal numbers. *Ecology*, **51**: 823–825.

Steel, R. G. D. & J. H. Torrie, 1980. *Principles and procedures of statistics: a biometrical approach*, second edition. McGraw-Hill, New York.

Stewart-Oaten, A., W. M. Murdoch & K. R. Parker, 1986. Environmental impact assessment: "pseudoreplication" in time? *Ecology*, **67**: 929–940.

Swihart, R. K. & N. A. Slade, 1985. Testing for independence in animal movements. *Ecology*, **66**: 1176–1184.

Swihart, R. K. & N. A. Slade, 1986. The importance of statistical power when testing for independence in animal movements. *Ecology*, **67**: 255–258.

Thompson, W. A. & J. R. Moore, 1963. Non-negative estimates of variance components. *Technometrics*, **5**: 441–449.

Trenbath, B.R., 1974. Biomass productivity of mixtures. *Advances in Agronomy*, **26**: 177–210.

Tukey, J. W., 1949a. Comparing individual means in the analysis of variance. *Biometrics*, **5**: 99–114.

Tukey, J. W., 1949b. One degree of freedom for non-additivity. *Biometrics*, **5**: 232–242.

Tukey, J. W., 1957. The comparative anatomy of transformations. *Annals of Mathematical Statistics*, **33**: 1–67.

Underwood, A. J., 1977. Movements of intertidal gastropods. *Journal of Experimental Marine Biology and Ecology*, **26**: 191–201.

Underwood, A. J., 1978. An experimental evaluation of competition between three species of intertidal prosobranch gastropods. *Oecologia*, **33**: 185–208.

Underwood, A. J., 1981. Techniques of analysis of variance in experimental marine biology and ecology. *Annual Reviews of Oceanography and Marine Biology*, **19**: 513–605.

Underwood, A. J., 1984. Vertical and seasonal patterns in competition for microalgae between intertidal gastropods. *Oecologia*, **64**: 211–222.

Underwood, A. J., 1986. The analysis of competition by field experiments. In *Community ecology: pattern and process*, edited by J. Kikkawa & D. J. Anderson, Blackwells, Melbourne, pp. 240–268.

Underwood, A. J., 1988. Design and analysis of field experiments on competitive interactions affecting behaviour of intertidal animals. In *Behavioral adaptation to intertidal life*, edited by G. Chelazzi & M. Vannini, Plenum Press, New York, pp. 333–350.

Underwood, A. J., 1990. Experiments in ecology and management: their logics, functions and interpretations. *Australian Journal of Ecology*, **15**: 365–389.
Underwood, A. J., 1991a. The logic of ecological experiments: a case history from studies of the distribution of macro-algae on rocky intertidal shores. *Journal of the Marine Biological Association of the United Kingdom*, **71**: 841–866.
Underwood, A. J., 1991b. Beyond BACI: experimental designs for detecting human environmental impacts on temporal variations in natural populations. *Australian Journal of Marine and Freshwater Research*, **42**: 569–587.
Underwood, A. J., 1992a. Beyond BACI: the detection of environmental impact on populations in the real, but variable, world. *Journal of Experimental Marine Biology and Ecology*, **161**: 145–178.
Underwood, A. J., 1992b. Competition in marine plant–animal interactions. In, *Plant–animal interactions in the marine benthos*, edited by D. M. John, S. J. Hawkins & J. H. Price, Clarendon Press, Oxford, pp. 443–475.
Underwood, A. J., 1993. The mechanics of spatially replicated sampling programmes to detect environmental impacts in a variable world. *Australian Journal of Ecology*, **18**: 99–116.
Underwood, A. J., 1994. On beyond BACI: sampling designs that might reliably detect environmental disturbances. *Ecological Applications*, **4**: 3–15.
Underwood, A. J. & M. G. Chapman, 1992. Experiments on topographic influences on density and dispersion of *Littorina unifasciata* in New South Wales. In *Proceedings of the third international symposium on littorinid biology*, edited by J. Grahame, P. J. Mill & D. G. Reid, The Malacological Society of London, London, pp. 181–195.
Underwood, A. J. & M. G. Chapman, 1996. Scales of spatial patterns of distribution of intertidal invertebrates, *Oecologia*, in press.
Underwood, A. J. & E. J. Denley, 1984. Paradigms, explanations and generalizations in models for the structure of intertidal communities on rocky shores. In *Ecological communities: conceptual issues and the evidence*, edited by D. R. Strong, D. Simberloff, L. G. Abele & A. B. Thistle, Princeton University Press, Princeton, NJ, pp. 151–180.
Underwood, A. J. & P. G. Fairweather, 1989. Supply-side ecology and benthic marine assemblages. *Trends in Ecology and Evolution*, **4**: 16–20.
Underwood, A. J. & S. J. Kennelly, 1990. Pilot studies for designs of surveys of human disturbance of intertidal habitats in New South Wales. *Australian Journal of Marine and Freshwater Research*, **41**: 165–173.
Underwood, A. J. & P. S. Petraitis, 1993. Structure of intertidal assemblages in different locations: how can local processes be compared? In *Species diversity in ecological communities: historical and geographical perspectives*, edited by R. E. Ricklefs & D. Schluter, University of Chicago Press, Chicago, pp. 38–51.
Ury, H. K., 1976. A comparison of four procedures for multiple comparisons among means – pairwise contrasts for arbitrary sample sizes. *Technometrics*, **18**: 89–97.
Weerahandi, S., 1995. Anova under unequal error variances. *Biometrics*, **51**: 589–599.
Welden, C. W. & W. L. Slauson, 1986. The intensity of competition versus its importance: an overlooked distinction and some implications. *Quarterly Review of Biology*, **61**: 23–44.
Wilk, M. B. & O. Kempthorne, 1955. Fixed, mixed, and random models. *Journal of the American Statistical Association*, **50**: 1144–1167.

Winer, B. J., D. R. Brown & K. M. Michels, 1991. *Statistical principles in experimental design*, third edition. McGraw-Hill, New York.
Wonnacott, T., 1987. Confidence intervals or hypothesis tests? *Journal of Applied Statistics*, **14**: 185–201.
Zar, J. H., 1974. *Biostatistical analysis*. Prentice-Hall, Englewood Cliffs, NJ.

Author index

Subject index